Catalysis in Organic Syntheses 1976

ACADEMIC PRESS RAPID MANUSCRIPT REPRODUCTION

Catalysis in Organic Syntheses 1976

EDITED BY

Paul N. Rylander
Engelhard Industries Division
Engelhard Minerals and Chemicals Corp.
Menlo Park, New Jersey

Harold Greenfield
Uniroyal Chemical
Division of Uniroyal Inc.
Naugatuck, Connecticut

ACADEMIC PRESS, INC. New York San Francisco London 1976

A Subsidiary of Harcourt Brace Jovanovich, Publishers

ACADEMIC PRESS, INC.
111 Fifth Avenue, New York, New York 10003

United Kingdom Edition published by
ACADEMIC PRESS, INC. (LONDON) LTD.
24/28 Oval Road, London NW1

Library of Congress Cataloging in Publication Data

Conference on Catalysis in Organic Syntheses, 5th, Boston, 1975.
Catalysis in organic syntheses.

Includes bibliographical references and index.
1. Chemistry, Organic–Synthesis–Congresses.
2. Catalysis–Congresses. I. Rylander, Paul Nels, Date II. Greenfield, Harold, Date III. Title.
QD262.C56 1975 547'.2 76-4507
ISBN 0–12–605340–5

PRINTED IN THE UNITED STATES OF AMERICA

Contents

LIST OF CONTRIBUTORS vii
PREFACE ix

ENGINEERING ASPECTS AND SURFACE ANALYSIS

Influence of Mass and Heat Transfer on Performance of Heterogeneous Catalysts in Gas/Liquid/Solid Systems
George W. Roberts 1

Auger Electron Spectroscopy. A Passing Curiosity or a Practical Surface Analytical Tool for Heterogeneous Catalysis
Madan M. Bhasin 49

SPECIAL TOPICS

Catalytic Tritiation in the Preparation of Labeled Compounds
Dale W. Blackburn and Sidney H. Levinson 75

Enzymatic Epoxidation and Oxygen Activation
Sheldon W. May 101

Examples of the Use of DTA for the Study of Catalyst Activity and Related Phenomena
Patrick K. Gallagher, David W. Johnson, Jr. and Eva M. Vogel 113

Catalyst Activity via Thermal Analysis
John R. Kosak 137

POLYMER-ATTACHED HOMOGENEOUS CATALYSIS

Polymer-Attached Catalysts
Robert H. Grubbs, E. M. Sweet, and Sawit Phisanbut 153

Polymer-Anchored Homogeneous Hydrogenation Catalysts and Their Use in Multistep Synthetic Reactions
Charles U. Pittman, Jr., Stephan Jacobson, Larry R. Smith, Willie Clements, and Harlin Hiramoto 161

CATALYTIC PROCESSES

Synthesis by Carbon Monoxide Insertion at Carbon-Chlorine Bonds
John A. Scheben and Irving L. Mador 181

Asymmetric Homogeneous Hydrogenation with Chiral Rhodium-Phosphine Catalysts
James D. Morrison, William F. Masler, and Susan Hathaway 203

Stereochemical Aspects of Metal-Catalyzed Formation and Hydrogenolysis of Epoxy Alcohols
James E. Lyons 235

Some Aspects of Nitroalkane Hydrogenation via Homogeneous Catalysis
John F. Knifton 257

Use of Hydrazine-Ruthenium on Carbon for Selective Synthesis of Nitro-Substituted *o*-Phenylenediamines
John L. Miesel, George O. P. O'Doherty, and John M. Owen 273

HYDROGENATION

Selective Hydrogenations Catalyzed by Dichlorodicarbonylbis (triphenylphosphine) Ruthenium (II)
Darryl R. Fahey 287

Polymer Hydrogenations with Soluble Lithium/Cobalt and Aluminum/Cobalt Catalysts
John Carl Falk 305

Hydrogeneolysis of Polyalkylcyclopropanes
Robert L. Augustine and Babubhai A. Patel 325

Preparation of N-Alicyclic Amides by Nuclear Hydrogenation of N-Aryl Amides
Russell E. Malz, Jr. and Harold Greenfield 343

SUBJECT INDEX 361

List of Contributors

Robert L. Augustine, Seton Hall University, South Orange, N.J. 07079

Madan M. Bhasin, Union Carbide Corp., Chemicals and Plastics, South Charleston, W. Va. 25303

Dale W. Blackburn, Smith Kline & French Laboratories, Philadelphia, Pa. 19101

Willie Clements, University of Alabama, University, Ala. 35486

Darryl R. Fahey, Phillips Petroleum Co., Bartlesville, Okla. 74004

John Carl Falk, Borg-Warner Corp., Des Plaines, Ill. 60018

Patrick K. Gallagher, Bell Laboratories, Murray Hill, N.J. 17974

Harold Greenfield, Uniroyal Chemical, Naugatuck, Conn. 06770

Robert H. Grubbs, Michigan State University, East Lansing, Mich. 48823

Susan Hathaway, University of New Hampshire, Durham, N.H. 03824

Harlin Hiramoto, University of Alabama, University, Ala. 35486

Stephan Jacobson, University of Alabama, University, Ala. 35486

David W. Johnson, Jr., Bell Laboratories, Murray Hill, N.J. 07974

John F. Knifton, Texaco Inc., Beacon, N.Y. 12508

John R. Kosak, E.I. duPont de Nemours and Co., Deepwater, N.J. 08023

Sidney H. Levinson, Smith Kline & French Laboratories, Philadelphia, Pa. 19101

James E. Lyons, Sun Oil Co., Marcus Hook, Pa. 19061

Irving L. Mador, Allied Chemical Co., Morristown, N.J. 07960

Russell E. Malz, Jr., Uniroyal Chemical, Naugatuck, Conn. 06770

William F. Masler, University of New Hampshire, Durham, N.H. 03824

Sheldon W. May, Georgia Institute of Technology, Atlanta, Ga. 30332

John L. Miesel, Lilly Research Laboratories, Greenfield, Ind. 46140

James D. Morrison, University of New Hampshire, Durham, N.H. 03824

George O. P. O'Doherty, Lilly Research Laboratories, Greenfield, Ind. 46140

John M. Owen, Lilly Research Laboratories, Greenfield, Ind. 46140

Babubhai A. Patel, Seton Hall University, South Orange, N.J. 07079

Sawit Phisanbut, Michigan State University, East Lansing, Mich. 48823

Charles U. Pittman, Jr., University of Alabama, University, Ala. 35486

George W. Roberts, Engelhard Industries, Edison, N.J. 08817

John A. Scheben, U.S. Industrial Chemicals Co., Cincinnati, Ohio 45237

Larry R. Smith, University of Alabama, University, Ala. 35486

E.M. Sweet, Michigan State University, East Lansing, Mich. 48823

Eva M. Vogel, Bell Laboratories, Murray Hill, N.J. 07974

Preface

The contents of this book reflect an idea born in the early 1960's. At that time, a group of men, whose daily work involved application of catalysis to organic syntheses, felt the need for a forum where those with similar interests could meet and exchange views. The New York Academy of Sciences was approached and asked to sponsor a conference dealing entirely with practical applications of catalysis. The Academy proved enthusiastic about the suggestion, and in 1966 the first meeting was held in New York under the joint chairmenship of Joseph M. O'Connor, Morris Freifelder, and Melvin A. Rebenstorf.

The conference was given the awkward but embracing title of "Conference on Catalytic Hydrogenation and Analogous Pressure Reactions." It was realized at the time the group would grow to have other major interests in addition to hydrogenation, but just what they would be was not then clear. The uncertainty about future events was anticipated in the flexible phrase, "Analogous Pressure Reactions." This first conference was a success, and was followed by the second, third, and fourth conferences held under the chairmanships of Joseph M. O'Connor, Melvin A. Rebenstorf, and Paul N. Rylander, respectively. The value of these conferences is attested to by the fact that each was sponsored by the New York Academy of Sciences despite the Academy's firm rule not to sponsor more than one conference in the same area. The proceedings of the four conferences have been published in the *Annals of the New York Academy of Sciences*, Volumes 148, 158, 172, and 214, respectively.

Gradually, it became clear that the original name of the conference did not accurately describe the groups' changing interests, and, when the fifth conference was held in Boston in April, 1975, it was given the name "Fifth Conference on Catalysis in Organic Syntheses." This conference was produced independently by a small group, whose members choose to call themselves The Organic Reactions Catalysis

Society. A salutary development arising from these collective conferences has been a meshing of the interests of organic chemists with those of catalytic chemists, and in recognition of this development The Organic Reactions Catalysis Society became affiliated in 1975 with the North American Catalysis Society.

Catalysis in Organic Syntheses is a collection of papers given at the Fifth Conference. Their diversity reflects the scope of the subject. Most of the papers are about both organic chemistry and catalysis simultaneously, and it is difficult to determine where one discipline begins and the other ends. A few of the papers attempt to look at the shadowy world of heterogeneous catalysts' surfaces and to relate catalyst properties with catalytic performance. In another paper a visitor from the field of chemical engineering examines the often overlooked effects of mass and heat transfer on catalyst performance. The breadth of interest evidenced in this collection of papers makes it clear that it will often be difficult to ascertain from what field the key will come that will open the lock to the solution of a perplexing problem in organic syntheses by catalysis.

It is a pleasure to express our thanks to Engelhard Industries, Uniroyal, Merck, Sharpe and Dohme, Strem Chemical, and the Catalysis Society of New England for their moral and material support of the Fifth Conference. We are indebted and grateful to Mrs. Rebeca Trautner, Ms. Mariya Bower, and Mrs. Irene Tafaro for their skill and care in preparing this often difficult material for camera-ready copy. The formulas reflect Mrs. Trautner's artistic talents.

THE INFLUENCE OF MASS AND HEAT TRANSFER ON THE PERFORMANCE OF HETEROGENEOUS CATALYSTS IN GAS/LIQUID/SOLID SYSTEMS

GEORGE W. ROBERTS

Engelhard Minerals and Chemicals Corporation
Engelhard Industries Division
Research and Development Department
Menlo Park
Edison, New Jersey 08817

ABSTRACT

The influence of mass and heat transport on the performance of heterogeneous catalysts is reviewed, with particular emphasis on systems where the catalyst is suspended as discrete particles in a liquid, and where at least one reactant is gaseous and the other occurs primarily in the liquid phase. The transport processes which necessarily accompany the catalytic reaction are defined, and the existence of concentration and temperature profiles throughout the three-phase system is established. These concentration and temperature differences can change the apparent catalyst performance; catalyst activity, selectivity and life are briefly reviewed in this light.

Rate expressions are defined for the various transport and reaction steps occurring in the catalytic system, and these are combined to give an overall rate expression. From this, the various possible rate-limiting steps are analyzed. Finally, methods for determining the rate-limiting step from laboratory experiments are developed. These methods fall into two categories: 1) diagnostic experiments, and; 2) calculative approaches. Certain rate-limiting steps, such as gas/liquid mass transport, are best approached experimentally, whereas others, such as pore diffusion, are most amenable to a calculative diagnosis.

INTRODUCTION

The ultimate objective of virtually all research in heterogeneous catalysis is to understand how the formulation of the catalyst and the selection of operating conditions influence the chemistry of the reactions which take place. Unfortunately, the physical processes of heat and mass transport, which necessarily accompany all heterogeneous catalytic reactions, can distort, or even totally obscure the instrinsic chemical relationships between the reaction and the catalyst. Under improper experimental conditions, the apparent reaction kinetics, the product distribution and the catalyst life will be determined much more by the nature of these transport processes than by the chemistry of the system.

Many of the classical works on the interaction between heat transport, mass transport and chemical reaction were published over three decades ago [Damkohler (1938), Frank-Kamenetskii (1939), Thiele (1939), Zeldovitch (1939)]. The literature in this area is now voluminous as well as being highly mathematical. With two notable exceptions, [Satterfield and Sherwood (1963), Satterfield (1970)], no real attempt has been made to reduce this literature to a form that is useful to the catalytic chemist.

The present paper is an attempt to fill the need for a very basic treatment of the interactions between transport processes and chemical reaction. The specific objectives are: 1) to provide a sound qualitative basis for understanding why the processes of heat and mass transport alter the behavior of heterogeneous catalytic reactions, and how these alterations are exhibited; 2) to present some easily-applied, quantitative tools for determining the importance of reaction-transport interactions in the laboratory.

The following five sections proceed gradually from a qualitative to a quantitative description of reaction-transport interactions. The fundamentals of transport processes are reviewed in Section II, and the existence of concentration and temperature gradients in all heterogeneous catalytic systems is established. Section III is devoted to explaining how these concentration and temperature gradients effect the apparent activity, selectivity and life of the catalyst. The concept of rate-limiting steps is reviewed in Section IV, leading into Section V, which deals with the quantitative analysis of various rate-limiting steps. Finally, in Section VI, some simple procedures for determining the rate-limiting step from experimental data are presented.

This paper will be concerned primarily with three-phase (gas/liquid/solid) systems, where the catalyst particles are suspended in a liquid, and where a gas phase is in intimate contact with the liquid. The liquid may contain one or more of the reactants in solution, but at least one of the reactants occurs mainly in the gas phase. Many important industrial reactions fall into this category, including a wide variety of hydrogenation reactions. Despite the practical importance of this type of system, comprehensive discussions of the special transport problems that occur in these systems are rather rare.

Any one of several different types of reactors may be used with the type of gas/liquid/solid system described above, including batch reactors, continuous slurry reactors and the so-called "ebullating bed", Fixed-bed reactors will not be treated in this paper. However, an excellent review of methods for analyzing transport limitations in fixed beds has been presented by Mears (1971).

A MECHANISTIC PICTURE OF REACTION/TRANSPORT INTERACTIONS

System Definition

Figure 1 is a physical picture of the catalytic system which will be treated. Porous catalyst particles are assumed to be suspended in a liquid phase. These catalyst particles almost always have a higher density than the liquid, so that some form of agitation is required to prevent the catalyst from settling to the bottom of the reactor. Agitation also prevents the discrete catalyst particles from forming larger aggregates, which would settle more rapidly and which would generally be less efficient catalysts than the original particles. The required agitation is frequently provided by a mechanical stirrer, although it can also be provided by bubbling gas rapidly through the liquid, or by circulating the liquid continuously through the reactor at a high velocity.

The liquid is assumed to be in contact with a gas phase, which may be present either as bubbles embedded in the liquid, as a continuous phase on top of the liquid, or in some combination of these two forms.

The important behavioral features of the system shown in Figure 1 can be illustrated by considering a simple, catalytic reaction, say $A + B \rightarrow C$. In order to make the following discussion as general as possible, it is assumed that reactant A is supplied to the reactor as a gas, and remains primarily in the gas phase. This assumption does not preclude the possibility that some A will dissolve in the liquid, which, in fact, always occurs to some extent. Reactant A is analogous to H_2 in most hydrogenation reactions.

Reactant B is supplied to the reactor in the liquid phase, and remains primarily in the liquid. Thus, B is analogous to the organic molecule that is hydrogenated in any of a large variety of organic hydrogenation reactions, e.g., the hydrogenation of benzene to cyclohexane.

In general, heat is either liberated or absorbed due to the occurrence of a chemical reaction. To sustain the analogy with hydrogenation, it is assumed that the reaction is exothermic.

FIGURE 1

SCHEMATIC DIAGRAM OF GAS/LIQUID/SOLID CATALYTIC SYSTEM

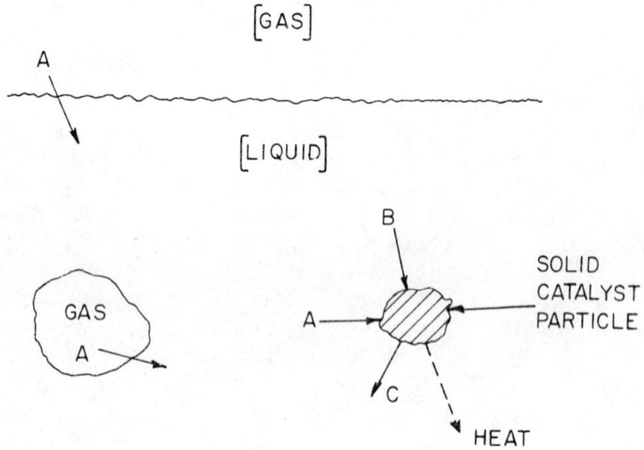

GENERALIZED REACTION:

$$\text{A (GAS)} + \text{B (LIQUID)} \longrightarrow \text{C (LIQUID)}$$

[EXOTHERMIC]

To simplify the analysis which follows, the reaction is considered to be essentially irreversible at reaction conditions. Therefore, there is no need to specify whether the product C is a liquid or a gas at reaction conditions. However, in most hydrogenations, the product will accumulate primarily in the liquid phase.

The next step in the analysis is to identify the transport processes, i.e., the mass and heat transfer steps, which must occur in order for the chemical reaction to proceed. Considering Reactant A first, it is evident from Figure 1 that A must be transported from the gase phase to the liquid phase, across the gas/liquid interface. After dissolving in the liquid, A must be further transported through the liquid to the external (geometric) surface of the catalyst particles. Finally, since the active sites in the catalyst are generally distributed throughout the interior of the particle, Reactant A must diffuse through the network of pores which makes up the particle. Once Reactant A has arrived in the interior of the particle, it can adsorb at an active site and react with some form of adsorbed B to form C. Some molecules of A will diffuse quite far into the particle interior before they react, but others will undergo reaction close to the particle surface.

This paper is not concerned with the exact pathway or mechanism by which A and B react. All of the elementary reaction steps, beginning with the adsorption of A and B onto the surface of the catalyst through the desorption of the product C, are referred to simply as "chemical reaction".

Thus, for Reactant A, three transport steps are absolutely necessary for the chemical reaction to occur: 1) transport of A from the bulk gas phase into the bulk liquid; 2) transport from the bulk liquid to the external (geometric) surface of the catalyst particle, and; 3) diffusion of A into the porous structure of the catalyst. The first two steps, gas/liquid mass transport and liquid/solid mass transport are in series with the chemical reaction itself. Every molecule of A that reacts in the catalyst must first be transported from the gas to the liquid, and then from the liquid to the external catalyst surface before reaction can occur. The third step, pore diffusion, occurs in parallel with the chemical reaction. This parallel relationship can be rationalized by focusing on a position in the interior of the catalyst. Many molecules of A arrive at this position over a given period of time. Some of these molecules will react at the specified position, but others will diffuse past, to react at other sites located deeper in the particle interior. Thus, at any position within the catalyst, a molecule has two parallel "pathways" open to it: 1) reaction, or; 2) diffusion deeper into the particle.

The situation for Reactant B is similar to that for A, but simpler. Since B is supplied in the liquid phase, no gas/liquid transport step is necessary. The only required transport steps are: 1) transport of B from the bulk liquid to the external surface of the catalyst, and; 2) diffusion of B into the interior of the catalyst. As with Reactant A, the former step is in series with the chemical reaction and the latter is in parallel.

The product C, if it is confined primarily to the liquid phase, will undergo transport processes which are exactly the reverse of those for Reactant B. After it is formed by the reaction, C must diffuse from the interior of the catalyst to the external surface, and then be transported from the external surface to the bulk liquid.

The heat that is liberated by the exothermic reaction must also find its way out of the catalyst particle. At steady-state, the rate at which heat leaves the particle must exactly equal the rate of heat generation by reaction. Since the main reservoir for heat is the bulk liquid phase, only two heat-transport steps are necessary: 1) conduction of heat from the particle interior to the external particle surface, and; 2) heat transport from the particle surface to the bulk liquid.

Rate Laws for Mass and Heat Transport

The fundamental reason that heat and mass transport processes influence the performance of heterogeneous catalysts is that these processes necessarily cause temperature and concentration gradients throughout the catalytic system. These gradients alter the reactant concentrations and the temperature within the catalyst, and these alterations, in turn, change the rate and the selectivity of the reaction, and can also affect the rate at which the catalyst appears to age.

In order to understand this situation more completely, the rate laws which govern mass and heat transport must be briefly considered. As shown in Figure 2, the fundamental mathematical relationship which governs mass transport is Fick's Law:

$$N_i = -D_i A \frac{dC_i}{dx} \qquad (1)$$

In this equation, N_i is the rate (gm.moles/sec.) at which species i diffuses through a plane in the normal direction. According to Fick's Law, this rate is equal to the product of the diffusion coefficient of species i, D_i(cm.^{2}/sec.), the area of the plane, A(cm.^{2}), and the concentration gradient of species i, dC_i/dx(gm.moles/cm.^{4}). The minus sign on the right-hand side of Equation 1 accounts for the fact that the direction of mass transport is opposite to the direction of the concentration gradient, i.e., the minus sign expresses the fact that mass is transported from regions of high concentration to regions of low concentration.

It is frequently necessary to consider mass transport across an area of finite thickness, rather than across a plane. In this case, Fick's Law may be written in the alternate form:

$$N_i = k_i A(C_{i,1} - C_{i,2}) \qquad (2)$$

where k_i is the mass-transfer coefficient for species i (cm./sec.) and $C_{i,1}$ and $C_{i,2}$ are the concentrations of species i on

FIGURE 2

FUNDAMENTAL RELATIONSHIPS FOR MASS TRANSPORT

FICK'S LAW

RIGOROUS FORM:

PLANE

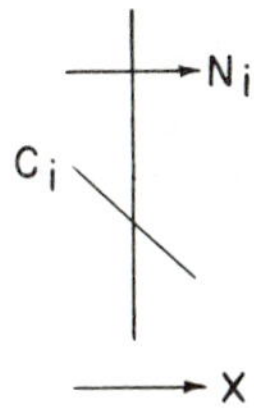

$$N_i \left(\frac{\text{moles } i}{\text{time}}\right) = -D_i \; A \frac{dC_i}{dx}$$

ALTERNATE FORM:

VOLUME ELEMENT

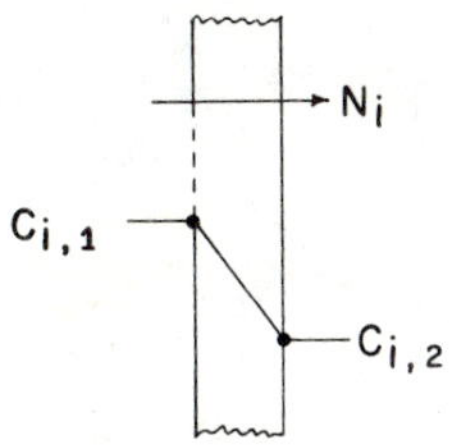

$$N_i \left(\frac{\text{moles } i}{\text{time}}\right) = k_i \, A(C_{i,1} - C_{i,2})$$

either side of the element. This situation is shown in the second part of Figure 2.

Equations 1 and 2 contain three important facts: 1) In order for species i to move from one point to another, a concentration difference or gradient must exist. This concentration difference or gradient is the "driving force" which causes mass transport to take place. 2) The magnitude of this concentration difference or gradient is directly proportional to the rate of mass transport. 3) For a given transport rate, the magnitude of the gradient or difference is inversely proportional to the area through which transport takes place.

Very similar relationships also apply to heat transport. As shown in Figure 3, the analog to Fick's Law is Fourier's Law:

$$q = -k_t A \frac{dT}{dx} \tag{3}$$

where q is the rate of heat transport (cal./sec.), k_t is the thermal conductivity (cal./sec., cm., °C), and dT/dx is the temperature gradient (°C/cm.). For a plane of finite thickness, Fourier's Law can be written as:

FIGURE 3

FUNDAMENTAL RELATIONSHIPS FOR HEAT TRANSPORT

FOURIER'S LAW

RIGOROUS FORM:

PLANE

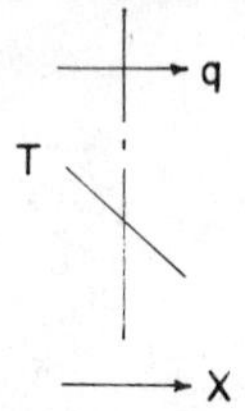

$$q\left(\frac{\text{calories}}{\text{time}}\right) = -k_t A \frac{dT}{dx}$$

ALTERNATE FORM:

VOLUME ELEMENT

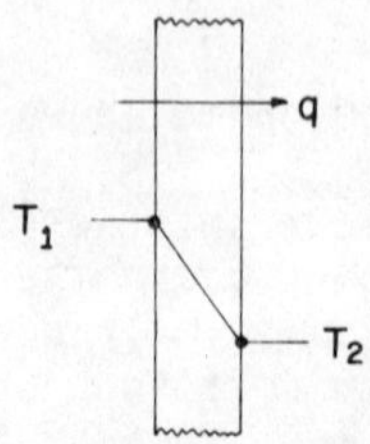

$$q\left(\frac{\text{calories}}{\text{time}}\right) = h A (T_1 - T_2)$$

$$q = hA(T_1 - T_2) \tag{4}$$

where h is the heat-transfer coefficient (cal./sec.,cm.^{2},°C) and T is the temperature (°C).

Heat transport is analogous to mass transport in the three major respects listed previously. Specifically, a temperature difference or gradient "driving force" is always required for heat transport to occur. The size of this temperature difference (gradient) is directly proportional to the rate of heat transport, and inversely proportional to the area through which transport occurs.

Concentration and Temperature Gradients in the Catalytic System

Once the transport steps which are necessary to a given catalytic process have been identified, the above rate laws can be applied to synthesize a qualitative picture of the concentration and temperature profiles that must exist in the catalytic system.

The top portion of Figure 4 is a more detailed schematic drawing of the gas/liquid/solid catalytic system. A relatively-stagnant gas layer exists on the gas side of the gas-liquid interface, and relatively-stagnant liquid layers exist: 1) on the liquid side of the gas-liquid interface and; 2) at the surface of the catalyst particles. These three stagnant films or "boundary layers" constitute the major resistance to mass transport between the bulk gas and the bulk liquid, and between the bulk liquid and the catalyst particle surface.

Essentially all of the "driving force" that is required to supply reactants to the surface of the catalyst occurs across these stagnant films. This situation is depicted in the lower half of Figure 4. Reactant A, the gaseous reactant, experiences a concentration drop as it moves across the gas film at the gas/liquid interface, and a further concentration drop as it moves across the adjacent liquid film. Mass transport through the bulk liquid is relatively rapid due to intensive mixing and high turbulence, so that the concentration decline through this portion of the liquid is negligible. However, as Reactant A moves through the stagnant film at the catalyst particle surface, it undergoes a further concentration drop.

The other major resistance to mass transport, the pore-diffusion resistance is distributed throughout the interior of the catalyst particle. Therefore, the concentration of Reactant A declines continuously as the reactant diffuses deeper and deeper into the particle. However, the concentration gradient becomes progressively smaller towards the interior of the particle.

FIGURE 4

CONCENTRATION GRADIENTS IN GAS/LIQUID/SOLID CATALYTIC SYSTEM

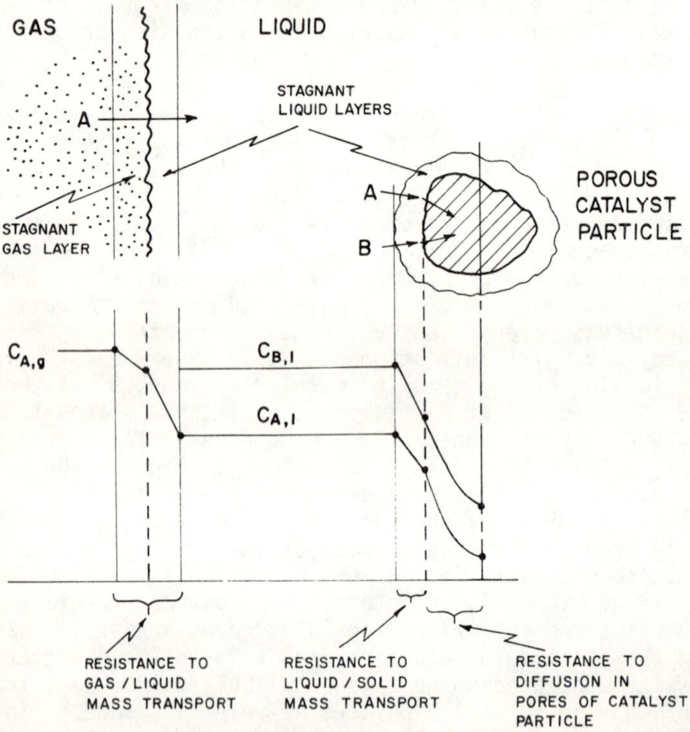

Since no gas/liquid transport step is necessary for Reactant B, the liquid-phase reactant, the only resistances which B encounters are the liquid/solid resistance and the pore-diffusion resistance. These resistances cause concentration declines, as shown in Figure 4.

A picture of the temperature profile in the gas/liquid/solid catalytic system can be constructed in a similar fashion. The only significant heat-transport resistances are the resistance to heat conduction inside the particle and the resistance in the liquid film at the particle surface. Therefore, for an exothermic reaction, the temperature gradients are as shown in Figure 5.

FIGURE 5

TEMPERATURE GRADIENTS IN GAS/LIQUID/SOLID CATALYTIC SYSTEM

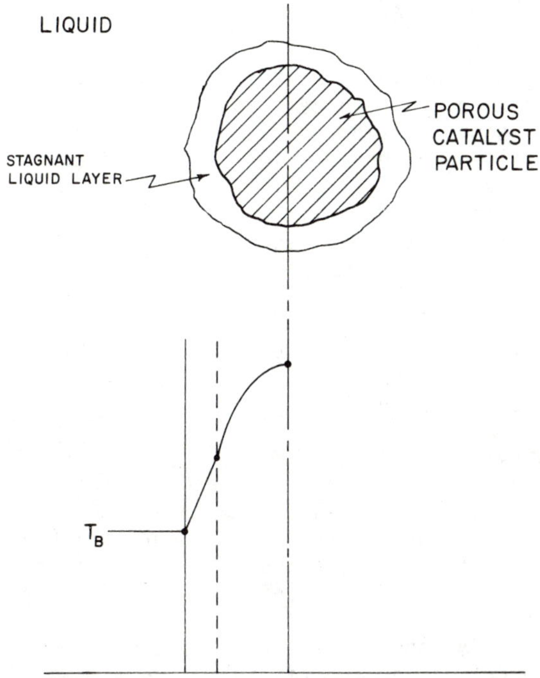

For an endothermic reaction, the temperature profiles are exactly reversed, with the external surface of the pellet being cooler than the bulk liquid, and the center of the particle being even cooler than the external surface.

It is important to remember that concentration and temperature profiles such as those shown in Figures 4 and 5 always exist in gas/liquid/solid catalytic systems. However, the magnitude of these gradients can vary widely, from totally negligible to very severe. The most important factor in determining the magnitude of the concentration and temperature gradients is the rate of the chemical reaction. If the observed rate of reaction is very high, either because a highly-active catalyst is employed or because operating conditions are severe, the transport rates will also have to be high in order to keep up with the reaction. These high transport rates will in turn create large concentration and temperature gradients, as is readily apparent from Equations 1 through 4. This argument substantiates the old adage that "transport processes never interfere with the performance of poor catalysts, only good ones".

THE INFLUENCE OF CONCENTRATION AND TEMPERATURE GRADIENTS ON OBSERVED CATALYST PERFORMANCE

There is an extensive literature dealing with the influence of transport on observed catalyst performance. It is well documented that concentration and temperature gradients within the catalytic system can have a significant effect on the apparent catalyst activity, selectivity and life. Some of the key results from these areas will be reviewed in the paragraphs which follow. It has also been established that transport effects can alter the apparent kinetics of the catalytic reaction. For instance, Weisz and Prater (1954) have shown that both the apparent activation energy and the apparent reaction order can be substantially changed by transport effects. Unfortunately, it is impossible to do justice to the subject of "disguised kinetics" in the present paper.

Catalyst Activity

In order to appreciate some of the quantitative developments which follow, it is necessary to have a qualitative picture of how and why transport-related concentration and temperature gradients alter the observed catalyst performance. In the case of catalyst activity, the explanations are straightforward.

With respect to concentration, it is clear from Figure 4 that the concentrations of the reactants are <u>always</u> lower in the catalyst particles than in the bulk gas or liquid. In general, the intrinsic rate of a chemical reaction can be correlated with a rate equation of the form:

$$-r_A \propto C_A^a C_B^b$$

where $a+b = n$ is the overall reaction order. For $n>0$, the reaction rate declines as the reactant concentrations are reduced. In this case, if the various mass-transport resistances are significant, the reduced reactant concentrations within the catalyst particle will cause the apparent activity to be lower than for the ideal case, where no gradients exist. This "ideal case" is depicted schematically in Figure 6. When the actual

FIGURE 6

DEFINITION OF "IDEAL" CATALYTIC SYSTEM

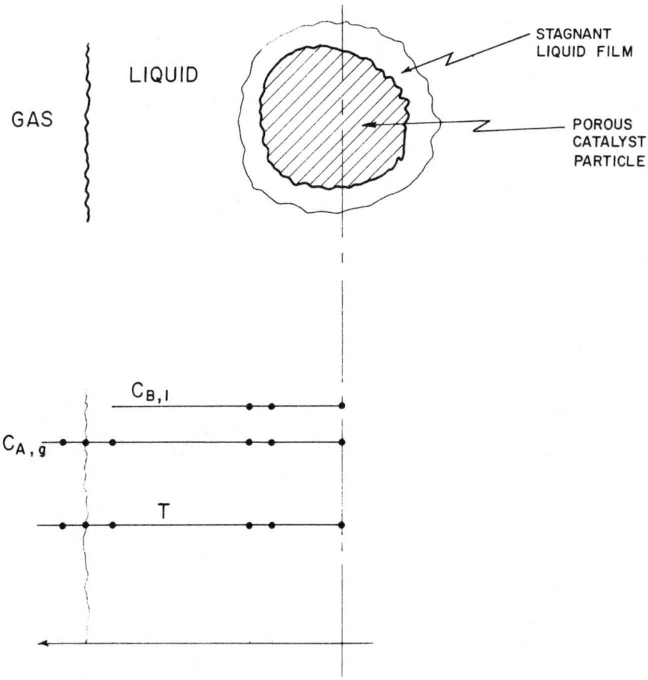

experimental concentration and temperature profiles are as shown in Figure 6, transport does not effect the catalyst performance, and intrinsic catalyst behavior is observed.

For the very unusual case of $n<0$, decreased reactant concentrations in the catalyst cause an increased reaction rate. This situation can occur when one reactant is very strongly adsorbed by the catalyst, and effectively "squeezes" the other reactant off the surface. In such a case, decreasing the concentration of the strongly-adsorbed reactant can actually increase the reaction rate, and make the observed catalyst activity greater than the instrinsic activity. However, this situation is rare enough to exclude from further consideration.

The effect of temperature on apparent catalyst activity is also easily rationalized. For exothermic reactions, the temperature within the pellet is always higher than in the bulk liquid. Since reaction rates increase with increasing temperature, the temperature gradients tend to make the actual rate higher than the rate for the "ideal" case. This temperature effect is directionally opposite to the previously-described concentration effect.

For systems where the catalyst particles are surrounded by a liquid phase, the concentration gradients generally have a stronger influence on the reaction rate than the temperature gradients. In such cases, the overall effect of the various transport processes is to make the observed catalyst activity less than the intrinsic activity.

For systems where the catalyst particles are surrounded by a gas, the situation is more complex. Depending on conditions, either the temperature or the concentration effect can predominate, and the apparent catalyst activity can be either higher or lower than the intrinsic value.

If the reaction is endothermic, the temperature within the catalyst particle is always less than that of the bulk liquid, making the actual reaction rate less than the "ideal" rate. For an endothermic reaction, the effects of the temperature and concentration gradients are reinforcing. The observed reaction rate is always less than the "ideal" rate, and the apparent catalyst activity is always less than the intrinsic activity.

Selectivity

A definitive analysis of the influence of mass transport on catalyst selectivity was originally presented by Wheeler (1951), who defined three types of selectivity, corresponding to different configurations of the reactions which take place. Type I selectivity corresponds to the two independent simultaneous reactions.

$$A \rightarrow \text{Products}_1 \qquad \text{(rapid)}$$

$$B \rightarrow \text{Products}_2 \qquad \text{(slow)}$$

An example of this reaction sequence occurs in the selective hydrogenation of C_4 acetylenes in the presence of butadiene. For these reactions, proper catalyst selection permits a very high rate of C_4 acetylenes hydrogenation compared to the rate of butadiene hydrogenation. This process is used commercially in purification of the butadienes that are produced as a byproduct of ethylene manufacture.

In the following discussion, it is assumed that $A \rightarrow Products_1$ is rapid compared to $B \rightarrow Products_2$. To simplify the analysis, it is further assumed that both A and B are supplied as liquids, so that gas/liquid mass transport can be neglected. However, all of the important conclusions which derive from the following analysis would remain unchanged if gas/liquid mass transport were included.

Figure 7 is a schematic drawing of the concentration gradients that exist in a Type I system. Considering first the

FIGURE 7

CONCENTRATION GRADIENTS FOR TYPE I SELECTIVITY

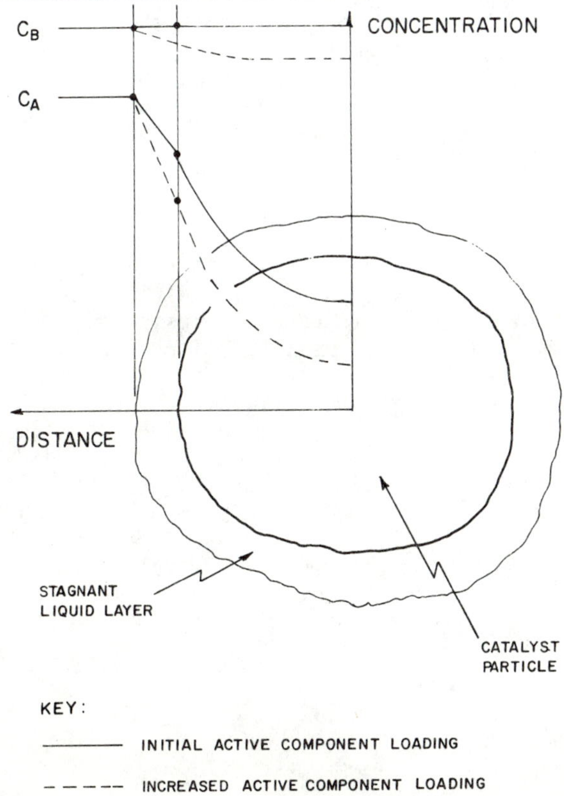

concentration profiles shown as solid lines, it can be seen that the concentration decline due to mass transport is much larger for Reactant A than for Reactant B. This is a simple consequence of the fact that A is consumed in the rapid reaction. Therefore, a higher driving force (concentration drop) is required to support this reaction, compared to the slower reaction, in which B is consumed.

If the two reactions are the same order, e.g., the fast reaction is first order in A and the slow reaction is first order in B, then the concentration profiles in Figure 7 show that the rate of the fast reaction will be decreased by the mass transport resistances to a much greater extent than that of the slow reaction. Therefore, the apparent selectivity of the catalyst will be altered so as to make the fast reaction appear less favored than it is under intrinsic conditions. If the fast reaction is desired, mass transport effects will cause a lowering of the apparent selectivity.

Now consider the effect of increasing the concentration of the active component in the catalyst. In general, this will increase the instrinsic rates of both reactions, and will lead to larger concentration declines for both A and B. However, the concentration profile for Reactant A will undergo a larger change than that for Reactant B, and the apparent selectivity wil decrease even further.

This simple analysis is consistent with the old adage that lower loadings of active component lead to improved selectivity. Transport effects are not the only way to rationalize this statement, but they certainly can be a contributing factor.

The effect that the heat transport resistances will have on apparent selectivity for Type I reactions depends on whether the reactions are exothermic or endothermic, and on which of the two reactions has the higher activation energy. In general, temperature gradients favor the reaction with the highest activation energy when the reactions are exothermic, and favor the reaction with the lowest activation energy when the reactions are endothermic. Further details are presented in Table I. In the event that the effect of heat transport on apparent selectivity is directionally opposite to the effect of mass transport, the mass transport effect will usually predominate, at least for systems where the catalyst is surrounded by liquid.

Type II selectivity corresponds to the parallel reactions

$$A \begin{matrix} \nearrow B \\ \searrow C \end{matrix}$$

This type of reaction configuration occurs in the hydrogenation of chlorobenzene, where benzene and HCl are formed in one reaction, and cyclohexyl chloride in the other.

If the reactions $A \rightarrow B$ and $A \rightarrow C$ are of the same order, the concentration gradients in the catalytic system will not change the apparent selectivity, i.e., both reactions will be slowed down to exactly the same extent by the lower concentration of A in the catalyst particle. However, if the reactions are of different orders, the reaction with the higher order is slowed down by the concentration decline more than the reaction with the lower order. Therefore, the apparent selectivity of the catalyst is altered by transport limitations so that the yield of the product formed in the lower-order reaction is increased relative to the yield of the product from the higher-order reaction.

Increasing the concentration of the active component(s) on the catalyst tends to make the concentration gradients steeper. This in turn slows down the higher-order reaction relative to the lower-order reaction. Thus, if the higher-order reaction is desired, the apparent selectivity of the catalyst is decreased by increasing the active concentration. This effect is again in line with the previously-stated rule-of-thumb. However, if the lower-order reaction is desired, increasing the concentration of active component increases the apparent catalyst selectivity, contradicting the normal effect.

As with Type I selectivity, the influence of the temperature gradients on catalyst selectivity depends on the relative activation energies of the two reactions, and on whether the reactions are exothermic or endothermic.

Type III selectivity corresponds to the series reactions

$$A \rightarrow B \rightarrow C$$

An example of this reaction sequence is the hydrogenation of nitrobenzene, where the initial product, aniline, can be further hydrogenated to cyclohexylamine.

Figure 8 is a schematic representation of the concentration gradients for a Type III system. Note that the concentration of Reactant A is lower in the catalyst particle than for the ideal case, and that the concentration of the intermediate, B, is higher in the particle than for the ideal case. The decreased concentration of A decreases the rate of formation of B and A, whereas the increased concentration of B increases the rate at which B reacts away to form C. Thus, the net effect of the concentration gradients is to decrease the apparent selectivity to B by increasing the net rate of formation of C relative to the rate of disappearance of A, i.e., more of the A that is consumed goes to form C, and less to B, as the transport limitations become more severe.

FIGURE 8

CONCENTRATION GRADIENTS FOR TYPE III SELECTIVITY

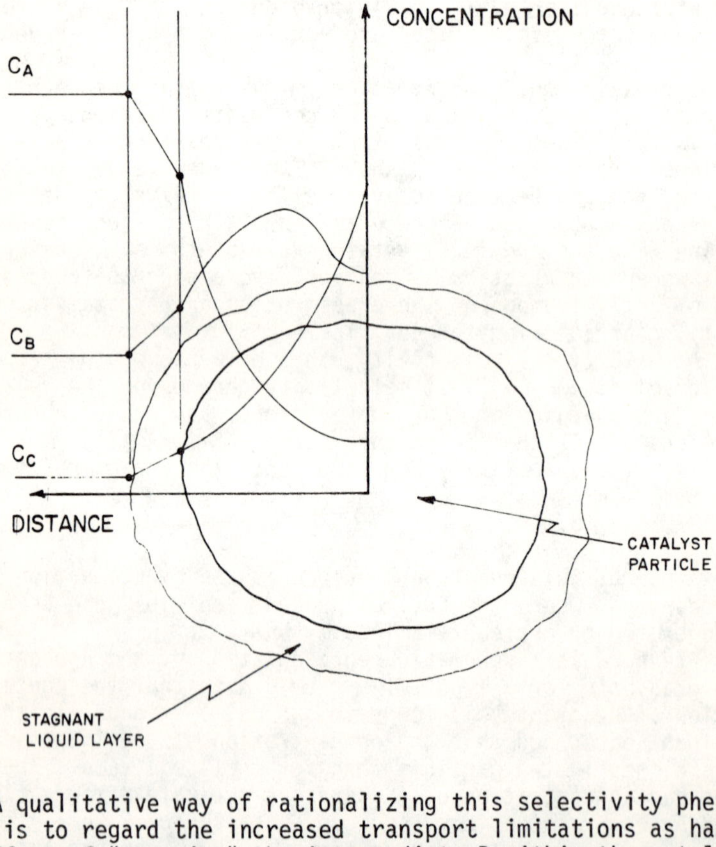

A qualitative way of rationalizing this selectivity phenomenon is to regard the increased transport limitations as having the effect of "trapping" the intermediate B within the catalyst particle, making it increasingly difficult for B to diffuse back to the bulk liquid, and increasing the probability that B will react further to form C, before it can escape from the interior of the catalyst.

Increasing the loading of the active component in the

TABLE I. *Effect of Temperature Gradients on Type I Selectivity*

Reaction Heat Effect	Reaction with Greater Activation Energy	Effect on Selectivity
Exothermic	$A \rightarrow X$	Selectivity to X Improved
Exothermic	$B \rightarrow Y$	Selectivity to X Decreased
Endothermic	$A \rightarrow X$	Selectivity to X Decreased
Endothermic	$B \rightarrow Y$	Selectivity to X Improved

Conclusion: Selectivity for reaction with higher activation energy improved by temperature gradients when reaction is exothermic. Selectivity for reaction with lower activation energy favored when reaction is endothermic.

catalyst makes the concentration gradients steeper, and this tends to make the apparent catalyst selectivity for B even poorer. Once again, the rule-of-thumb that increasing the loading of active component in the catalyst decreases catalyst selectivity is consistent with the effect predicted by analyzing the interaction between chemical reaction and transport phenomena.

The effect of temperature gradients within the catalytic system on apparent catalyst selectivity is very similar to that occurring in Type I and Type II systems. The exact effect depends on the relative activation energies of the two reactions, and on whether the reactions are exothermic or endothermic. In the event of a directional conflict between the mass and heat transport effects, the mass transport effect will usually predominate when the catalyst is surrounded by liquid.

<u>Catalyst Life</u>

The loss of catalyst activity with time is a very complex phenomenon. Many mechanisms exist for activity loss, including support sintering, coke laydown, chemisorption of a "poison", physical blocking of the catalyst surface, and many others.

It is impossible in this paper to comprehensively review all of the various types of activity loss, and how each of these interact with the transport processes. In order to illustrate the type of interaction that can occur, as well as the possible pitfalls that await the casual researcher, one simple case will be analyzed.

Consider the reaction $A \rightarrow B$, taking place on a catalyst which is assumed to be slowly poisoned by irreversible chemisorption of a poison, P. Figure 9A shows the situation that occurs in the "ideal" case, where essentially no mass-transport limitations exist. Since there are no temperature or concentration gradients in the system, each site on the catalyst contributes equally to the reaction rate, and the overall rate decreases in direct proportion to the number of sites which are poisoned. This decline in reaction rate is shown in the lower half of Figure 9A. The graph is not influenced by the way in which the poison deposits in the catalyst, i.e., whether it deposits uniformly throughout the pellet, or concentrates at the external surface or at the center of the particle.

Contrast this behavior with the situation shown in Figure 9B, which corresponds to a catalyst where the initial reaction rate is severely influenced by liquid/solid mass transport. As illustrated in the top half of Figure 9B, the intrinsic reaction rate on the initially unpoisoned catalyst is so rapid that all of the concentration driving force is used up in getting Reactant A through the stagnant liquid film at the particle surface, so that the concentration of A in particle interior is negligibly small. The observed reaction rate is determined solely by how fast A can diffuse through the stagnant film. Under these conditions, the reaction is said to be "controlled" by liquid/solid mass transport.

Consider the case where the poison, P, distributes uniformly throughout the catalyst particle. Even after a moderate amount of poison has deposited, the intrinsic reaction rate is

FIGURE 9

EFFECT OF TRANSPORT LIMITATIONS ON CATALYST LIFE

A. IDEAL CASE (NEGLIGIBLE INITIAL TRANSPORT LIMITATIONS)

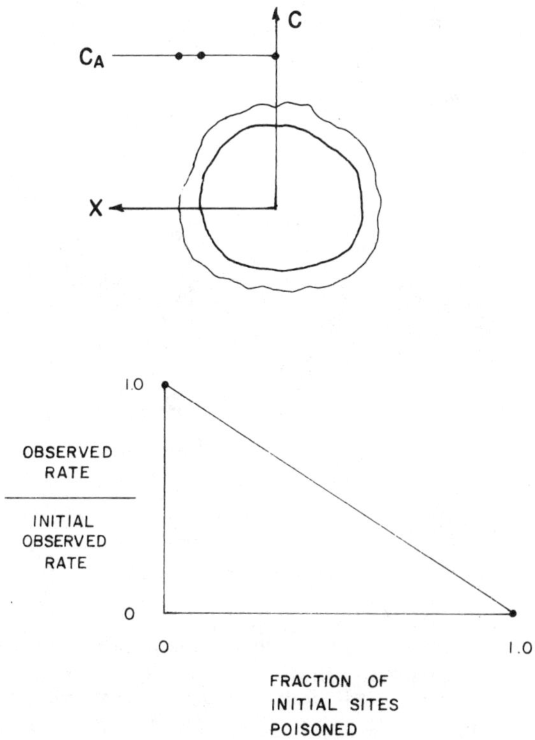

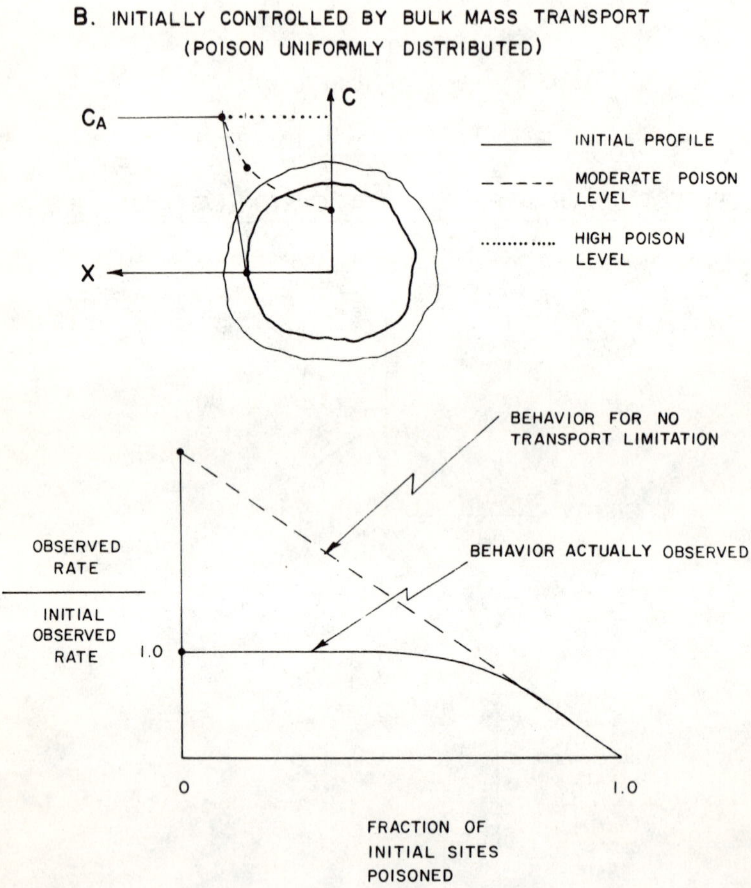

still so high that liquid/solid mass transport of A completely determines the reaction rate. Even though the laydown of poison does reduce the intrinsic reaction rate, as shown by the dashed line in the lower part of Figure 9B, the observed reaction rate is essentially unchanged.

This behavior can theoretically persist up to high levels of poisoning. Thus, the rate may appear unchanged over a very long period of time, despite the fact that more and more sites are being rendered ineffective by chemisorption of the poison.

Eventually however, a point will be reached where so many sites are poisoned that the intrinsic reaction rate is no longer high enough to require a very severe concentration difference across the liquid film. At this point, the concentration gradient across the film will decline, and the concentration within the catalyst particle will begin to rise. This situation is shown by the dashed line, labelled "moderate poison level", in the top half of Figure 9B. When this situation occurs, the observed reaction rate will begin to fall off because the reaction rate is no longer solely determined by liquid/solid mass transport.

At very high levels of poison, the intrinsic rate will be so low that concentration gradients become negligible, and "ideal" deactivation behavior is observed once again. The actual deactivation curve will then approach the "ideal" line, as shown in Figure 9B.

Obviously, this type of aging behavior is very easy to misinterpret. For long periods of time, the actual rate shows little change, suggesting that catalyst deactivation is not a serious problem. However, extrapolation of the observed low aging rate to longer times could lead to very erroneous conclusions, and to very unpalatable consequences if the extrapolation were applied commercially.

The relatively sudden change in aging rate that occurs as the rate-limiting step shifts from liquid-solid mass transport to intrinsic kinetics is sometimes misinterpreted as an experimental problem, such as an upset in temperature control, the introduction of an extraneous contaminant, a feedstock change, etc.

The above discussion has concentrated on a single, relatively-straightforward example of how transport effects interact with chemical kinetics to influence the apparent aging behavior of a heterogeneous catalyst. Many other types of behavior are possible, leading to different curves of the actual rate versus the fraction of sites poisoned. The exact nature of the deactivation curve depends in a complex way on such factors as: 1) the severity of the initial transport limitation; 2) the distribution of the poison within the catalyst particle; 3) the intrinsic kinetics of the reaction; 4) the heat effect (exothermic or endothermic), and many other factors. The most recent detailed treatment is given by Butt (1972).

RATE-LIMITING STEPS

The concept of a rate-limiting step was used rather loosely in the preceeding discussion of catalyst aging. At this point, a more quantitative approach is needed, in order to set the stage for the final section of this paper, dealing with the analysis of experimental data.

Figure 4 shows all of the individual mass transport steps that take place in the catalytic system. Rate expressions will be developed for each of these steps, and then the individual equations for each step will be combined to give an expression for the overall, observed rate in the catalytic system.

For Reactant A, the rate of gas/liquid transport is given by:

$$-r_A = k_{gl}A_{gl}(C^*_{A,l} - C_{A,l}) \qquad (5)$$

In this equation, $-r_A$ is the rate of transport of A from the bulk gas to the bulk liquid (gm.moles/sec.), k_{gl} is the gas/liquid mass-transfer coefficient (cm./sec.), A_{gl} is the interfacial area between the gas and liquid phases, and C_A is the concentration of A in the liquid phase. $C^*_{A,g}$ is the (hypothetical) concentration of A in the liquid which is in equilibrium with A in the bulk gas phase, and $C_{A,l}$ is the actual concentration of A in the bulk liquid.

The rate of transport of A from the bulk liquid to the external surface of the catalyst particles is given by:

$$-r_A = k_{ls}A_{ls}(C_{A,l} - C_{A,s}) \qquad (6)$$

where: k_{ls} is the liquid/solid mass-transfer coefficient (cm./sec.), A_{ls} is the total external surface area of the catalyst particles, and $C_{A,s}$ is the concentration of A at the external surface of the catalyst particles.

At steady-state, the rate of gas/liquid transport must be equal to the rate of liquid/solid transport, and both of these rates must be equal to the rate at which A is consumed by chemical reaction within the pellet. Therefore, the same symbol, $-r_A$, has been used to designate the rates of these two transport steps.

The last rate expression that is necessary for Reactant A is an equation giving the rate at which A is consumed by chemical reaction within the catalyst particle. However, this rate may be strongly influenced by pore diffusion in the particle, and by the intrinsic chemical kinetics, which have not been specified.

For simplicity, it will be assumed that the intrinsic reaction rate is first-order in A. If pore diffusion effects are negligible, the rate of chemical reaction would be given by:

$$-r_A = k_rWC_{A,s} \qquad \text{(negligible pore diffusion)}$$

where k_r is the intrinsic first-order rate constant (cm.^{3}/sec., gm.) and W is the total weight of catalyst particles. To account for the effect of pore diffusion, the above equation is multiplied by an "effectiveness factor", denoted by the symbol η , i.e.,

$$-r_A = \eta k_r W C_{A,s} \tag{7}$$

The effectiveness factor is defined as the actual reaction rate divided by the rate that would have existed in the absence of pore-diffusion effects.

The value of the effectiveness factor is usually between 0 and 1. Values close to zero occur when the pore-diffusion resistance is significant, and concentration gradients within the catalyst particle are very steep. When the effectiveness factor is close to 1, pore diffusion has an insignificant effect. Under certain rare circumstances, the effectiveness factor can exceed unity, but this situation is sufficiently unusual so that it will be neglected for present purposes.

The rate of transfer of Reactant B from the bulk liquid to the external catalyst surface is given by:

$$-r_B = -r_A = k'_{ls} A_{ls} (C_{B,l} - C_{B,s}) \tag{8}$$

The prime on the mass-transfer coefficient, k'_{ls}, signifies that this parameter does not usually have the same value as the mass-transfer coefficient for Reactant A. Since it has been assumed that 1 mole of B reacts with one mole of A, the rate of liquid/solid transport of Reactant B, $-r_B$, is equal to the rate of reaction of A, $-r_A$.

The intrinsic chemical reaction rate is assumed to be zero-order in Reactant B. This assumption is not as restrictive as it might seem and does allow a significant simplication of the mathematics.

Many of the concentrations in Equations 5 through 8 are virtually impossible to measure directly, e.g., $C_{A,s}$, $C_{B,s}$ and frequently, $C_{A,l}$. These concentrations can be eliminated by combining Equations 5,6 and 7 to give an expression for the overall rate of reaction:

$$-r_A = \frac{C_{A,g}^*}{\dfrac{1}{\eta k_r W} + \dfrac{1}{k_{ls} A_{ls}} + \dfrac{1}{k_{gl} A_{gl}}} \tag{9}$$

The only concentration in this expression is $C_{A,g}^*$, which is easily calculated from the known partial pressure of A in the gas phase.

Equation 9 is of the form:

$$\text{Rate} = \frac{\text{Driving Force}}{\text{Resistance}}$$

From the denominator of Equation 9, it can be seen that the total resistance is made up of three separate resistances in series, i.e.,

Total Resistance = Resistance to Gas/Liquid Mass Transport $\left(\frac{1}{k_{gl}A_{gl}}\right)$

+ Resistance to liquid/Solid Mass Transport $\left(\frac{1}{k_{ls}A_{ls}}\right)$

+ Resistance to Chemical Reaction and Pore Diffusion $\left(\frac{1}{\eta\, k_r W}\right)$

For spherical catalyst particles, the total external particle area, A_{ls}, can be expressed as a function of particle size and total catalyst weight, i.e.,

$$A_{ls} = \frac{6W}{d_p\,\rho_p}$$

where ρ_p is the particle density (gm./cm.^{3}) and d_p is the particle diameter (cm.). Substitution of this relationship into Equation 9 gives:

$$-r_A = C^*_{A,g}\;\frac{1}{\dfrac{1}{\eta k_r W} + \dfrac{d_p\rho_p}{6k_{ls}W} + \dfrac{1}{k_{gl}A_{gl}}} \qquad (10)$$

Four special cases of Equation 10 will now be considered.

Case A ($k_{gl}A_{gl} \ll \eta\, k_r W$ and $d_p\rho_p/6k_{ls}W$)

If it is assumed that the product $k_{gl}A_{gl}$ is much smaller than either $\eta k_r W$ or $d_p\rho_p/6k_{ls}W$, then $1/k_{gl}A_{gl}$ will be much larger than either of the other terms in Equation 10, and the other terms may be neglected in comparison. In this case, Equation 10 reduces to:

$$-r_A \cong k_{gl}A_{gl}C^*_{A,g} \qquad (11)$$

Note that the overall reaction rate does not depend at all on the parameters which effect the chemical reaction (η, k_r or W), or the parameters which effect liquid/solid transport (d_p, ρ_p, k_{ls}, k'_{ls}, W, or C_B). The only variables which influence the rate are those which influence gas/liquid mass transport, e.g., stirring rate, gas distributor design, etc. In physical terms, the gas/liquid transport resistance is much larger than either of the other resistances, and the observed rate is completely determined by how rapidly A can be transferred from the gas to the liquid.

The concentration gradients for this case are shown in Figure 10A. Note that all of the gradient occurs across the gas-liquid films, i.e., all of the available concentration driving force is expended in getting the reactant from the bulk gas phase to the bulk liquid phase. The reactant concentration is essentially zero in the bulk liquid and throughout the catalyst particles. Gas/liquid mass transfer is not nearly rapid enough to keep the liquid saturated with A.

For a situation such as that described above, the reaction is said to be controlled by gas/liquid mass transport.

Case B ($6k_{ls}W/d_p\rho_p \ll \eta\, k_rW$ and $k_{gl}A_{gl}$)

If $6k_{ls}W/d_p\rho_p$ is much less than either ηk_rW or $k_{gl}A_{gl}$, then $d_p\rho_p/6k_{ls}W$ will be much larger than either of the other resistance terms in Equation 10. In physical terms, this corresponds to a situation where the liquid/solid transport resistance is much larger than either the gas/liquid resistance or the resistance to chemical reaction/pore diffusion.

In this case, Equation 10 reduces to:

$$-r_A = \frac{6k_{ls}W}{d_p\rho_p} \quad C^*_{A,g} \tag{12}$$

This expression shows that the only variables which affect the overall rate are those which influence liquid/solid transport, such as the catalyst weight, W, the external particle area (via d_p), and the transport coefficient, k_{ls}. The rate does not depend on parameters such as k_r, k_{gl}, C_B or A_{gl}.

The concentration gradients for this case are shown in Figure 10B. All of the gradients occur across the liquid film at the catalyst surface, and all of the available concentration driving force is expended in getting the reactant from the bulk liquid to the particle surface. The reactant concentration is essentially zero throughout the catalyst particle.

For this situation, the overall rate is completely determined by the rate of liquid/solid transport, and the reaction rate is said to be controlled by mass transport of Reactant A from the bulk liquid to the external surface of the catalyst

FIGURE 10

CONCENTRATION PROFILES FOR VARIOUS RATE-CONTROLLING STEPS

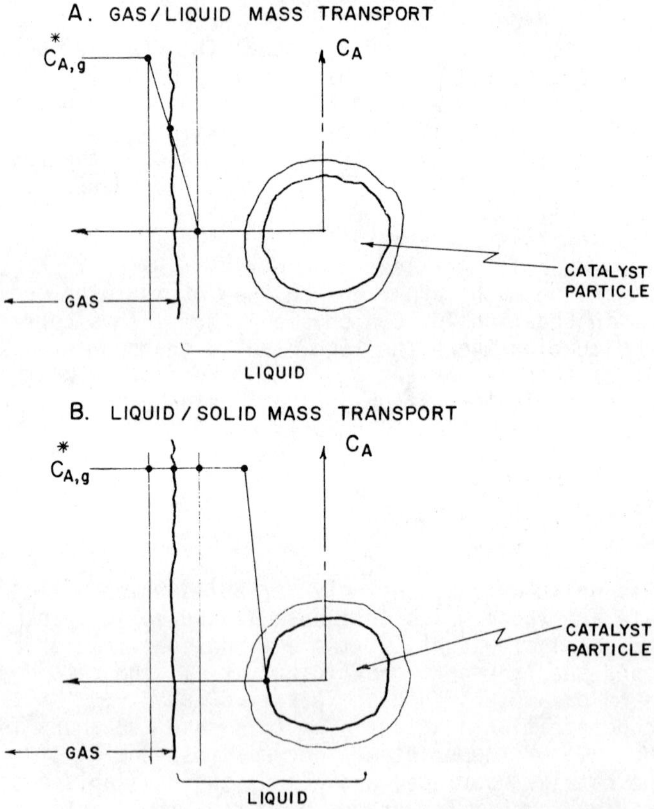

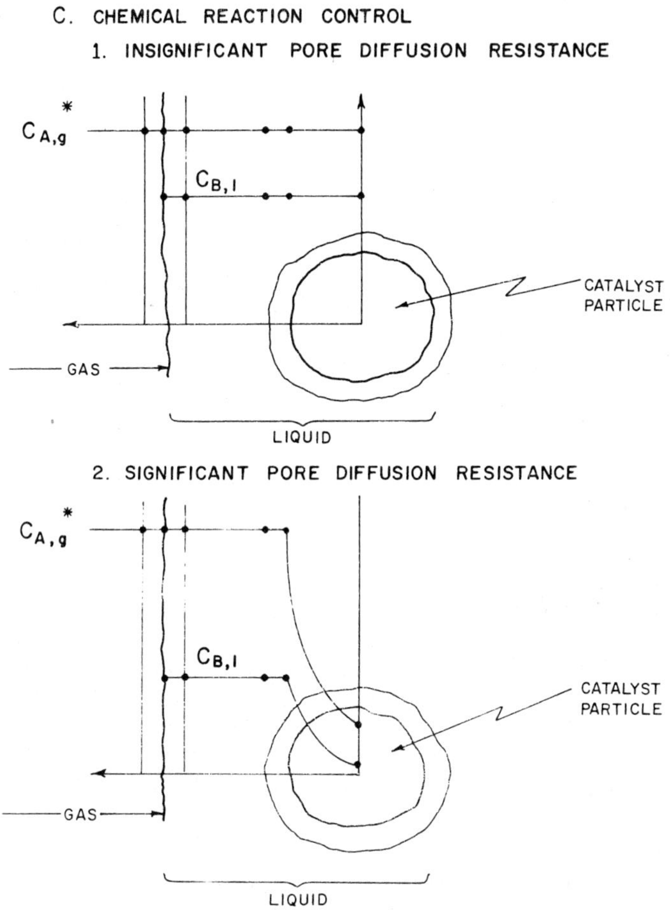

particles.

Case C

The overall reaction can also be controlled by liquid/solid transport of Reactant B. The gradient picture is very similar to that shown in Figure 10, except that Reactant B is involved instead of Reactant A. The overall rate equation is:

$$-r_A = \frac{6k'_{ls}W}{d_p\rho_p} \; C_{B,l} \qquad (13)$$

In order for this situation to occur, the maximum rate of liquid/solid transport of B would have to be significantly less than the maximum rates of the three individual steps for Reactant A, i.e., $k_{ls}A_{ls}C_{B,l} << k_{ls}A_{ls}C^*_{A,g}$ or $k_{gl}A_{gl}C^*_{A,g}$ or $k_rWC^*_{A,g}$.

Case D ($\eta k_rW << 6k_{ls}W/d_p\rho_p$ and $k_{gl}A_{gl}$)

If $(1/\eta k_rW)$ is significantly greater than the other resistances, then Equation 10 reduces to:

$$-r_A = \eta k_rWC^*_{A,g} \qquad (14)$$

In physical terms, the chemical reaction/pore diffusion resistance is much larger than either the gas/liquid or liquid/solid resistances, and the observed rate is determined by the rate of reaction inside the catalyst particles.

In order to develop a picture of the concentration gradients for this case, it is necessary to distinguish between two sub-cases: 1) insignificant pore-diffusion resistance ($\eta \cong 1.0$), and; 2) significant pore diffusion resistance ($\eta \leq 0.50$).

Figure 10C.1 shows the concentration profiles for the case where the resistance to pore diffusion is insignificant. This corresponds to the "ideal" case defined earlier in Figure 6, where none of the driving force is used up by the transport steps, and the full reactant concentration is available at all points within the catalyst particle.

The situation shown in Figure 10C.1 is an example of a system where the overall reaction is controlled by the intrinsic chemical kinetics. None of the various transport steps have any significant influence on the observed rate.

Figure 10C.2 shows the concentration gradients in a system where the overall rate is controlled by chemical reaction within the catalyst particle, but where the pore diffusion resistance is significant. All of the concentration decline occurs within the catalyst particle, and the driving force for reaction is significantly lower in the center of the particle than at the catalyst surface.

The controlling mechanism for this case is not easy to describe in a few words. To be strictly correct, one should say that the overall reaction rate is controlled by a combination of pore diffusion and chemical reaction, since these two processes occur in parallel. However, this case is frequently referred to as "pore diffusion control", even though pore diffusion does not strictly control the reaction, in the same sense as gas/liquid and liquid/solid mass transport.

The observed rate of reaction will be sensitive to different experimental variables, depending on which of the above steps control the overall rate. A summary of how the rate can be expected to respond to the primary experimental variables, for various controlling steps, is presented in Table II.

Several of these effects deserve emphasis:

1) The primary effect of improved agitation is to increase the gas/liquid interfacial area, A_{gl}. Consequently, the rate given by Equation 11 will depend strongly on agitation. The parameters in Equations 12 through 14 are at best weak functions of agitation, so that the rates given by these equations will not vary significantly with agitation. In short, improved agitation will increase the observed reaction rate only when gas/liquid mass transport is at least partially rate controlling.

2) Decreasing the diameter (d_p) of the catalyst particles increases the effectiveness factor, η, and the liquid/solid mass-transfer coefficient, k_{ls}. Equations 12 through 14 show that reducing the particle size will increase the observed rate when the reaction is controlled by either liquid/solid mass transport or by pore diffusion. Contrary to popular opinion, reducing the particle size is not a definitive test for pore diffusion alone.

3) Varying the pore structure of the catalyst will generally have an effect on the pore-diffusion resistance, but this effect is very complex. Alteration of pore structure can be very useful as a problem-solving tool, but it is not a reliable diagnostic test.

4) Temperature has a strong effect on only one parameter, the intrinsic rate constant, k_r. The transport coefficient k_{gl} and k_{ls} are generally weak functions of temperature. Therefore, temperature has a major influence on the observed rate only when chemical reaction is at least partially controlling.

5) If gas/liquid mass transport is controlling, the observed rate will not depend on the amount of catalyst used. However, the rates of the other possible controlling steps are all directly proportional to the catalyst charge.

These facts form the basis for several quantitative diagnostic tests, which are treated in the following section.

QUANTITATIVE ANALYSIS OF CONTROLLING STEPS

Since different controlling steps can cause dramatic differences in observed catalyst behavior, it is important to quantitatively determine the rate-controlling step(s) as early as possible in any experimental program. Such a determination usually involves a combination of experimental and calculative tests.

Experimental Approaches

Stirring Rate

A very direct test of the importance of gas/liquid transport can be made by running several experiments at different agitation rates, with all other variables constant. If the observed reaction rate depends on agitation, this constitutes strong evidence that gas/liquid mass transport controls, or at least influences, the reaction. If no effect is observed, the resistance to gas/liquid mass transport is probably insignificant compared to at least one other resistance.

Figure 11 shows such a series of experiments for nitrobenzene hydrogenation over a 3% palladium on carbon catalyst. At low stirring speeds, the agitation rate has a pronounced effect on the observed reaction rate. Clearly, the rate of nitrobenzene disappearance at 800 rpm. is significantly less than at, say, 2000 rpm. Since the quantity $k_{gl}A_{gl}$ increases with increased stirring rate, the increase in observed nitrobenzene reaction rate with increasing agitation is in line with theory.

At stirring speeds above about 2000 rpm., the observed reaction rate is no longer sensitive to agitation. This strongly suggests that the gas/liquid mass-transport resistance at this point has been reduced to the point where some other resistance is controlling.

To avoid a strong influence by gas/liquid mass transport in this system, experiments must be conducted at stirring speeds in excess of 2000 rpm. However, if the catalyst or the reaction conditions are changed, these agitation experiments must be repeated, since the "critical" stirring rate will depend on the overall reaction rate.

Amount of Catalyst

Consider a series of experiments where the reaction rate is measured as a function of the amount of catalyst charged to the reactor, with all other conditions held constant. The analysis of the data from such a series of experiments is based on Equation 10, which can be rearranged to give:

$$\frac{1}{-r_A} = \frac{1}{W} \quad \frac{1}{\eta k_r C^*_{A,g}} + \frac{d_p \rho_p}{6 k_{ls} C^*_{A,g}} + \frac{1}{k_{gl} A_{gl} C^*_{A,g}} \qquad (15)$$

If the model represented by this equation is valid, a plot of $1/-r_A$ versus $1/W$ should be a straight line with:

$$\text{Intercept} = 1/k_{gl}A_{gl}C^*_{A,g}$$

$$\text{Slope} = \frac{1}{\eta k_r C^*_{A,g}} + \frac{d_p \rho_p}{6 k_{ls} C^*_{A,g}}$$

FIGURE 11

EFFECT OF AGITATION RATE ON OBSERVED REACTION RATE FOR NITROBENZENE HYDROGENATION

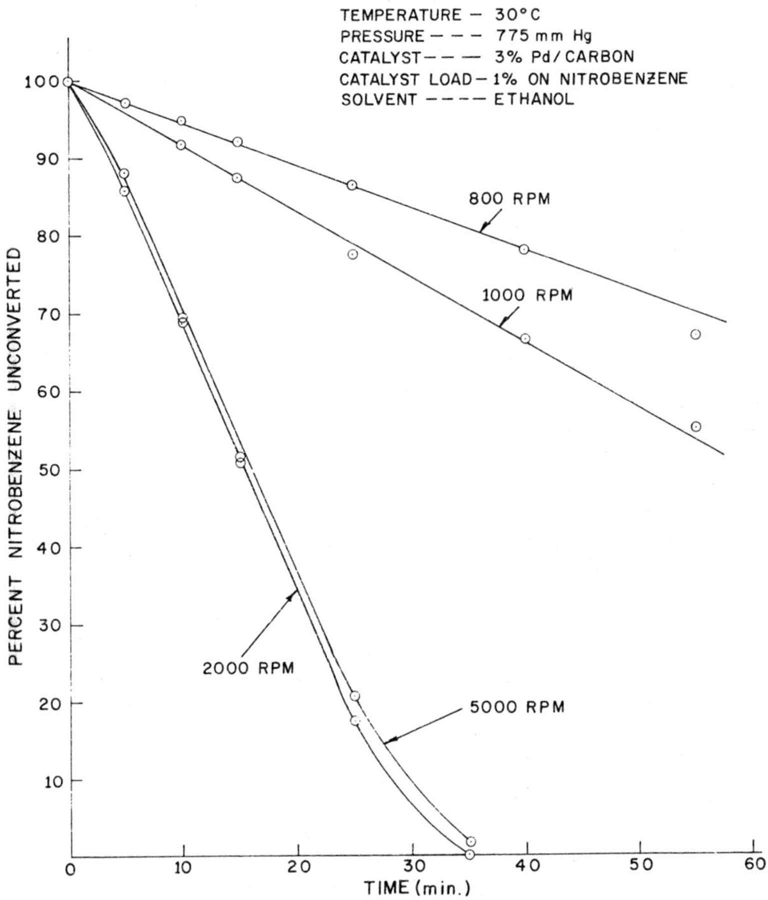

It can be seen from Equation 11 that the intercept is exactly equal to the rate that would exist if the reaction were controlled by gas/liquid mass transport. To be certain that gas/liquid mass transport is not controlling, the actual experimental value of $1/-r_A$ should be much larger than the intercept.

Figure 12 is a plot based on Equation 14 for nitrobenzene

FIGURE 12

EFFECT OF CATALYST LOADING ON OBSERVED REACTION RATE FOR NITROBENZENE HYDROGENATION

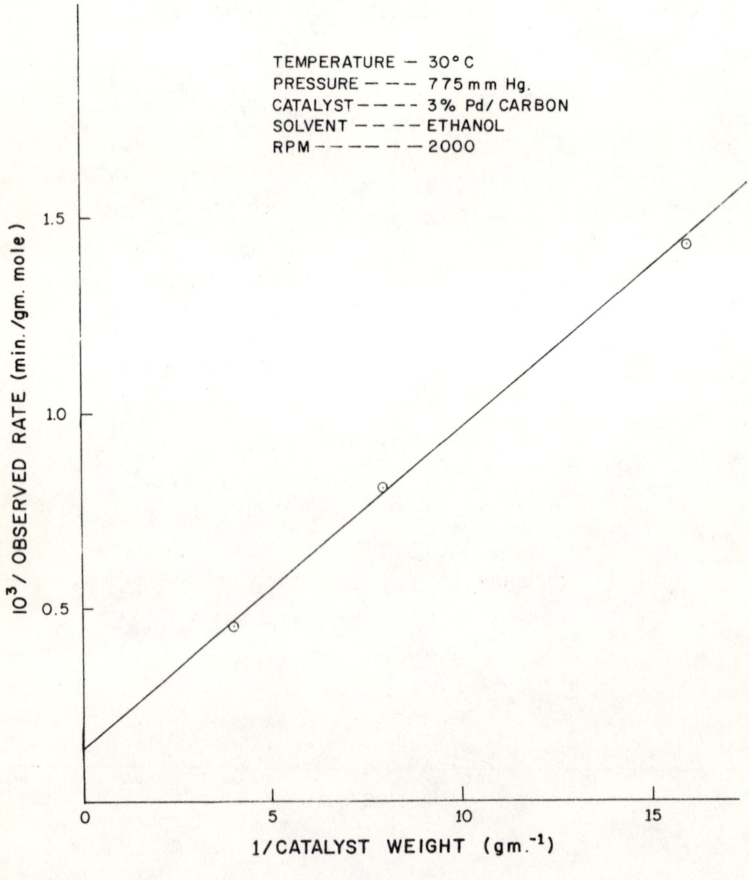

hydrogenation. Note that the intercept of this plot is small compared to the value of $1/-r_A$ for the run with the lowest catalyst loading. The actual values are 0.14×10^3 min./gm. mole for the intercept and 1.43×10^3 min./gm.mole for the experiment. These numbers show that the gas/liquid resistance for this run was only about 10% of the total resistance. Therefore, transfer of hydrogen from the gas to the bulk liquid did not control the reaction rate.

For the run with the highest catalyst loading (the point farthest to the left on Figure 12), the intercept is about 30% of the actual value of the ordinate, 0.14×10^3 versus 0.45×10^3 min./gm.mole. Although transfer of hydrogen from the gas to the liquid is not truly controlling in this case, it nevertheless constitutes an important resistance, which probably does influence the observed reaction rate.

Figure 13, which is a similar plot for the hydrogenation of thiophene over a modified cobalt catalyst, shows a situation which is much closer to a true case of gas/liquid mass transport control. Note that the value of the intercept is more than 70% of the actual value of $1/-r_A$ for the run with the highest catalyst loading.

The slope of a plot based on Equation 14 is proportional to the sum of the resistances to liquid/solid mass transport and chemical reaction/pore diffusion. However, there is no obvious way at present to separate these two terms. Even if the intercept of a plot of $1/-r_A$ versus $1/W$ is sufficiently small, it is still not possible to determine whether liquid/solid mass transport or chemical reaction/pore diffusion, or some combination of the two, controls the overall rate. Additional information is required for this judgment to be made.

Temperature Effect

The mass-transfer coefficients k_{gl} and k_{ls} in Equation 11 through 13 are weak functions of temperature; the apparent "activation energy" for these coefficients is usually less than 5 kcal./mole. When the overall reaction is controlled by either gas/liquid or liquid/solid mass transport, the apparent activation energy for the reaction will be the same as for the appropriate mass-transfer coefficient, provided that the temperature dependence of $C^*_{A,g}$ is properly taken into account.

By way of contrast, the intrinsic rate constant, k_r, generally has an activation energy of 10 kcal/mole or greater. If the observed activation energy exceeds this value, the reaction is almost certainly limited by chemical reaction itself, either with or without pore diffusion.

When η is small ($<\sim 0.5$), the apparent activation energy for a reaction which is controlled by chemical reaction within

FIGURE 13

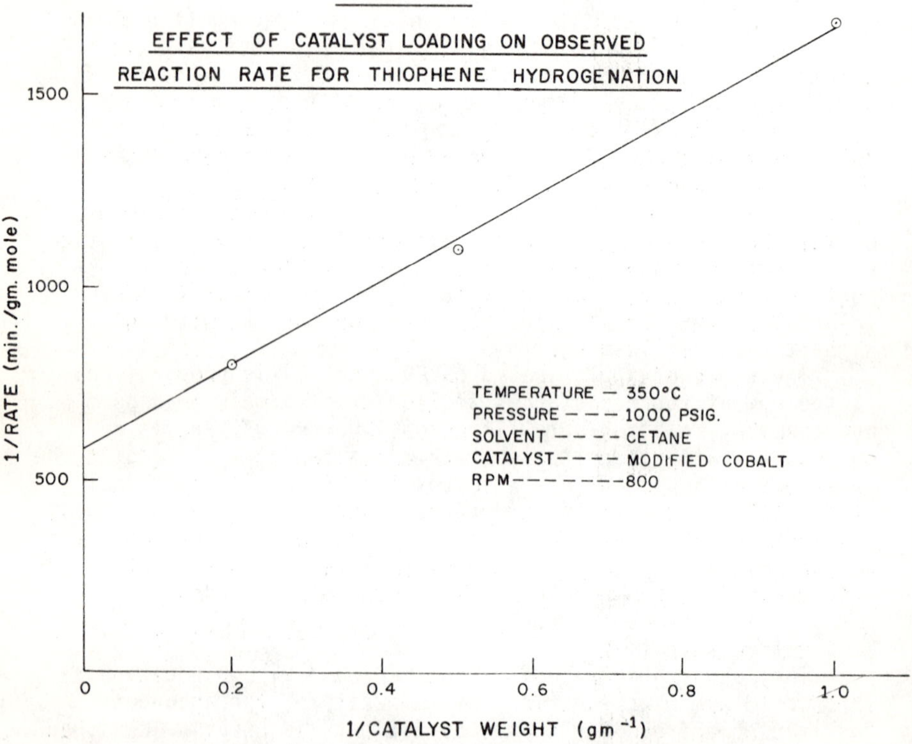

the particle is reduced to about half of its intrinsic value. Thus, an observed activation energy of 10 kcal/mole could arise from a case where $\eta \cong 1.0$ with an intrinsic activation energy of 10 kcal mole, or from a case where $\eta \lesssim 0.5$ with an intrinsic activation energy of 20 kcal./mole.

In summary, if the observed activation energy is less than 5 kcal/mole, this is rather strong evidence that the reaction is controlled by either gas/liquid or liquid/solid mass transfer, or some combination of the two. If the observed activation energy is greater than 10 kcal/mole, the reaction is almost certainly controlled by chemical reaction within the particle. However, the importance of pore diffusion cannot be ascertained solely from the observed temperature effect.

Measurement of the reaction rate at various temperatures is an effective tool for determining the relative importance of the two terms, $1/\eta k_r C^*_{A,g}$ and $d_p\rho_p/6k_{ls}C^*_{A,g}$, which make up the value of the slope in Figures 12 and 13. If the slope is a strong function of temperature, $1/\eta k_r$ is undoubtedly much larger than $d_p\rho_p/6k_{ls}$. However, if the slope is a very weak function of temperature, the reverse is true.

Calculative Approaches

In order to quantitatively determine the importance of: 1) liquid/solid mass transport, and; 2) pore diffusion, it is frequently useful to perform some simple calculations, as described below:

Liquid/Solid Mass Transport

The theoretical rate of reaction for a situation where the reaction is completely controlled by liquid/solid mass transport is first calculated from Equation 12. The theoretical reaction rate is then compared with the experimentally-measured rate. If the two values are in close agreement, this is strong evidence that the reaction is controlled by the assumed step. However, if the two values are seriously divergent, some other step must be controlling.

a. Estimation of k_{ls}

The rate of mass transport between suspended solid particles and the surrounding fluid has been studied by a number of researchers, and values of the mass-transfer coefficient have been measured and correlated. Perhaps the most reliable correlation (Brian and Hales, 1969) is:

$$\left(\frac{k_{ls}d_p}{D_A}\right)^2 = 4.0 + 1.21\ N_{Pe}^{2/3} \tag{16}$$

In this equation, N_{Pe} is a dimensionless group called the Peclet Number, which is defined by

$$N_{Pe} = \frac{d_pV}{D_A} \tag{17}$$

In Equations 16 and 17, D_A is the diffusivity of Reactant A in the liquid and V is the velocity of the solid particle <u>relative to the surrounding liquid</u>. A value of D_A can usually be esti-

TABLE II

VARIABLES AFFECTING OBSERVED REACTION RATE

Controlling Step	Variables With: Major Influence	Minor Influence	Insignificant Influence
Gas/Liquid Mass Transport	Stirring Rate Reactor Design (impellor, gas distributor, baffling, etc.) Concentration of reactant in gas phase	Temperature	Concentration of Liquid-Phase Reactant Amount of Catalyst Catalyst Particle Size Concentration of Active Component(s) on Catalyst
Liquid/Solid Mass Transport (Gaseous Reactant)	Amount of Catalyst Catalyst Particle Size Concentration of reactant in gas phase	Temperature Stirring Rate Reactor Design Viscosity Relative Densities	Concentration of Liquid-Phase Reactant Concentration of Active Component(s) on Catalyst
Liquid/Solid Mass Transport (Liquid Reactant)	Amount of Catalyst Catalyst Particle Size Concentration of reactant in Liquid Phase	Temperature Stirring Rate Reactor Design Viscosity Relative Densities	Concentration of Gas-Phase Reactant Concentration of Active Component(s) on Catalyst

TABLE II (Cont d.)

Controlling Step	Variables With: Major Influence	Minor Influence	Insignificant Influence
Chemical Reaction (Insignificant Pore Diffusion Resistance)	Temperature Amount of Catalyst Reactant Concentrations Concentration of Active Component(s) on Catalyst		Stirring Rate Reactor Design Catalyst Particle Size
Chemical Reaction (Significant Pore Diffusion Resistance)	Amount of Catalyst Reactant Concentrations Temperature* Catalyst Particle Size Concentration of Active Component(s) on Catalyst*	Pore Structure	Stirring Rate Reactor Design

*These variables do not exert as strong an influence as when pore diffusion resistance is negligible.

mated with fair accuracy via the methods presented by Reid and Sherwood (1966).

The most difficult task in applying Equation 16 is estimating the value of V in Equation 17. Most powdered catalysts are quite small ($<50\mu$), and particles of this size tend to move with the liquid. Even with intense agitation, it is difficult to increase the liquid velocity relative to the particle. Increasing the agitation rate only serves to make the liquid move faster.

Gravity is the principal force which causes the catalyst particles to move relative to the liquid. Since the catalyst generally has a higher density than the surrounding liquid, the particles tend to settle down through the fluid. This settling is periodically interrupted when the agitator circulates fluid from the bottom of the reactor to the top, so that at any point in time the catalyst particles are uniformly distributed throughout the liquid phase.

The terminal velocity of a particle settling through a fluid is given by Stokes' Law

$$V = \frac{g d_p^2 \, (\rho_a - \rho_1)}{18\, \mu} \tag{18}$$

where ρ_1 is the liquid density, ρ_a is the apparent density of the catalyst particle in the liquid, and μ is the viscosity of the liquid. Combining Equations 16,17, and 18 gives:

$$\left(\frac{k_{1s} d_p}{D_A}\right)^2 = 4.0 + 1.21 \left(\frac{g d_p^3 (\rho_a - \rho_1)}{18 \mu D_A}\right)^{2/3}$$

In actuality, mechanical agitation and/or the gas bubbles rising through the liquid tend to make the value of the mass-transfer coefficient somewhat higher than predicted by the above equation. To adjust for this effect, it is reasonable to use a value of k_{1s} about twice the predicted value. Building this factor of two into the above equation gives:

$$\left(\frac{k_{1s} d_p}{D_A}\right)^2 = 16 + 4.84 \left(\frac{g d_p^3 (\rho_a - \rho_1)}{18 \mu D_A}\right)^{2/3}$$

b. Estimation of reaction rate for liquid/solid mass transfer control.

If the overall reaction is controlled by liquid/solid mass transport, the observed rate will be given by Equation 12. If the superscript $\sim$ is used to denote a predicted value, Equation

12 becomes:

$$-\tilde{r}_A = \frac{6\tilde{k}_{ls}W}{d_p\rho_p} C^*_{A,g} \tag{20}$$

In this equation, $\tilde{k}_{ls}$ is the value of the mass-transfer coefficient predicted from Equation 19.

The next step is to compare the predicted rate, $-\tilde{r}_A$, with the experimentally-observed rate, $-r_A$. If these two quantities are close, say within a factor of 2, it is likely that the overall reaction rate is controlled by liquid/solid transport of Reactant A.

If the two rates, $-\tilde{r}_A$ and $-r_A$, are not close, another calculation must be made before the hypothesis of liquid/solid transport control can be rejected. Specifically, the possibility of control by liquid/solid transport of Reactant B must be checked, in a manner exactly analogous to that described above. First, k'_{ls} is calculated for Species B from Equation 19. Next, $-\tilde{r}_B$ is calculated from the analog of Equation 13, i.e.,

$$-\tilde{r}_B = \frac{6\tilde{k}'_{ls}W}{d_p\rho_p} C_{B,l}$$

Finally, the value of $-\tilde{r}_B$ is compared to the experimentally-observed rate, $-r_B$. If the values compare to within a factor of about 2, it is likely that liquid/solid transport of Reactant B controls the reaction rate. However, if these rates are more than a factor of 10 different, some other step is almost certainly controlling.

An example of the application of this calculational approach is given in Appendix A.

Pore Diffusion

It is impossible in this paper to even begin to review the extensive literature on pore diffusion which has developed over approximately the last two decades. A comprehensive survey of the subject as it existed prior to 1970 has been presented by Satterfield (1970). In the following treatment, attention is focused on developing a single, simple approach which is a sufficient starting point for the analysis of most data from gas/liquid/solid catalytic systems.

For most reactions where the pores of the catalyst are filled with liquid, temperature gradients inside the catalyst particle are negligible. For this case, the effectiveness factor, η, is a function of only two dimensionless variables, the reaction order, n, and a dimensionless modulus, Φ, which is defined by:

$$\Phi = \frac{d_p^2}{4D_{A,e}} \left(\frac{-r_A}{V_c}\right) \frac{1}{C_{A,1}}$$

In this equation, $D_{A,e}$ is the effective diffusivity of Reactant A inside the porous structure of the catalyst, and V_c is the total geometric volume of catalyst particles. Therefore, the quantity $(-r_A/V_c)$ is the observed rate of reaction per unit volume of catalyst particles.

As a first approximation, the effective diffusivity may be estimated from the equation:

$$D_{A,e} = \frac{\theta D_A}{\tau}$$

where θ is the porosity of the catalyst particle and τ is the "tortuosity factor", which typically has a value of about 4. Substituting the above equation, with $\tau = 4$, into the defining equation for Φ gives

$$\Phi = \frac{d_p^2}{D_A\theta} \left(\frac{-r_A}{V_c}\right) \frac{1}{C_{A,1}} \qquad (21)$$

All of the quantities on the right-hand side are either known from the experimental data, i.e., d_p, θ, $-r_A$, V_c, $C_{A,1}$, or can be estimated from physical-property correlations, i.e., D_A. If gas/liquid mass transport is rapid, $C_{A,1}$ can be taken to be equal to $C^*_{A,g}$. However, if the assumption of rapid gas/liquid transport is not valid, the value of $C_{A,1}$ can be estimated from Equation 6.

Figure 14 is a plot of N versus Φ for a first-order reaction (n=1). The curves for different reaction orders are slightly displaced from the one shown on Figure 14, but for the purposes of a first approximation, the differences are not significant.

Once Φ has been has been calculated, Figure 14 can be used directly to estimate the value of η. Note that if the value of Φ is less than about 1, the effectiveness factor is greater than about 0.95, and pore diffusion effects are insignificant. For $\Phi > 10, \eta < 0.5$ and pore diffusion exerts a very significant influence on the reaction.

An example of the use of Figure 14 to estimate the importance of pore diffusion is given in Appendix B.

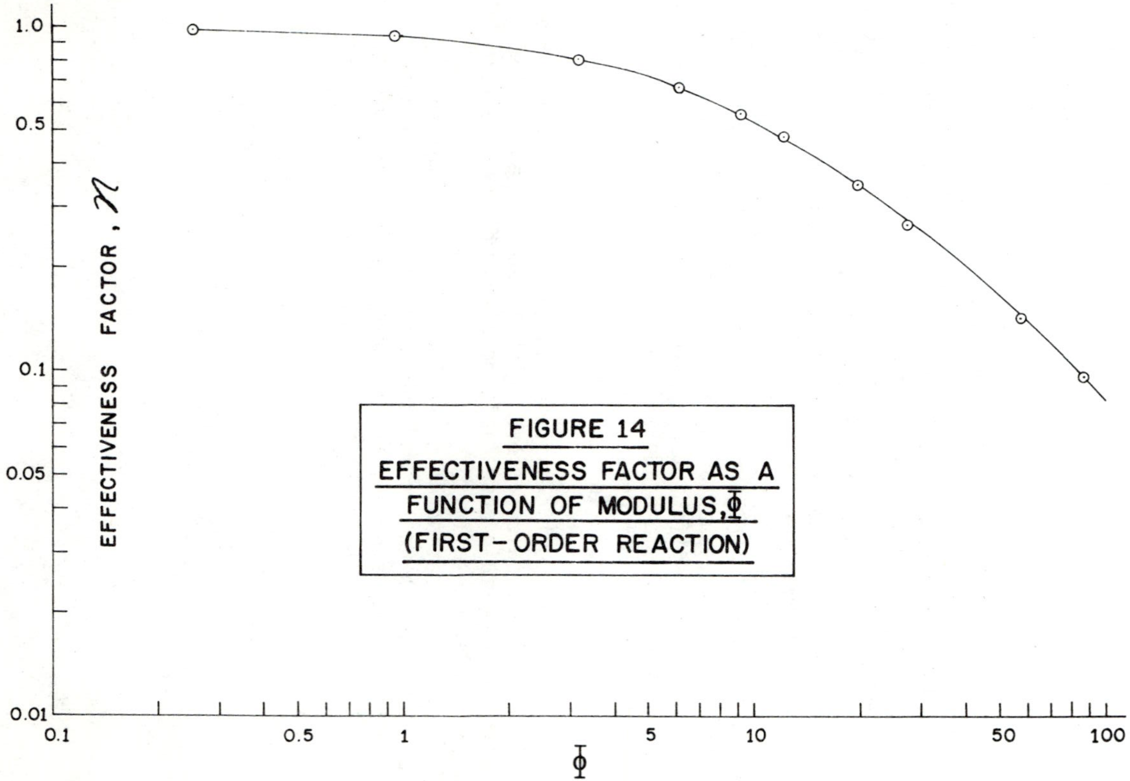

FIGURE 14

EFFECTIVENESS FACTOR AS A FUNCTION OF MODULUS, Φ (FIRST-ORDER REACTION)

CONCLUSION

This paper is an attempt to condense the very complex subject of chemical reaction-transport phenomena interactions into an elementary form that can be easily utilized by the catalytic chemist. Some broad approximations are necessary in any such attempt, and there are important exceptions to some of the generalizations which have been made. Nevertheless, the value of having a simple, easily-applied set of procedures for analysis of reaction/transport interactions is believed to far outweigh any deficiencies caused by the need to generalize.

Some basic truths have been discussed in this paper, which should be kept in mind as an aid to understanding the many complex phenomena involved in multiphase catalysis:

1) Temperature and concentration gradients will always exist in multiphase catalytic systems. The magnitude of these gradients depends on the reaction rate, with faster reactions causing steeper gradients.

2) These gradients can influence the apparent activity, selectivity, and life of heterogeneous catalysts; and changes in physical factors, such as agitation rate, which affect the gradients, can also affect the apparent behavior of the catalyst.

3) Several simple diagnostic experiments can help the catalytic researcher to understand which, if any, transport steps play a major role in influencing the apparent behavior of the catalyst.

4) These diagnostic experiments can be effectively supplemented with simple calculations which allow a more detailed exploration of certain rate-limiting steps which are difficult to attack experimentally.

Careful application of these tools should allow the catalytic researcher to understand the factors which influence the reaction being studied, and to design experiments which will reflect true catalyst behavior.

Appendix A - Calculative Test for Liquid/Solid Mass Transport Control

Consider the data point farthest to the right on Figure 12. For this point, the experimentally-observed rate was 0.70 x 10^{-3} gm. moles of nitrobenzene per minute, or 2.10 x 10^{-3} gm. moles of H_2/min.

The amount of catalyst used was 0.0615 grams, with an average particle size of 16μ. The particle density (in air) (ρ_p) was about 0.95 gm./cm.^{3} and the porosity or fractional pore volume (θ) was about 0.50. Since the density of ethanol (ρ_l) at reaction conditions is about 0.78 gm./cm.^{3}, the apparent density (ρ_a) of the catalyst when its pores are filled with ethanol is about 1.3 gm./cm.^{3} (0.95 + 0.5 x 0.78).

The viscosity of ethanol (μ) at reaction conditions is about 1.0 centipoises, or 0.01 gm./cm.sec. Since the concentration of nitrobenzene in the liquid is relatively small (∿4 mole %), the density and viscosity of the liquid will be assumed to be the same as those of pure ethanol.

Consider that H_2 is the species being transported. The diffusion coefficient of H_2 in ethanol (D_A) is about 5.0 x 10^{-5} cm.^{2}/sec. and the equilibrium solubility of H_2 in ethanol ($C^*_{A,g}$) is about 3.5 x 10^{-7} gm. moles/cm.^{3}. These values will be assumed to apply to the actual liquid in the reactor.

The mass-transfer coefficient is estimated from Equation 19

$$\left(\frac{k_{ls} d_p}{D_A}\right)^2 = 16 + 4.84 \left(\frac{g d_p^3 (\rho_a - \rho_l)}{18 \mu D_A}\right)^{2/3} \qquad (19)$$

Substituting the above values into this equation gives:

$$\left(\frac{k_{ls} \times 16 \times 10^{-4}}{5 \times 10^{-5}}\right)^2 = 16 + 4.84 \left(\frac{981 \times 16^3 \times 10^{-12} \times (1.30 - 0.78)}{18 \times 0.01 \times 5 \times 10^{-5}}\right)^{2/3}$$

which reduces to

$$\tilde{k}_{ls} = 0.132 \text{ cm./sec.}$$

The rate for this reaction, if it were controlled by liquid/solid mass transport of H_2 would be:

$$-\tilde{r}_A = \frac{6 \tilde{k}_{ls} W C^*_{A,g}}{d_p \rho_p} \qquad (20)$$

Substituting the appropriate values gives

$$-\tilde{r}_A = \frac{6 \times 0.132 \times 0.0615 \times 3.5 \times 10^{-7}}{16 \times 10^{-4} \times 0.95} \text{gm.moles } H_2\text{/sec.}$$

$$-\tilde{r}_A = 0.67 \times 10^{-3} \text{ gm. moles } H_2 \text{ min.}$$

Comparing the theoretical and experimental rates,

$$-\tilde{r}_A \text{ (calculated)} = 0.67 \times 10^{-3} \text{ gm.moles } H_2\text{/min.}$$
$$-r_A \text{ (experimental)} = 2.1 \times 10^{-3} \text{ gm.moles } H_2\text{/min.}$$

These values are reasonably close, considering the approximations involved in the calculations. This agreement suggests that the reaction rate for this run was significantly influenced, and possibly controlled, by transport of dissolved H_2 from the bulk liquid to the external surface of the catalyst particles.

In an actual situation, the above calculation should be repeated for the assumption that the reaction is controlled by liquid/solid transport of nitrobenzene. Such a calculation is exactly analogous to the one shown above, except that D_A is replaced by D_B, the diffusion coefficient for nitrobenzene in ethanol, and $C^*_{A,g}$ is replaced by $C_{A,1}$, the initial concentration of nitrobenzene in ethanol.

Appendix B - Estimation of Effectiveness Factor

Consider the same data point treated in Appendix B. The modulus Φ is calculated from Equation 20

$$\Phi = \frac{d_p^2}{D_A\theta} \left(\frac{-r_A}{V_c}\right) \frac{1}{C_{A,1}} \tag{21}$$

The quantity $-r_A/V_c$ is given by $-r_A\rho_p/W$ so that

$$\Phi = \frac{d_p{}^2}{D_A\theta} \left(\frac{-r_A\rho_p}{W}\right) \frac{1}{C_{A,1}}$$

Substituting appropriate values from Appendix A gives

$$\Phi = \frac{(16 \times 10^{-4})^2 \times (2.10 \times 10^{-3}/60) \times 0.95}{5 \times 10^{-5} \times 0.50 \times 0.0615 \times 3.5 \times 10^{-7}} = 160$$

From Figure 14, for this value of Φ, $\eta \cong 0.06$. Therefore, the resistance to pore diffusion is very significant.

The above calculation assumes that the concentration of hydrogen at the surface of the pellet is the same as that in the bulk liquid. However, Appendix A suggests that the hydrogen concentration does drop significantly between the bulk liquid and the catalyst surface. If so, a lower value should have been used for $C_{A,l}$ in Equation 21. This would not affect the overall conclusion, however, since reducing $C_{A,l}$ would cause Φ to increase and η to decrease.

ACKNOWLEDGEMENT

The author wishes to thank Engelhard Industries for permission to publish this paper, and gratefully acknowledges the contributions of his colleagues, Dr. Miron Abramovici and Dr. Eric W. Stern, whose experimental skill made possible the examples used in this paper.

NOMENCLATURE

English letters:

A - area (cm.^{2})
a - individual reaction order (dimensionless)
b - individual reaction order (dimensionless)
C - concentration (gm.moles/cm.^{3})
C^* - concentration of solute in liquid in equilibrium with concentration in gas (gm.moles,cm.^{3})
D - bulk diffusion coefficient in liquid (cm.^{2}/sec.)
D_e - diffusion coefficient in pores of catalyst (cm.^{2}/sec.)
d_p - diameter of catalyst particle (cm.)
g - acceleration due to gravity (cm./sec.^{2})
h - heat-transfer coefficient (cal/sec.,cm.,$^\circ$C)
k - mass-transfer coefficient (cm./sec.)
k_r - first-order reaction rate constant (cm.^{3}/sec.,gm.)
k_t - thermal conductivity (cal./sec.,cm.,$^\circ$C)
N - rate of mass transport (gm.moles/sec.)
N_{Pe} - Peclet Number ($=d_pV/D_A$)(dimensionless)
n - overall reaction order (dimensionless)
q - rate of heat transport (cal./sec.)
$-r$ - rate of chemical reaction (gm.moles/sec.)
T - temperature ($^\circ$C)
V - liquid velocity (cm./sec.)
V_c - volume of catalyst (cm.^{3})
W - weight of catalyst (gm.)
x - distance (cm.)

Greek Letters:
η - effectiveness factor (dimensionless)
θ - porosity of catalyst particle (dimensionless)
μ - liquid viscosity (gm./cm.,sec.)
ρ_a - apparent density of catalyst particle (liquid-filled pores)(gm./cm.^{3})
ρ_l - density of liquid (gm./cm.^{3})
ρ_p - density of catalyst particle (air-filled pores)(gm./cm.^{3})
τ - tortuosity (dimensionless)
Φ - modulus (dimensionless)

Subscripts:
A - denotes species A
B - denotes Species B
g - refers to gas
gl - refers to gas/liquid interface
l - refers to liquid
ls - refers to liquid/solid interface
s - refers to solid

Superscripts:
' - denotes species B
$\sim$ - denotes a calculated value

REFERENCES

1. Brian, P.L.T., and Hales, H.B., Am. Inst. Chem. Eng. J. 15, 419 (1969)
2. Butt, J.B., in "Chemical Reaction Engineering", Advances in Chemistry Series, No. 109, American Chemical Society, Washington, D.C., 259 (1972)
3. Damkohler, G., Chem. Ing., 3, 430 (1937)
4. Frank-Kamenetskii, D.A., Zhur. Tekh. Fiz., 9, 1457 (1939)
5. Mears, D.E., J. Cat., 20, 127 (1971)
6. Reid, R.C., Sherwood, T.K., "Properties of Gases and Liquids", 2nd. ed., McGraw-Hill, N.Y. (1966)
7. Satterfield, C.N., "Mass Transfer in Heterogeneous Catalysis", MIT Press, Cambridge, Mass. (1970)
8. Satterfield, C.N., and Sherwood, T.K. "The Role of Diffusion in Catalysis", Addison-Wesley, Reading, Mass. (1963)
9. Weisz, P.B., and Prater, D.C., in "Advances in Catalysis", VI, Academic Press, N.Y., 143 (1954)
10. Wheeler, A., in "Advances in Catalysis", III, Academic Press, N.Y., 249 (1951)
11. Zeldovitsch, J.B., Acta Physicochim. U.R.S.S., 10, 583 (1939)

AUGER ELECTRON SPECTROSCOPY
A PASSING CURIOSITY OR A PRACTICAL SURFACE ANALYTICAL TOOL FOR HETEROGENEOUS CATALYSIS

MADAN M. BHASIN

Union Carbide Corporation
Chemicals and Plastics Research and Development Department
Technical Center, P. O. Box 8361
South Charleston, West Virginia 25303

ABSTRACT

In recent years, Auger† electron spectroscopy (AES) has been demonstrated to be a powerful tool for the analysis of surface and sub-surface elemental composition in the fields of thin films, semiconductors and metallurgy. In heterogeneous catalysis, surface analysis by AES has found little application, if any, although AES analysis has been used in determining surface purity in LEED-Auger studies of surface structures of single crystals and the nature of chemisorption processes on such surfaces. However, the high sensitivity (0.1 atom percent) to analysis of top few (1 - 5) atomic layers of surfaces and high speed of analysis (typically ~30 minutes for complete scan of the periodic table) of AES, made possible by recent advances in instrumentation, should make AES a very powerful tool for qualitative and quantitative analysis of heterogeneous catalysts.

Practical applications of AES to surface poisoning of two heterogeneous catalyst of commercial interest are described. One of these applications involved surface poisoning by lead of copper catalysts used in the synthesis of methylchlorosilanes (by reaction of methylchloride and silicon). The other application involved surface poisoning by iron of a commercial palladium-gamma-alumina catalyst used in the selective hydrogenation of diolefins in the presence of olefins and aromatics. Classical bulk analytical procedures had previously failed to explain the catalyst poisoning observed.

---

†Pronounced as "OJ".

INTRODUCTION

General Background

Auger electron spectroscopy (AES) is named after Pierre Auger (1925) who first discovered the tracks of Auger electrons in a Wilson Cloud Chamber while studying the "Photoelectric Effect". Several excellent reviews on the theory and applications of AES have been published recently, for example, Palmberg (1972 and 1973), Chang (1971), Pentenero (1971), Weber (1972) and Karasek (1974).

In a recent review of Electron Spectroscopy by Hercules and Carver (1974) covering the period of 1972 and 1973 there are 465 papers on ESCA† (Electron Spectroscopy for Chemical Analysis) as compared to 80 on AES. Furthermore, there are about 44 papers covering "Surfaces and Catalyst", ESCA studies and several papers dealing with the application of ESCA to commercial catalysts. In contrast, surface studies using AES deal mainly with adsorption on clean, single crystal surfaces, surface diffusion, surface segregation on grain boundaries, thin films, and semiconductors. This paper will attempt to show that AES is not just a passing curiousity but a very practical surface analytical tool for heterogeneous catalysis and that, in the near future, AES can potentially become as basic and as important a physical measurement as B.E.T. surface area in catalysis and adsorption. In this paper, the principles of AES will be briefly described with emphasis on the applicability of AES to heterogeneous catalysis. In later sections, application of AES to poisoning studies of two commercial, heterogeneous catalysts will be discussed.

What Is Auger Electron Spectroscopy?

Auger Electron Spectroscopy is a highly sensitive method for the chemical analysis of surface atoms. AES analysis is accomplished by energy analysis of the Auger electrons emitted from a sample that has been excited by an energetic beam of electrons having energies of 1-10 Kev. An Auger electron is generated when an atom that has lost one of its core electrons, say a K shell electron (ionization by the incident electron beam), relaxes to its doubly ionized equilibrium state by letting an L_1 electron fall into the K shell and simultaneously releasing an L_3 electron of kinetic energy equal approximately to $E_K - E_{L_1} - E_{L_3}$ (see Figure 1). The initially ionized state

†The term ESCA has become more popular although the term XPS more exactly represents the X-ray excited photoelectron spectroscopy.

FIGURE 1

Diagram Showing a KL_1L_3 Auger Transition

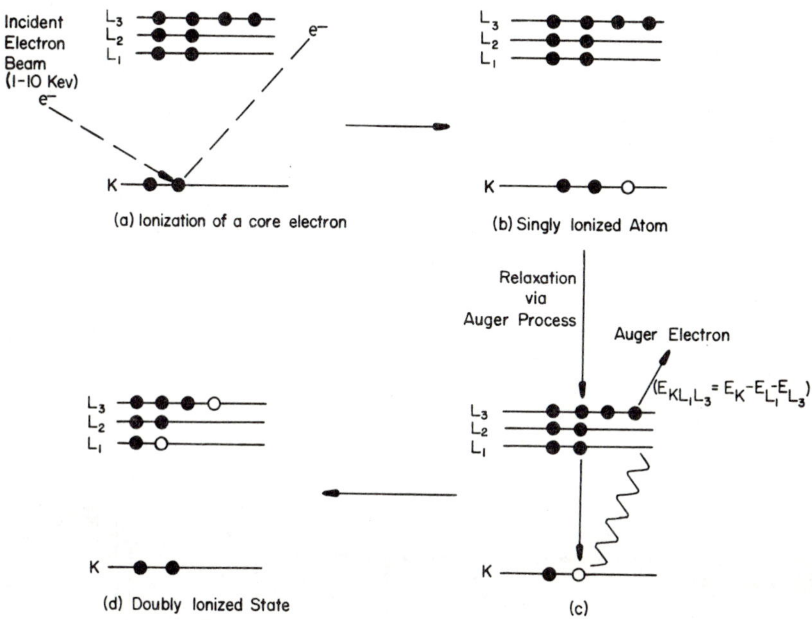

can also relax to its equilibrium state through the emission of an X-ray, thereby giving an X-ray emission spectrum (basis for electron probe analysis). By use of a low energy electron beam (≤4 Kev), the Auger transitions are made more favorable. The energy of an Auger electron is only dependent on electronic energy levels of the atom from which it comes, and is independent of the energy of the incident electron beam. An atom can exhibit several Auger transitions. Generally speaking, most of the elements exhibit one or two strong Auger transitions. The number of Auger transitions possessed by an atom increases with the atomic number of the element.

What is an Auger Spectrum?

An Auger spectrum is obtained when an energy analysis is performed on the secondary electrons emitted from a sample

FIGURE 2

SECONDARY ELECTRON ENERGY DISTRIBUTION CURVES,
N(E) and dN(E)dE from a Sample of Silver
Ep = 1 KV and Ip = 40 μA

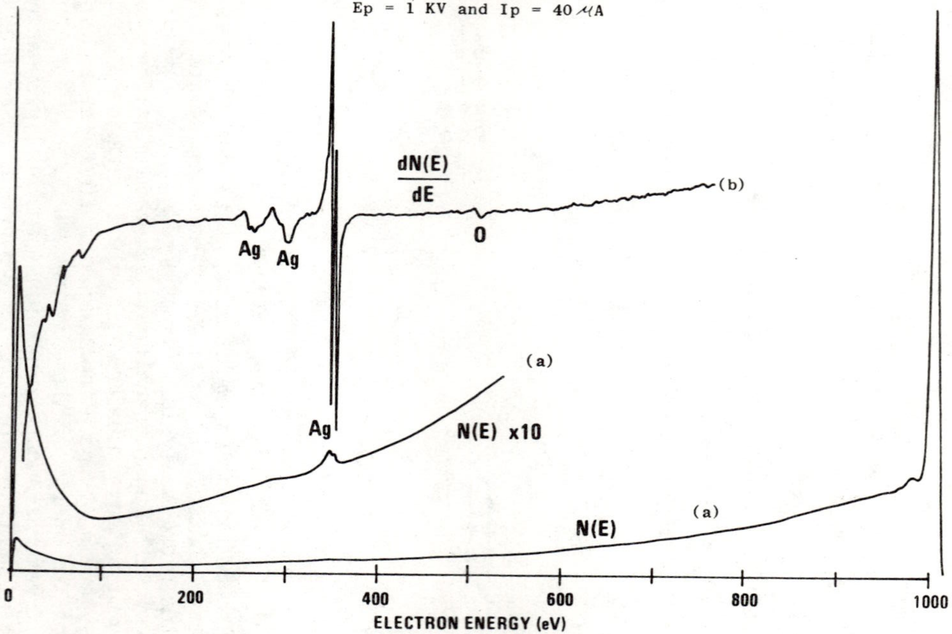

undergoing irradiation by a primary electron beam. A secondary electron energy distribution function, N(E), is plotted as a function of the energy(E) of the secondary electrons. Figure 2(a) shows such a spectrum for silver. Although Auger peaks of the transitions can be detected in such a plot of N (E) against E, their prominence is greatly enhanced in a plot of the differentiated function, $\frac{dN(E)}{d(E)}$, against E (Figure 2b). Differentiation primarily removes the large background caused by secondary electrons. The plot of $\frac{dN(E)}{d(E)}$ against E has now become known as the "Auger Spectrum". The negative peaks (or minima) in the Auger spectrum have been chosen to characterize the energies of the Auger transitions. These Auger peaks of individual elements are used as a fingerprint for characterizing an unknown sample in much the same manner as an infrared spectrum. Auger spectra of all of the elements (except H and He) have been obtained (Palmberg, et al, 1972b). The major Auger peaks of most of these elements have also been plotted in the form of a chart (see Figure 3), which serves as a quick reference guide for the Auger spectrum of each of the elements.

Sampling Depth

The sampling depth in AES is dependent on the escape depth of the Auger electrons which, in turn, is dependent on the electron-electron mean free path. The available literature indicates that for Auger peaks in the range of 50 to 2000 eV, the escape depth is in the range of 4 to 20 Å. Since a large number of Auger peaks occur at ≤1000 eV, the escape depth can be safely assumed to be below 10Å, which is approximately equivalent to only a few atomic layers. A compendium of the published data on relationship between escape depth and Auger transition energy has been made by Tracy and Palmberg (see Palmberg, 1973).

Analysis Time and Signal-to-Noise Ratio

In instruments equipped with a cylindrical mirror analyzer, typical analysis time for scanning all the elements (except H and He) of the periodic table is 15-30 minutes, without any substantial loss in signal-to-noise (S/N) ratio. Such a scan can detect elemental composition in the range of a few tenths of one atomic per cent. High sensitivity scans (to detect 0.1 per cent atomic composition) can also be made in 1-5 minutes in the limited regions of interest in the Auger spectrum. A S/N ratio of ~1000 is easily attainable under the scanning conditions described above. An example of the sensitivity of the technique is given in Figure 4. Phosphorous at a level of around 0.04 per cent is clearly present in the de-embrittled state of the steel.

FIGURE 3

A Chart of Principle Auger Electron Energies

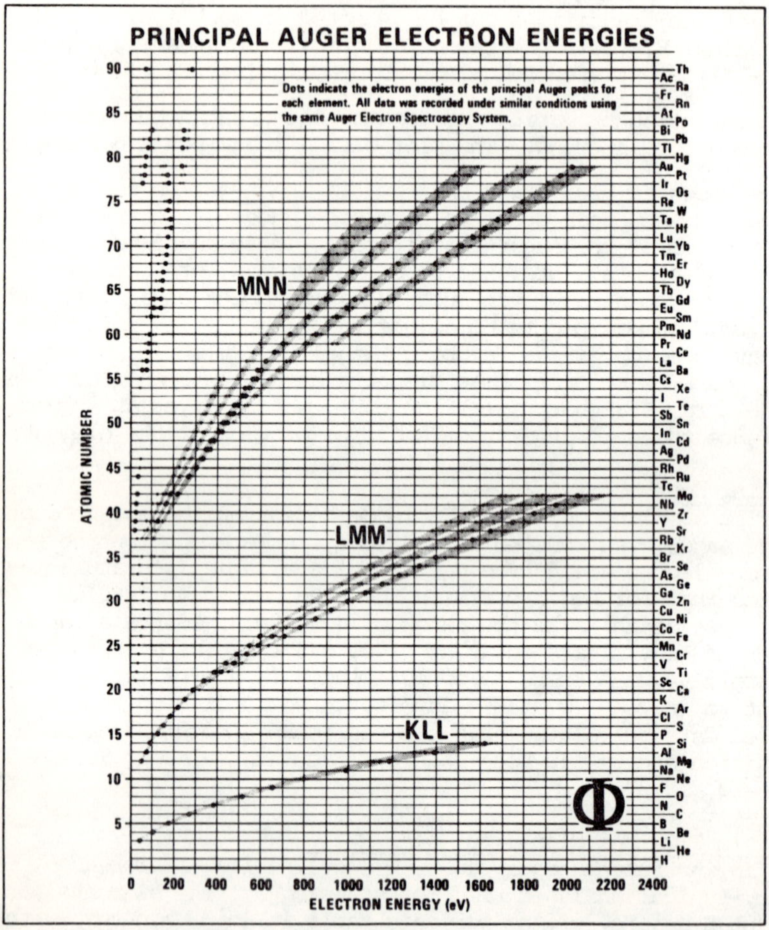

<u>Magnifying and Scanning Capabilities</u>

Low resolution SEM attachments capable of resolving 5-10μm particles are currently available on most of the instruments. An optical microscope (10-20X magnification) is also available for viewing and positioning of the sample. The future trends are towards better SEM/Auger combinations with even better resolving power. Scanning Auger capabilities have also been made available recently by Physical Electronics Industries, Inc. In this mode of analysis, surface maps of in-

FIGURE 4

AUGER SPECTRA FROM SURFACES EXPOSED BY FRACTURE OF EMBRITTLED AND DE-EMBRITTLED Ni-Cr-C-P STEELS

Ep = 3 KV and Ip = 50 μA

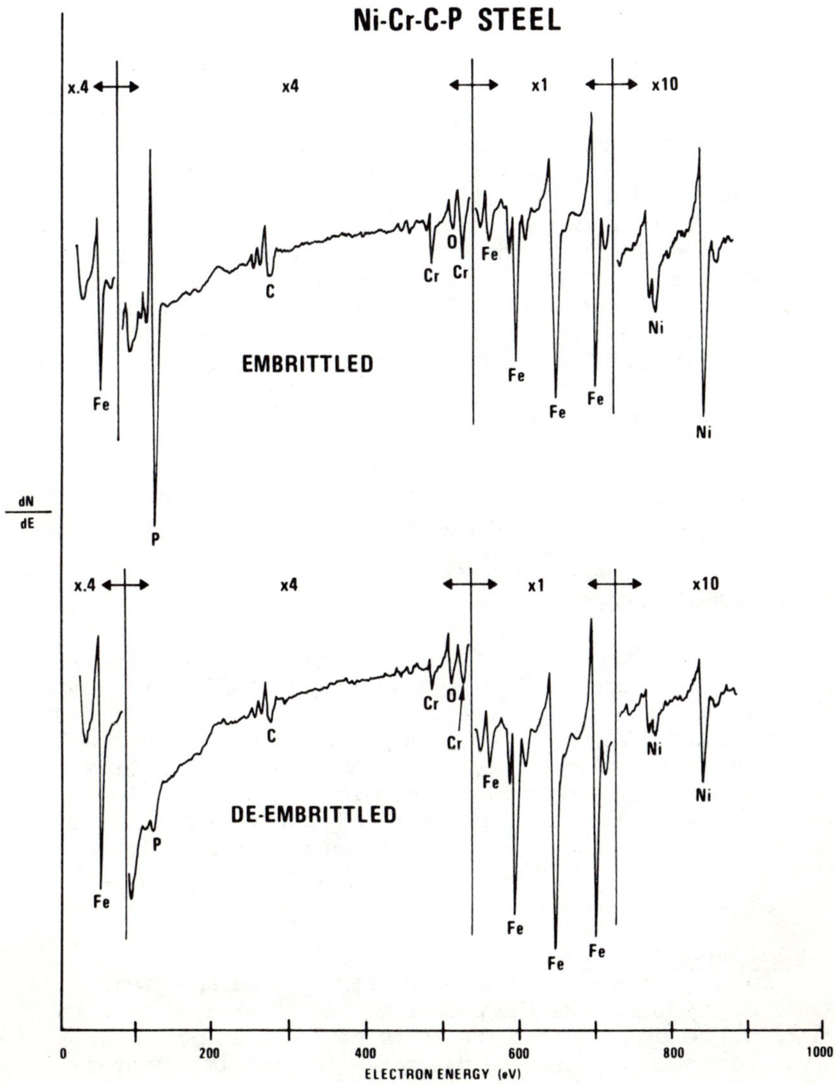

dividual elements present in a 200 x 100 μm^2 area of sample can be obtained using a 5 μm beam. Furthermore, Auger line scans over any region of interest can be obtained for each element present in the sample (see Karasek, 1974).

Chemical Shifts in AES

The chemical shifts in AES are very complex and not very well understood at present. However, chemical shifts are known for many light elements (Z <40) and these shifts in AES peaks can be used as fingerprints for particular oxidation states. For example: aluminum, manganese, magnesium, and silicon can each be distinguished in their zerovalent and oxidized states (spectra of $\overset{o}{Mn}$ and oxidized Mn is given in Figure 5). In the oxidized state, manganese has its characteristic three peak spectrum at 540, 587 and 636 eV (Figure 5, top spectrum). Signals from impurities like S, C and Fe are also seen. The main AES signal from oxygen is at 511 eV. As the surface is cleaned and reduced by argon ion sputtering (term used for etching or cleaning top surface atomic layers by high energy argon or xenon ion bombardment) the AES spectrum of $\overset{o}{Mn}$ shifts by +3 eV for 540 and 587 eV Auger peaks and only +1 eV for the 636 eV Auger peak (Figure 5 middle and bottom spectrum). Furthermore, the signals from S and C impurities are also reduced substantially. The shape of the carbon Auger peak in graphite and carbides is strikingly different, thus providing an easy means of determining the form of carbon contamination on surfaces (Palmberg a, 1972). The information on chemical shifts will expand in the future as more work is done in AES.

EXPERIMENTAL METHODS

The basic components of Auger electron spectrometer are an "Ultra High Vacuum" (UHV) System, an electron energy analyzer, an electron gun for sample excitation (Figure 6). For obtaining sub-surface composition profiles, sputter-etching (by argon or xenon ions) device can be very useful. Although the experimental methods given below apply to any equipment with a cylindrical mirror analyzer and having a large volume UHV system, the author's experience has been primarily with Physical Electronics Industries', instrument (Model 40-100). All spectra were obtained on this instrument.

Sample Considerations

Solid samples in any shape or form can be analyzed. Powdered samples, like the copper catalysts studied in this work, can be pressed into the shape of a pellet or a flat plate (for example, using a die normally used in infrared

FIGURE 5

AUGER SPECTRA OF PURE Mn AND OXIDIZED Mn SURFACE

Ep = 3 KV, Ip = 50 μA

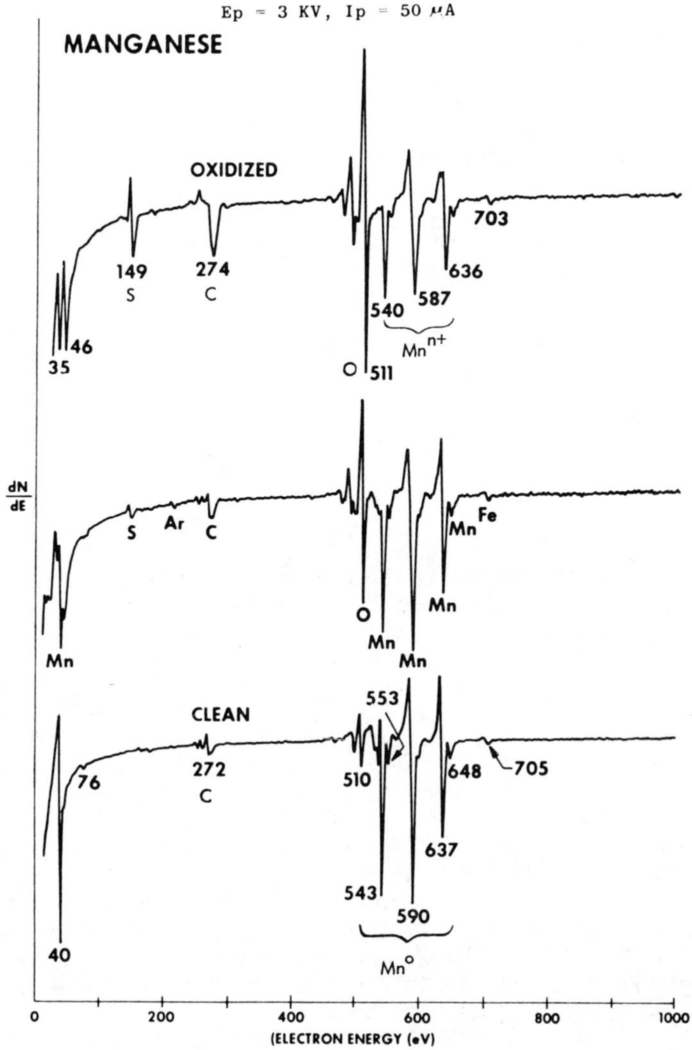

FIGURE 6

SCHEMATIC DIAGRAM OF AN AUGER ELECTRON SPECTROMETER

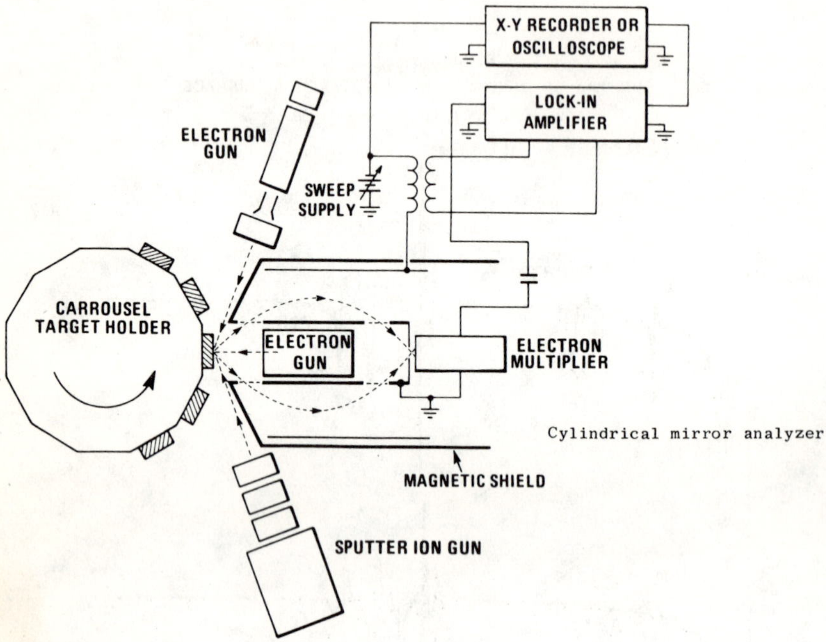

work) and then mounted into the holders. Sample holders can be easily made to handle samples of different sizes and shapes. Thus, samples as small as 1/64" to as large as 1" diameter can be mounted. By the use of "Specimen Manipulator", the sample can be positioned such that the primary electron beam can be focused on any particular spot of interest. Six degrees of freedom are provided; three linear translations, two tilt modes, and one rotation. The samples should be handled with extreme care so as not to contaminate them by touch (e.g. NaCl in sweat) or by friction with another surface.

About 24 samples of 1/8" size can be loaded into the system at one time. However, if sputtering each sample independent of the other is to be accomplished, the number of samples reduces to twelve. Only six samples can be loaded at a time if the sputtered atoms have to be collected on a plate and analyzed by AES. This is only necessary when dealing with some samples that charge very badly.

All work was done (with a primary electron beam energy of 5KeV) using a cylindrical mirror analyzer. Typical primary beam currents of 15 -30 ua were used in all Auger scans. Since some of the catalyst samples (like Pd on gamma-alumina or other catalysts on hydrophillic supports) did contain some adsorbed water, the pressure in the vacuum chamber occasionally increased from 10^{-8} to ~10^{-6} torr when the electron beam hit the surface. However, this loss of vacuum did not present any problem in recording high sensitivity scans. The problem of water desorption from such samples was solved later on by drying the samples at 150-400°C in dry nitrogen or air and then storing them under nitrogen in larger bottles containing silica gel. Minimal exposure to laboratory air during loading of such dried samples did not upset the high vacuum (10^{-8} - 10^{-9} torr) in the vacuum chamber during Auger measurements. In case of Pd on gamma-alumina catalyst, samples of fresh and used catalysts consisting of dark outer ringed areas (where most of the catalytic components were concentrated in order to minimize diffusional resistance) and light inside areas were analyzed by AES. Typically, the size of the circular spot analyzed was 0.1 mm in diameter. Although only the analysis of one spot is given in the Figures 8-13 several spots were generally scanned to gain better statistical confidence in the results.

Various Modes of Surface and Sub-Surface Analysis

DIRECT SURFACE AND SUB-SURFACE ANALYSIS

The outer as well as inner (by cleaving a solid pellet) surface of solids can be analyzed directly using 25-100 micron (0.025 to 0.1 mm) size electron beams. If the size of the sample is relatively large compared to the electron beam and the sample is heterogeneous, several spots can be analyzed to get a representative average analysis. If the surface is contaminated the top atomic layers (containing the contaminant) can be cleaned off by the use of an argon or xenon ion sputter gun for a few minutes. The surface can be monitored in the energy range of interest (for any element) on an oscilliscope while the sample is being sputtered. The size of the argon ion beam or the size of the area being sputtered is about 4-6 mm in dia., thus giving a uniform sputtering over the area being monitored. Sputtering for a short time can be used to remove, at least partially, any adsorbed carbon on the surface. Complete removal of adsorbed carbon would require heating the sample under <10^{-7} torr vacuum followed by sputtering. The analysis of subsurface atomic layers is obtained after sputtering for the required length of time and analyzing the sub-surface. The length of time required to remove a single layer (i.e. sputtering rate) is known for completely flat surfaces, though for rough surfaces like in a porous catalyst, sputtering rates are very approximate. Thus, concentration profiles in the sub-surface atomic layers can be obtained by the use of sputtering techniques (Weber, 1972).

Direct surface analysis by AES is possible with most of the conductor and insulator surfaces; however, in case of some insulator surfaces (generally having rough texture), the charging is so severe that the signal cannot be monitored at all. The analysis of such surfaces will be discussed later in another section.

The high sensitivity of AES to surface composition is shown by analysis of embrittled and de-embrittled steels in Figure 4. In the de-embrittled state, the amount of phosphorous is only ~0.1 per cent corresponding to 400 ppm in the bulk; however, in the embrittled state as much as a few per cent phosphorous is detected. Furthermore, the amount of nickel has doubled in the embrittled state.

INDIRECT ANALYSIS OF THE SURFACE LAYERS

The surface and sub-surface atoms can also be analyzed by collecting the sputtered atoms under argon ion bombardment onto a clean plate of say, aluminum, tantalum, etc. This is especially useful when surface charging problems are encountered. Experimentally, it is accomplished by collecting sputtered atoms onto a rotatable collection plate nearby, followed by Auger analysis of the collection plate (see Figure 7). The collection efficiency of the sputtered atoms is critically dependent on the configuration and hence the solid angle subtended by the collection plate. Other arrangements can be devised that will yield even higher collection efficiency.

FIGURE 7

Indirect Analysis of Surface and Sub-Surface Composition

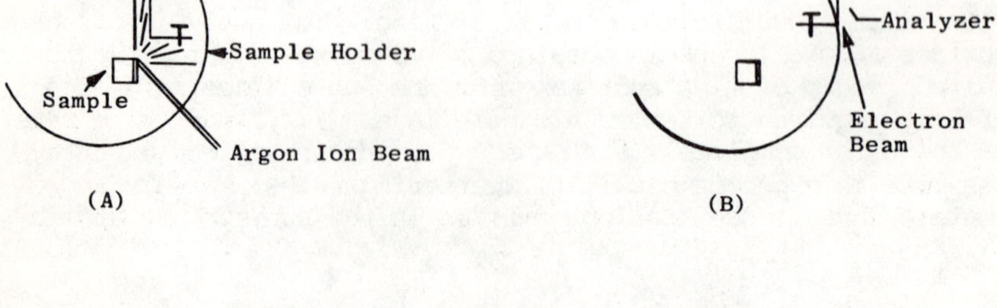

DIRECT ANALYSIS OF CESIUM-COATED SURFACE

Surface charging can also be prevented by coating the surface with cesium using a cesium ion gun. Auger spectrum is recorded after a monolayer (or less) of cesium has been deposited on the surface. This technique can be very useful in the study of some of insulator oxides that charge up under the incident electron beam.

Quantitative Analysis

Quantitative analysis can be done by the use of pure element standards under identical experimental conditions. The atomic concentration, C_X, of element X in the sample is given by:

$$C_X = C_{X,\ std.} \frac{I_X}{I_{X,\ std.}} \qquad (1)$$

where I_X and $I_{X,\ std.}$ are the Auger currents (peak-to-peak in the differential Auger spectrum) from the sample and the standard, respectively. The relation in Equation (1) assumes that the escape depths for Auger electrons in the sample and standard are equal. Thus, it is a very useful relationship when the sample and standards are similar, but large errors can result when the escape depths are different. Surface heterogeneity in the top few layers can further complicate the analysis. Surface topography of rough surfaces also affects the Auger yields; however, meaningful quantitative data can be obtained when comparing similar rough surfaces; for example, porous catalysts of the same type. An accuracy of ~5 per cent can be achieved if care is taken to minimize the problems cited above, and if sufficient number of sample points are taken to be statistically representative of the original sample population. With powdered samples, the problems of sample heterogeneity are not serious, however, with pelleted heterogeneous catalysts, the sampling problems are very serious. Hence, with heterogeneous samples one must analyze a sufficient number of spots (preferably on several pellets) which will be representative of the original sample.

Sensitivity factors for elements of interest were obtained from standard Auger spectra (at 3 KeV) published recently (Palmberg, b, 1972) assuming the sensitivity of oxygen to be unity. These factors may not be very accurate in themselves at 5 KeV, but the comparison of relative amounts from like samples is reasonably good.

Surface atomic per cents given in Tables I and II represent the normalized analyses of the top 2-10 layers of the catalyst. These analyses are not absolute but are self-consis-

TABLE I. *Conventional Chemical Analysis of Two Commercial Copper Catalysts*

	Lot A	Lot B
Copper Metal	14.6%	18.8%
Cu_2O	48.2	56.2
CuO	31.1	22.0
Fe	0.98	1.2
Mg	0.20	----
Pb	0.26	0.20
Soluble SO_4	1.52	N.D.
Part. Size	6.1 microns	5.5 microns
Surf. Area	0.5 m^2/g	0.65 m^2/g
Performance	Good	Poor

TABLE II. *Auger Surface Elemental Analysis of Copper Catalysts*

Element	Sensitivity Factor	Surface Atomic Percent	
		Lot A	Lot B
Sulfur	1.5	0.88	0.31
Chlorine	1.5	1.47	1.49
Lead	0.4	1.66	4.89
Carbon	0.3	N.D.	t
Tin	2.0	0.33	0.45
Oxygen	1.0	38.64	39.30
Iron	0.5	2.43	2.23
Copper	0.5	50.34	49.16
Magnesium	0.4	1.52	0.56
Arsenic	0.1	0.55	0.56
Aluminum	0.25	1.35	0.71
Silicon	0.20	0.83	0.33
Copper/Lead Ratio		23.3	8.04

tent and trends are believed to be significant. Actual sampling depth is not known and is very difficult to determine, but for like composition materials, it should be identical.

The Auger peaks of each of the elements found in the two catalysts have been labelled with the symbols of the elements near the negative peaks in the differential of the secondary electron energy distribution in most of the Figures. The main spectrum in each figure covers the energy range of 0-2000 eV. However, in the case of the copper catalyst sample lead had a pair of weak high-energy peaks at ~2170 eV and 2275 eV (MNN Auger transition), besides the characteristic strong low-energy peak at 92 eV (NOO Auger transition). The high-energy peaks are shown in the scan in the right hand corner in Figures 8 and 9.

Figure 8.

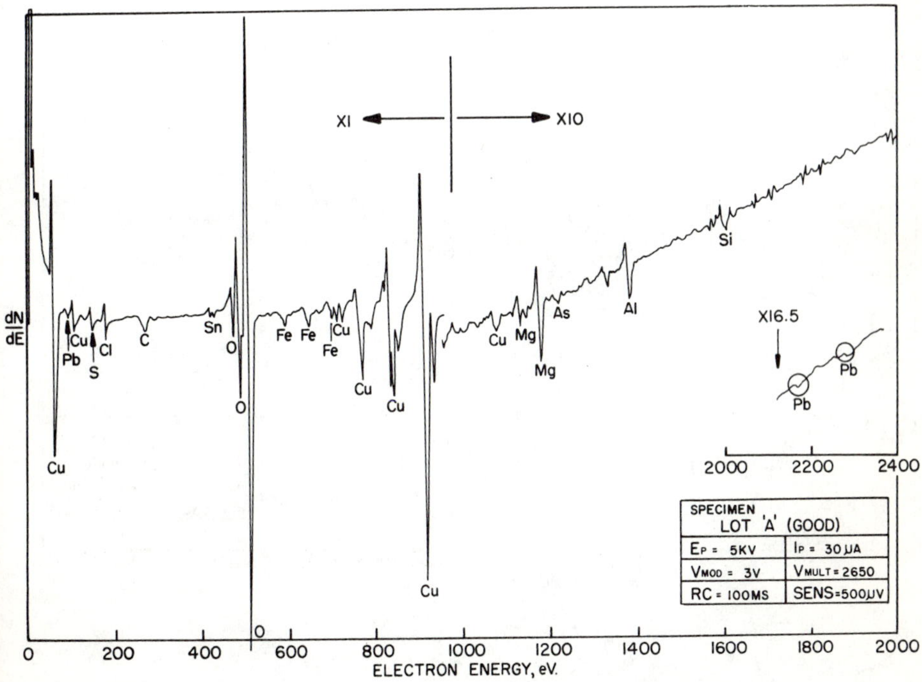

Vacuum Requirements

Because hydrocarbons are efficiently cracked to carbon in the incident electron beam, oil diffusion pumps cannot be used in AES. Ultra high vacuum of 10^{-7} to 10^{-9} torr necessary for routine work is obtained by use of ion pumps and sublimation pumping. For most metal and metal oxide surfaces, a vacuum of 10^{-7} torr is sufficient, but a vacuum of $\leq 10^{-9}$ torr is preferable. For sputter-etching work the system can be filled with an inert gas (Ar or Xe) up to 10^{-6} - 10^{-8} torr depending on sputtering rate described (Weber, 1972).

Figure 9.

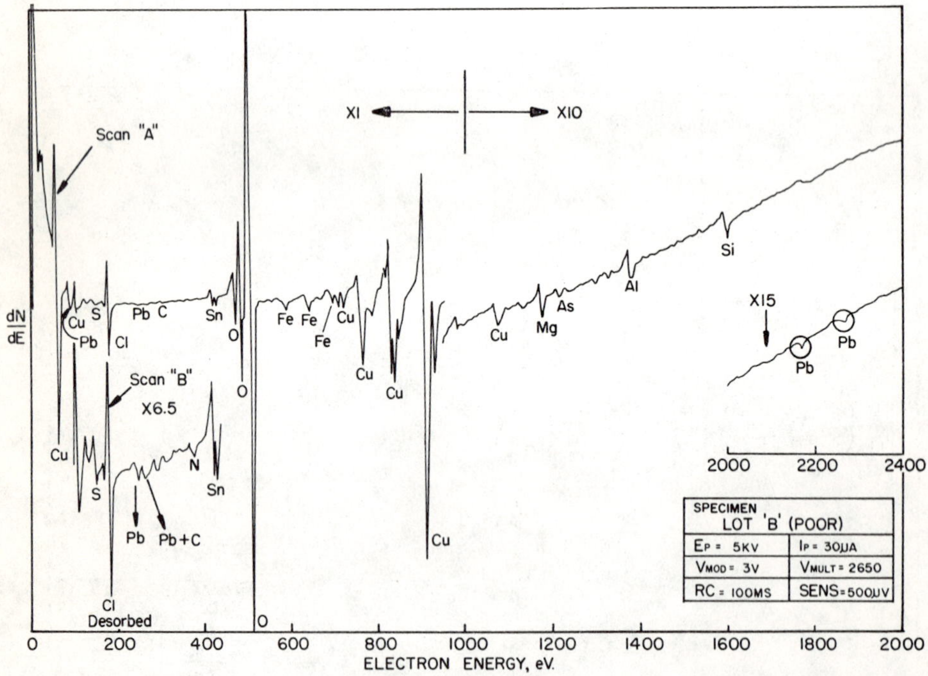

APPLICATIONS TO HETEROGENEOUS CATALYSIS

Surface Poisoning of a Copper Catalyst by Lead

Two copper catalysts of widely differing catalytic activity were analyzed by AES to explain the difference in their performances. Such copper catalysts are used to catalyze the reaction of methyl chloride with silicon to produce methyl chlorosilanes (Voorhoeve, a, 1967). Conventional chemical analysis failed to show any significant differences (see Table I). Known poisons for such copper catalysts are Pb, Sb, As and Bi (possibly Sn also) (Voorhoeve, b, 1967). However, Lot A catalyst (good) had 0.26% Pb as against 0.20% in Lot B catalyst (poor). Lead in much smaller amounts (~0.01%) can poison these catalysts (Voorhoeve, b, 1967). Chemical analysis of other elements and physical properties of the two catalysts appear to be the same (Table I). Analysis by scanning electron microscope and by X-ray analysis (EDAX) also failed to show any meaningful differences. Since AES is a highly sensitive tool for analyzing the top 2-10 surface atomic layers (Palmberg, 1972 and, a, 1973), such an analysis was performed on the two copper catalysts. AES scans of the good catalyst (Lot A) and poor catalyst (Lot B) are given in Figures 8 and 9, while quantitative surface analysis (using sensitivity factors explained in Experimental section) is given in Table II.

It is quite evident from Figures 8 and 9 and Table II that Lot A has about one-third as much lead as Lot B based on the low energy peak belonging to the NOO transition. However, the difference in Pb levels of the two samples is less pronounced when comparing the high energy peaks of the MNN transition. This is to be expected when the impurity is concentrated toward the top surface atomic layer. The escape depth for the high energy transition may be about tenfold deeper than that for the low energy NOO transition. The level of lead in Lot B is 4.9 surface atomic per cent of all elements including oxygen. In relation to copper, however, the lead level is about 13 per cent. It should be emphasized here that the Auger analysis given (using NOO transition) represents the analysis of the top few (1-2) layers of the surface (Palmberg, 1973). Thus, if lead is concentrated more towards the top surface atomic layer, the level of lead could be even higher. Furthermore, since all surface Cu atoms may not be catalytically active, 13 atoms of lead per 100 atoms of Cu can very well explain the inactivity of Lot B catalyst.

The fact that catalyst, Lot B, has a higher concentration of lead on the copper surface despite having a lower total lead content could be explained in a variety of ways. Extended roasting of the catalyst during its manufacture is known to result in an inactive product. This would suggest that the lead may be concentrating on the catalyst surface as a result of its migration to the grain boundaries during the roasting step.

Some of the other significant differences in the two catalysts are summarized below:

(1) Copper catalyst, Lot B, had much higher surface Cl, at start of Scan 'A' of the Auger spectrum; however, some of the chlorine desorbed under the incident electron beam. This phenomenon is quite typical of some of the well-known chlorides (Palmberg and Rhodin, 1968).

(2) The level of sulfur in Lot A was three times higher than in Lot B. This form and level (0.88 atom %) of sulfur is obviously not deleterious to the function of the good catalyst; namely, Lot A. Bulk analysis indicates that it is present as sulfate.

(3) The levels of tin, iron and arsenic are roughly the same in the two catalysts.

(4) The levels of aluminum and magnesium are 2-3 times higher in the good, active catalyst (Lot A) than in the inactive catalyst. What role these impurities play in the performance of such copper catalysts is open to speculation, although they are known to be promoters for the copper catalyst (Voorhoeve, a, 1967).

(5) No significant amounts (~0.1 atomic per cent) of Bi or Sb, the other poisons, were detected in the two catalysts analyzed.

Surface Poisoning by Iron of a Commercial Palladium-Gamma-Alumina Catalyst

A commercial selective hydrogenation catalyst (in the form 3/16" cylindrical pellets) lost most of its activity during normal operation. This catalyst consists of 0.5 per cent Pd, with chromium and molybdenum added as promoters, supported on a γ-alumina support. It is used for the selective hydrogenation of diolefins in an olefin-aromatic-paraffin containing stream. The loss in catalyst activity appeared to be associated with a change in feedstock that occurred a few days before. The catalyst was unloaded, the loose dust screened, and analyzed by X-ray fluorescence spectroscopy (XRF). The XRF analysis showed the presence of 1.7 per cent Fe, 0.16 per cent S and 0.01 per cent Cl. This level of iron is not necessarily a poison since iron oxide is a weak hydrogenation catalyst and moreover this amount of Fe is not very large for a ~200 m^2/g surface area support. As far as sulfur is concerned, about 0.1 per cent is also present in the fresh, active catalyst. The sulfur is very likely in sulfide or sufate form and is not deleterious to catalyst activity. The surface area of spent catalyst was also not changed significantly. Auger electron spectroscopic analysis was performed on the fresh and the spent catalyst to understand the cause or causes of catalyst deactivation.

The Auger spectrum of the outer surface of fresh catalyst is shown in Figure 10. This spectrum is shifted by ~10 eV and this energy should be subtracted from all peaks below 520 eV oxygen peak. The outer surface is rich in Pd, Cr and Mo, while Co is present in only small amounts. Some sulfur is also present on the fresh catalyst. The sulfur signal at 150 eV also contains one of the secondary peaks of Mo. The sulfur concentration is obtained by subtracting the proportional amount of Mo signal (relative to the 180 eV Mo major peak). The presence of Mo is further confirmed by the presence of high energy Mo

Figure 10.

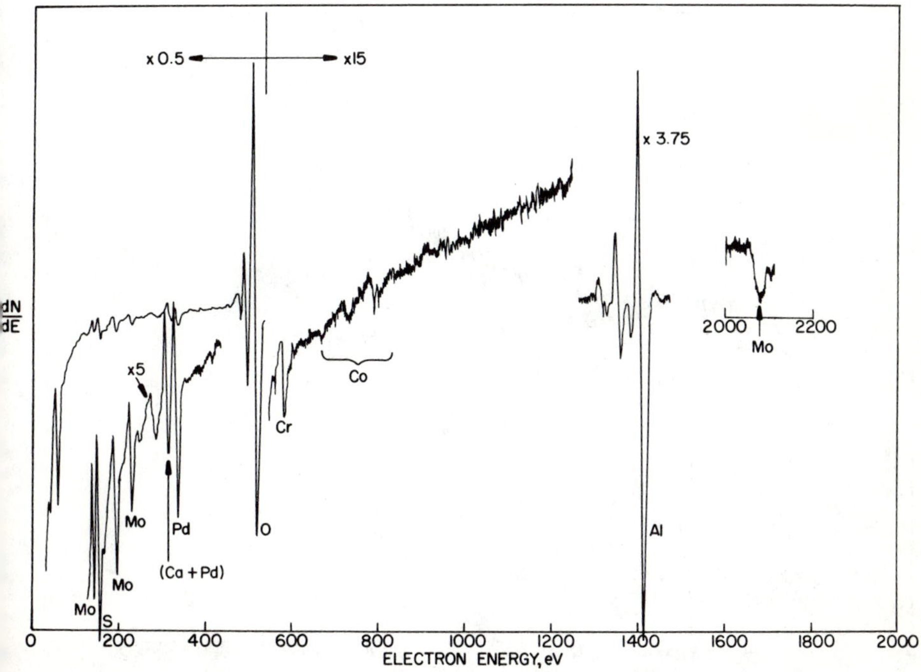

peak at ~2080 eV. As is evident in Figure 10, a substantial amount of Ca is also present. The relative surface atomic per cent of these elements shown in Figure 10 and Figures 11, 12 and 13 (for other catalysts) are given in Table III. In contrast to the richness of catalytically active components and the impurities (Ca and S) on outer dark surface, the interior surface (see Figure 11 and Table III) has only a trace of Mo and S and only a small amount of Pd while Cr and Co are present in about the same amount as the outer surface.

The outer surface of the spent catalyst (Figure 12 and Table III) contains so much Fe that it has just about completely (>74 per cent) suppressed the Pd, Mo, Cr signals and even Al

Figure 11.

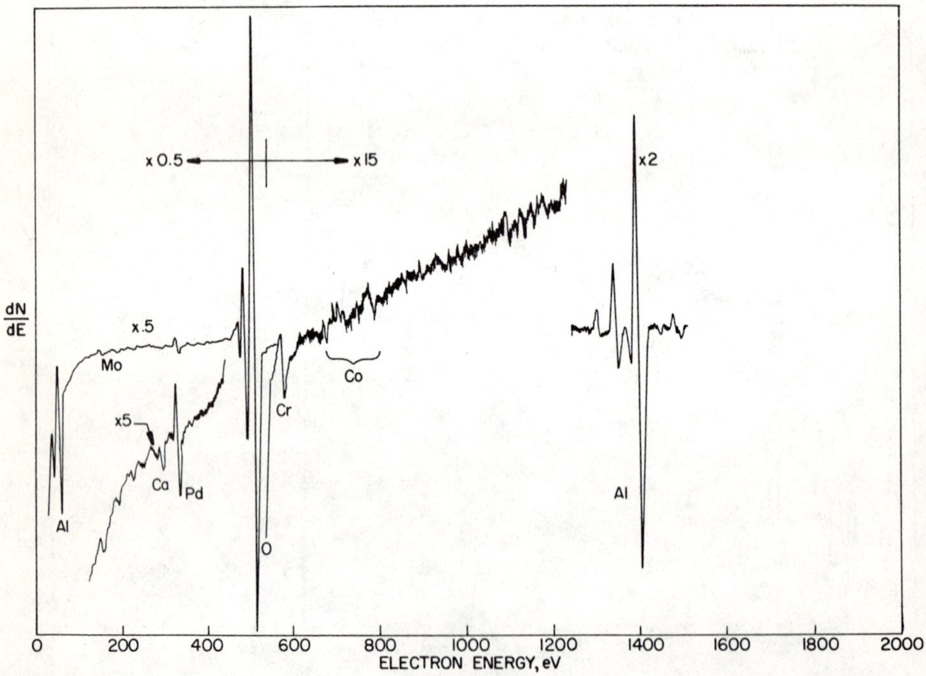

TABLE III. *Surface Atomic Percent*[a]

Elements	Sensitivity Factor[a]	Fresh Catalyst: Dark Area (Outer Surface)	Fresh Catalyst: Light Area (Interior Surface)	Spent Catalyst: Dark Area (Outer Surface)	Spent Catalyst: Light Area (Interior Surface)
Mo	0.5[b]	3.26	0.16	0.0	t
S	1.5[b]	1.51	0.11	1.40	t
Ca	1.5	0.70	0.11	0.93	0.60
Pd	2.0	1.26	0.52	0.32	0.54
O	1.0	55.79	56.88	70.16	57.94
Cr	0.4[c]	0.75	0.48	0.00	0.00
Fe	0.5	†	0.00	13.68	2.53
Co	0.5	0.24	0.19	0.35	t
Al	0.25	36.49	41.55	13.16	38.38(Est.)
		100.00	100.00	100.00	99.99

[a]Divide the peak heights of Auger peaks by their corresponding sensitivity factors to obtain relative atomic ratios, which on normalizing give the surface atomic percentages. [b]Sulfur peak at 150 eV interferes with Mo secondary peak at 148 eV. Sulfur peak height is obtained by subtracting 0.17 times the Mo peak height at 186 eV. [c]The sensitivity factor for Cr corresponds to full peak at 570 eV. Due to interference from the tail of large oxygen peak at 510 eV, sometimes only half of Cr - 570 eV peak is observed. If so, a sensitivity factor of 0.2 should be used.

signal arising from the support. Thus, iron is not only covering substantial (>90 per cent) amount of the catalytically active components (Pd, Cr, Mo) but is covering the surface of the alumina support also. Furthermore, since the Auger signal is known to originate in the first 2-5 atomic layers for most of the elements (Palmberg, a, 1972 and 1973), iron may very likely be masking greater than 80 or 90 per cent of the surface exposed atoms if iron is present in the top most layer only. Thus, even though the presence of iron, per se, in small amounts is not a poison, the masking of the highly active Pd surface atoms and of the Cr/Mo promoter atoms makes iron almost as deadly a poison as sulfur, mercury, lead, etc. Although analysis of only one spot of 0.1 mm is shown in Figures 10-13, several other spots were analyzed to gain statistical confidence. The amount of iron present on the interior surface (Figure 13 and Table III) was much less (a factor of 5 to 6) but the amount of Pd is much less also. The sharp gradient in iron from the outside to the interior of the pellet strongly suggests that iron came from an outside source. The feedstream to the catalyst was found to contain both powdered rust and dissolved† iron. The powdered

Figure 12.

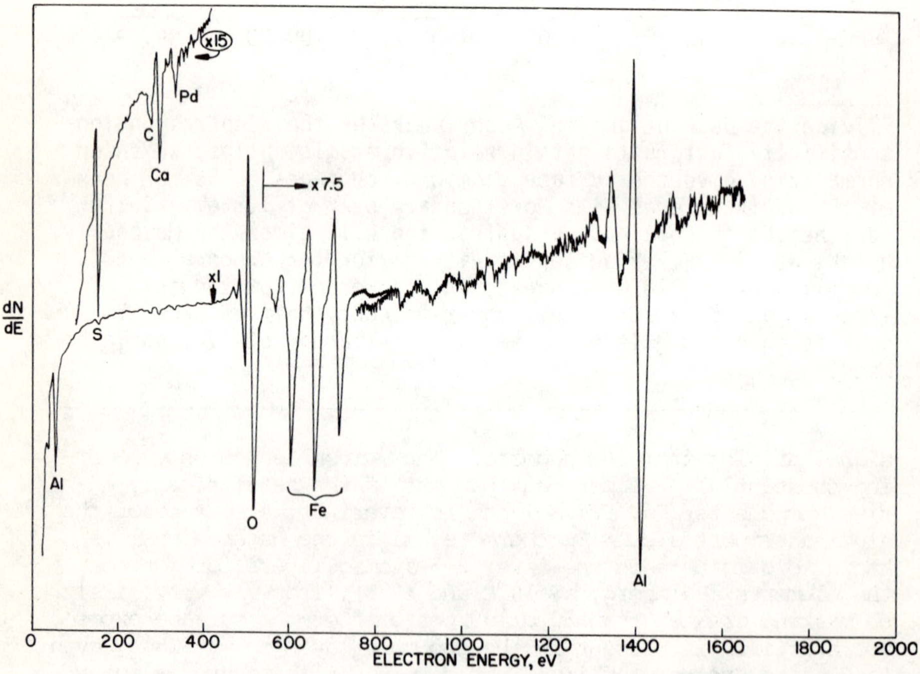

†The dissolved iron may have also contained fine particles of iron (<10 nm).

Figure 13.

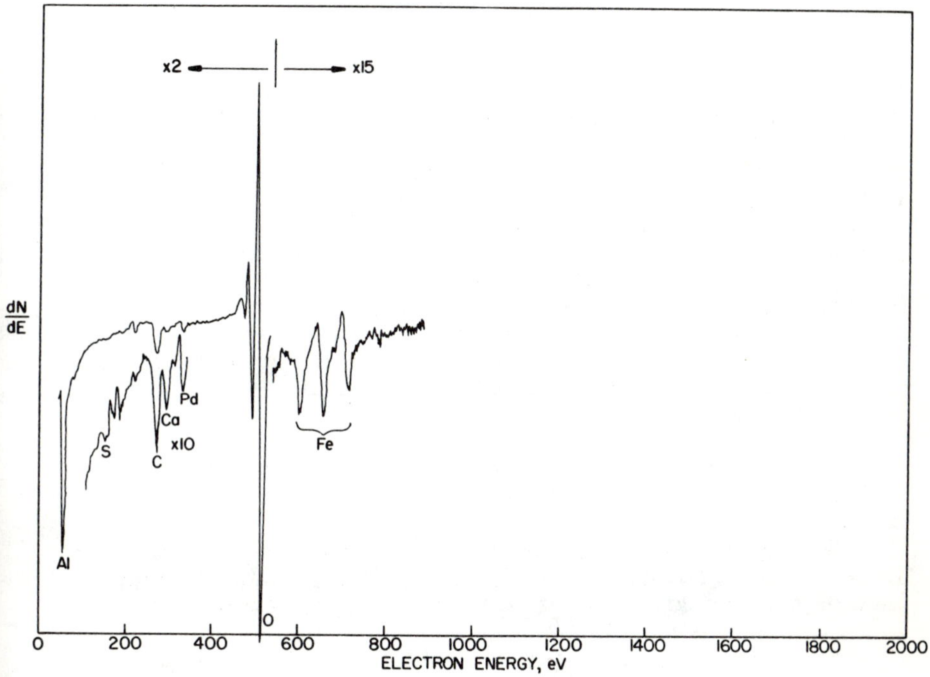

rust was localized towards the top of the catalyst bed but the dissolved iron was spread throughout the bed. Screening of the powdered rust did not increase catalyst activity significantly. The dissolved iron was the major source of poison for the active sites on the palladium surface as well as the sites on the promoters.

CONCLUSIONS

1. AES is a very useful tool in the analysis of trace a-

mounts (~0.1 surface atomic per cent) of elements on the surfaces of solids. All elements except hydrogen and helium can be analyzed by AES. In practical applications, AES offers high speed of analysis (15-30 minutes) with high sensitivity and yet it is as simple to use as infrared spectroscopy.

2. Since no sample preparation is necessary for porous, spherical or insulator surfaces, AES is ideally suited for the study of catalysts and catalytic behavior. The usefulness of AES analysis in studying the poisoning of two different catalysts (a powdered Cu catalyst and a gamma-alumina supported palladium catalyst), has been demonstrated. Furthermore, AES was shown to be helpful in detecting the presence of and determining the amount of surface impurities, and in establishing the concentration gradients of active components (and impurities) within the pellets. Thus, AES can be very useful in a wide variety of catalytic problems, for example, in establishing surface purity before and after use, understanding the roles of catalysts, promoters, etc. And, AES in combination with other surface analytical techniques can help, in years to come, make science out of the art of heterogeneous catalysis.

3. Moisture present in gamma-alumina supported catalyst did not present any serious problems in the high vacuum chamber of the Auger spectrometer.

4. The sampling depth in AES is in the range of 4-20 Å depending on the atomic number of the element, thus making it highly sensitive to most of the elements on the surfaces of solids.

5. By the use of sputtering techniques, sub-surface atomic layers can also be analyzed by AES. Thus, surface enrichment or surface depletion of elements in multicomponent catalysts can be studied.

6. A better estimate of top surface atomic layer and subsurface layer composition can be obtained by running in-depth concentration profiles by well-known inert ion sputtering techniques.

7. Typical analysis time per sample was less than one-hour. Complete high sensitivity, elemental analysis for most of the periodic table, in such a short time has not heretofore been possible.

ACKNOWLEDGEMENT

The author wishes to thank Physical Electronics Industries, Inc. for permission to use some of the figures used in the "Introduction" section of the paper. Thanks are also due to Union Carbide Corporation for the permission to publish this work.

REFERENCES

1. Bhasin, M. M., J. Cat., 34, 356 (1974).
2. Bhasin, M. M., J. Cat., 38, 218 (1975).
3. Chang, C. C., Surface Science, 24, 53 (1971).
4. Hercules, D. M., and Carver, J. C., Anal. Chem. 46, 133R (1974).
5. Karasek, F. W., Res. and Dev., p 48, Oct. (1974).
6. Pentenero, A., Catalysis Revs., 5(2), 199 (1972).
7. Palmberg, P. W., and Rhodin, T. N., J. Phys. Chem. Solids, 29, 1917 (1968).
8. Palmberg, P. W., "Electron Spectroscopy" ed. D. A. Shirley, pp 835-859, North Holland Publishing Company, Amsterdam, 1972.
9. Palmberg, P. W., Riach, G. E., Weber, R. E., and MacDonald, N. C., "Handbook of Auger Electron Spectroscopy", Physical Electronics Industries, Inc., Edina, Minn., 1972.
10. Palmberg, P. W., Anal. Chem., 45, 549A (1973).
11. Voorhoeve, R. J. H., "Organohalosilanes - Precursors to Silicones", Ch. 4., Elsevier Publishing Company, 1967(a).
12. (b) ibid., p 133 (1967b)
13. Weber, R. E., Res. and Dev., p 22, Oct. (1972).

CATALYTIC TRITIATION IN THE PREPARATION OF LABELED COMPOUNDS

DALE W. BLACKBURN and SIDNEY H. LEVINSON

SmithKline Corporation
Philadelphia, Pennsylvania 19101

Over the past decade our group has gained expertise in catalytic hydrogenation and radiosynthesis. We would like to combine these technologies in this paper, Catalytic Tritiation in the Preparation of Labeled Compounds. In the not so distant past several scientific meetings had as their goal the further development of the radioisotope method. A series of ten symposia on Advances in Tracer Methodology (Rothchild, 1963, 1965, 1966) were sponsored by several nuclear companies of the northeastern United States between the years of 1957 to 1965. Their goal was to stimulate the further use of radioisotopic techniques. Three international conferences on METHODS OF PREPARING AND STORING MARKED MOLECULES† were held from 1963 to 1966 to encourage the preparation of labeled compounds and to provide them as tools for research in Europe. Euratom, the sponsoring organization, was also instrumental in the formation of the JOURNAL OF LABELLED COMPOUNDS (Presses Européennes, Brussels, Belgium) to provide a forum for presenting isotope labeling technology.

We use tritium labeling for less than 10% of the compounds prepared. We like to think that the reason is our proficiency with carbon-14 synthesis but no doubt the major reason for the low utilization of tritium labeling is the lability mystique which surrounds hydrogen. We hope to discuss this to some degree and encourage greater use of the technology for specific tritium labeling. Tritium is a selective and sensitive probe for hydrogen and if positioned correctly can be used as an ancillary label for carbon in biological research. The recently published authoritative text, TRITIUM AND ITS COMPOUNDS, by Evans (1974) is of immense help in the search for relevant

†Euratom, Proceedings of the Conference on Methods of Preparing and Storing Marked Molecules. a. Eur 1625e, Brussels, Nov. 13, 1963; b. Eur 2200e, Venice, Aug. 23, 1964; c. Eur 3746d,f,e, Brussels, Nov. 28, 1966.

literature. In this paper we have drawn on much of our own tritium work to illustrate chemical technique and technology.

PROPERTIES OF TRITIUM

Tritium, a radioactive isotope of hydrogen with a half-life of 12.35 years, emits low energy beta radiation;

$$^{3}H(T) \longrightarrow {}^{3}He + \beta \quad (E_{max} = 18.6 \text{ Kev})$$

This compares with the maximum energy of 155 Kev for carbon-14. Because of its low energy beta, tritium is not an external hazard. However, detection of contamination is difficult and requires the liquid scintillation counting of smear samples. Tritium has a high maximum specific activity of 29.12 Ci†/milliatom which increases the ultimate sensitivity of detection. For example, biological mechanisms can be explored on a cellular level using an autoradiographic technique. Thymidine-^{3}H can be used to follow nuclei division.

Because of the greater mass of tritium compared to protium the zero-point energy of the carbon-tritium bond is lower than carbon-protium and the bond strength increased several fold. Because of this increased stability, tritium can be used as an ancillary tracer for carbon. Since the relative mass difference between tritium and protium is larger than any other element, the primary isotope effect in the formation and cleavage of hydrogen bonds is large.

PREPARATION OF TRITIUM LABELED COMPOUNDS

The methods for labeling compounds with tritium may be classified as shown in Table I. We prefer not to discuss exchange techniques because of the low specific activities (activity/unit weight) obtained and lack of specificity. The resulting compounds are usually general labeled (G). It is worth noting that the homogeneous catalyst reagent $H_2TPO_4 \cdot BF_3$ of Yavorski and Gorin (1962) can be used at ambient temperature to tritiate aromatic rings.

Biochemical methods are important in the stereoselective reactions in the resolution of labeled racemates and specific

---

†Curie, a unit of activity, is 3.7×10^{10} disintegrations/second.

TABLE I. *Methods of Labeling Compounds with Tritium*

I. Exchange Techniques
   A. Radiation Induced (Wilzbach)
   B. Heterogeneous Catalytic
   C. Homogeneous Catalytic
II. Biochemical Methods
III. Chemical Exchange (IC) With Specificity Followed by Reaction to Remove Lability
IV. Direct Chemical Synthesis

conversions of nucleosides. It is, however, outside the range of this paper.

The first method of labeling we will consider is a combination of selective exchange followed by chemical reaction which eliminates the possible back-exchange loss of the label. The first example involves the intramolecular conversion of labile to non-labile tritium. We have applied the method of Kalberer and Rutschmann (1963) in the specific labeling of the 9-position of phenothiazines shown in Table II. The substituted anthranilic acid, I, was exchange labeled by stirring in dioxane solution with tritium oxide. Decarboxylation fixed the tritium label in II. Part of the label was lost in the Bernthsen thionation to form the 9-^3H-phenothiazine, III. Sodium hydride in dimethylsulfoxide is the preferred reagent for the semi-micro alkylation to the indicated 10-alkylaminoalkylphenothiazine-9-^3H, IV. We could subsequently store the tritiated phenothiazines for future studies because of the longer half-life of tritium compared to sulfur-35. We have supplied these compounds for numerous research investigations. Interestingly, the only suggestion of lability of tritium in the phenothiazines was found by our biochemistry group when an in vitro chemical procedure using an ion exchange resin showed some loss of tritium label as tritium oxide.

TABLE II. *Preparation of Phenothiazines -9-^3H*

X	$\rangle N(CH_2)_3-$
H	$(CH_3)_2N(CH_2)_3-$
Cl	same
CF_3	

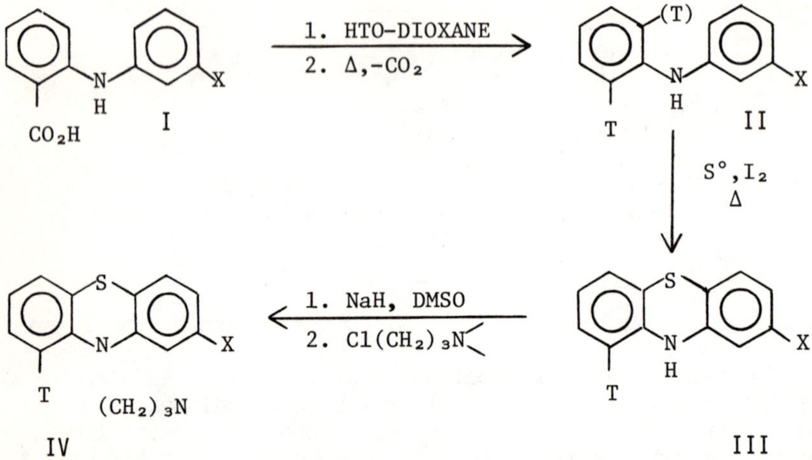

A semi-labile tritium can be converted to non-labile. In a recent paper, van Kordelaar and coworkers introduced tritium into the 16-position of estrenone, V, by base exchange (van Kordelaar, et al, 1973). The tritium was then stabilized in the lynestrenol, VI, by the subsequent ethynylation of the 17-keto group.

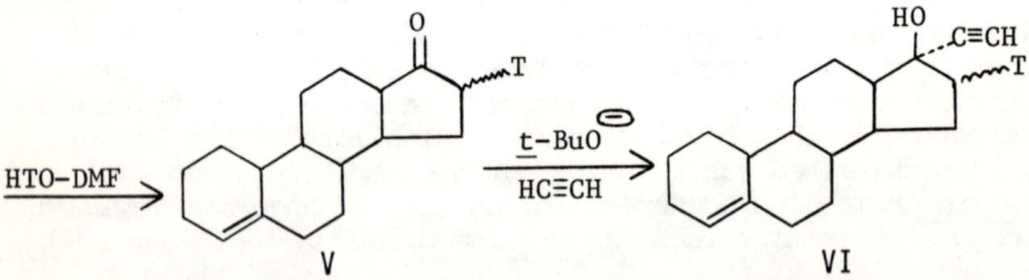

A general exchange labeled intermediate may be used to label a specific moiety within a complex molecule. For example a carbon-14 diphenidol, VII, was prepared for initial metabolic studies in dogs and humans (Kaiser, et al, 1972). The major metabolite was identified as the substituted aminovaleric acid VIII (^{14}C). The Grignard of bromobenzene-^{3}H (G) was reacted with the critical intermediate IX, followed by hydrolysis to obtain the tritium labeled metabolite VIII (^{3}H). This was used for subsequent <u>in vitro</u> assays.

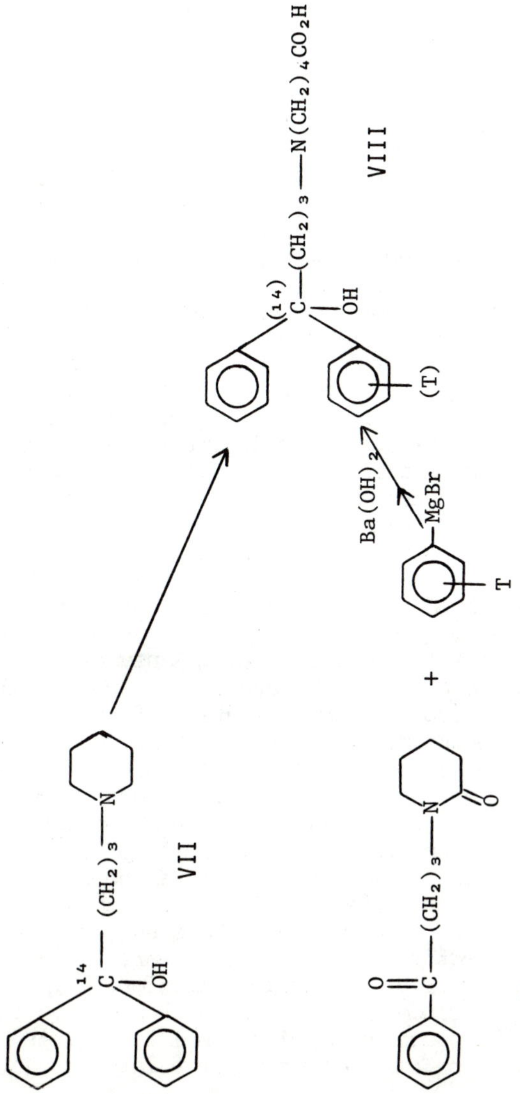
14
C
OH
(CH2)3
N
VII
(14)
C
OH
(CH2)3
N(CH2)4CO2H
VIII
(T)
Ba(OH)2
MgBr
T
+
O
C
(CH2)3
N
O
IX

Direct Chemical Synthesis

The direct addition of tritium to carbon-carbon unsaturated bonds is capable of giving very high specific activities. The catalyst of choice is palladium on carbon. More efficient incorporation of label effecting higher specific activities can be achieved by the following techniques:

- Minimizing amount of protium on the catalytic surface.
- Blocking of groups containing labile protium as for example esterification of carboxylic acids.
- Increasing the rate of reaction.
- Decreasing the polarity of the solvent system to reduce exchange dilution of tritium gas. The solvents useful in tritiation in decreasing order of polarity are the following: ethanol > isopropanol > dimethylformamide > ethyl acetate >> dioxane or tetrahydrofuran. Recently hexafluorobenzene has been used as a non-exchanging solvent to minimize solvent dilution of the tritium atmosphere (Gill and Jones, 1972).

Heterogeneous catalytic tritiation can cause tritium migration along an alkyl chain. For example, when oleic acid is reduced heterocatalytically only 45% of the tritium is on the 9,10-carbons of the resulting stearic-^{3}H acid (Koch, 1969). The homogeneous catalyzed tritiation of the same substrate with chlorotris (triphenylphosphine) rhodium (I) gives 99% of the tritium associated with the 9,10-carbons.

In a study in our laboratory a large quantity of tritiated chloropheniramine maleate, XI, was needed; catalytic tritiation of the unsaturated precursor X appeared to offer the most direct approach.

In contrast to the literature, Raney nickel catalyst in dioxane-ethyl acetate (1:1) gave the cleanest product in the exploratory atmospheric hydrogenation of the cis, trans isomers of X (Blackburn, et al, 1966). The custom catalytic tritiation by the radiochemical supplier gave 8.5 Ci of ^{3}H-chlorpheniramine. Although we expected to obtain some of the dechlorinated side-product XII, cold runs indicated the material could be removed by the crystallization of the maleate salt from isopropanol. The ^{3}H-chlorpheniramine was carrier-diluted but numerous crystallizations failed to remove the by-product XII. This and other experiences have suggested to us that it is best to affect purification at an early stage on the crude labeled product prior to carrier-dilution. Typical purification methods include preparative paper and thin layer chromatography. The presence of high specific activity impurities can not always be easily removed at a later stage.

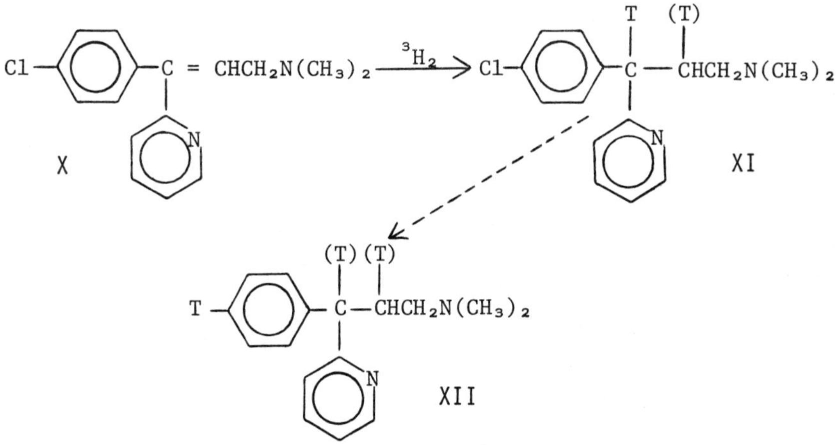

Purification by Phase Solubility Equilibria

In any tritiation reaction, high specific activity contaminants can form either by radiolytic self-decomposition or by undesired side-reactions such as addition or dehalogenation. Phase solubility equilibrium can effect removal of these radiochemical impurities. Solubility phase equilibrium analysis is a standard technique for determination of purity of complex chemical compounds. The technique can be adapted as a purification method by agitating the finely pulverized compound in a solvent in which it is very slightly soluble until equilibrium is reached. The impurities are gradually passed into the mother liquor. The superiority of this purification technique is shown with a comparative study of the purification of a recovered sample of ^{3}H-chlorpheniramine maleate (Table III). The method has been used routinely in our laboratory for purification.

The catalytic addition of tritium to the benzoyl carbonyl gives biostable 7-^{3}H-phenethanolamines. The label is of high specific activity and inexpensive. We have employed the technique in labeling several compounds where the classical carbon-14 synthesis would have been difficult.

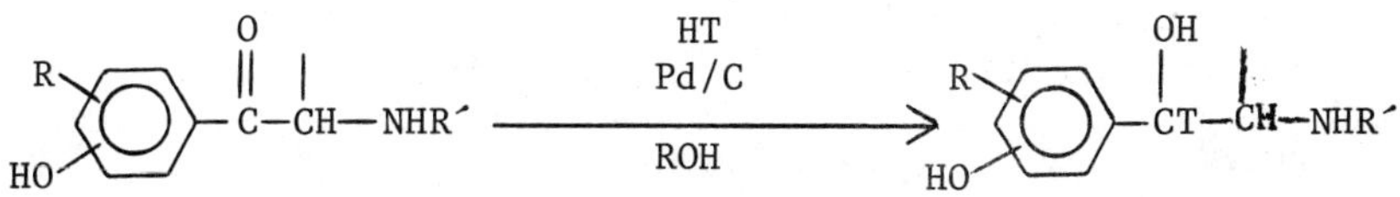

TABLE III. *Comparative Purification of ^{3}H-Chlorpheniramine Maleate by Recrystallization and Solubility Phase Equilibrium*

	Purity (%) Radiochemical	Purity (%) Chemical	Recovery (%)
Crude ^{3}H-Chlorpheniramine Maleate	92	95	
A. 1st Recrystallization			
Solid	94	96	
Mother Liquor	87		
2nd Recrystallization			
Solid	98	97-98	70
Mother Liquor	94		
B. Phase Solubility Equilibrium			
Solid	97-98	99	89-90
Mother Liquor	82		

Tritiated metal hydrides would appear to give higher specificity; however, the specific activity is limited by the tritium exchangeable into the metal hydride reagent.

Catalytic Dehalogenative Tritiation

The usual catalysts are palladium on carbon, barium sulfate, calcium carbonate, or Raney nickel. The halogen-tritium replacement occurs in a semipolar solvent in the presence of base such as triethylamine, alkalai metal hydroxide, or potassium acetate to neutralize the tritiated halogen acid produced.

We first recognized the unusual stability of a tritium label even under such drastic conditions as employed in the Freidel-Crafts reaction. The biphenyl-3-^{3}H, XIII, was reacted with dichloroacetyl chloride and the resulting intermediate XIV converted to the biphenylglyoxal, XV. The specific activity of starting material, intermediates and derived products were nearly identical.

Selective reductive dehalogenation with Raney nickel was employed to label 2-amino-5-bromobenzotrifluoride which was in turn used to label the second moiety of a molecule where

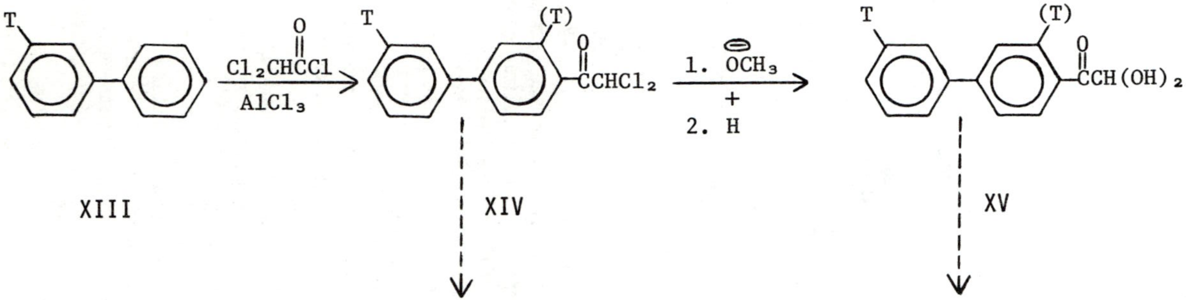

XIII XIV XV

carbon-14 labeling would have been difficult:

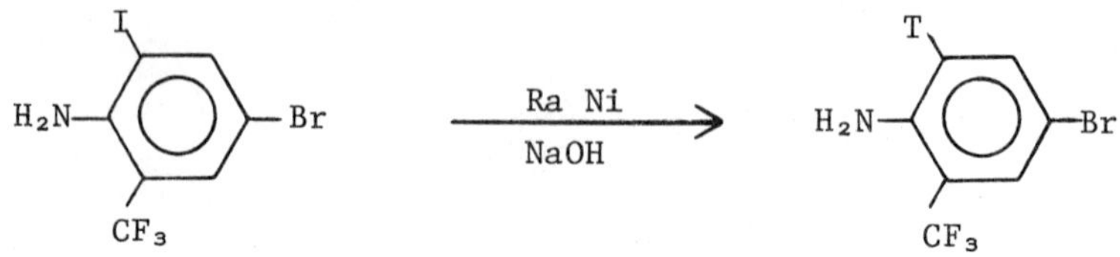

The technique is often used for double labeling a compound where metabolic cleavage of the two moieties is possible. The multiple labels are carried out as separate syntheses to provide flexibility in the subsequent metabolic studies.

Much of the synthetic effort may be involved in the synthesis of a halide precursor near the final step in order to reduce the number of radioactive steps. It was desirable to obtain labeled dipeptide L-histidyl-L-leucine in order to set up an assay procedure for 'converting enzyme' to form angiotensin II. The L-histidyl-L-leucine, XVI, was prepared by the activated ester coupling of the N-carbobenzoxy azide of L-histidine and methyl L-leucine (Burghard, 1967). After removal of the protective groups, XVI was iodinated to the diiododipeptide XVII by a method similar to that used by Brunings (Brunings, 1947) for making L-2,5-diiodohistidine.

The iodinated precursor XVII was added to a methanol solution containing prereduced palladium on calcium carbonate and an equivalent amount of sodium hydroxide. After tritiation, the labile tritium was removed by lyophilization from 80% ethanol, and the product was purified by preparative paper chromatography. The specific activity of the L-histidine-2,5-$^3H_2$-L-leucine, XVI-^3H, was 1.23 Ci/mM.

Tritium placed ortho to phenolic substituents is chemically semi-labile being slowly lost in acid solutions. However, this label can be biostable in certain peptides. Brundish and Wade (Brundish and Wade, 1973) specifically synthesized 46 Ci/mM [3,5-$^3H_2$-Tyr^{23}]-β-corticotrophin-(1-24)-tetracosapeptide, tetracosactin. Tyrosine of the (11-24)-peptide was iodinated, and the diiodo(11-24)-peptide was coupled to the protected 1-10 fragment. They tritiated the deprotected diiodo (1-24)-peptide using 5% palladium on carbon and rhodium on calcium carbonate in dimethylformamide.

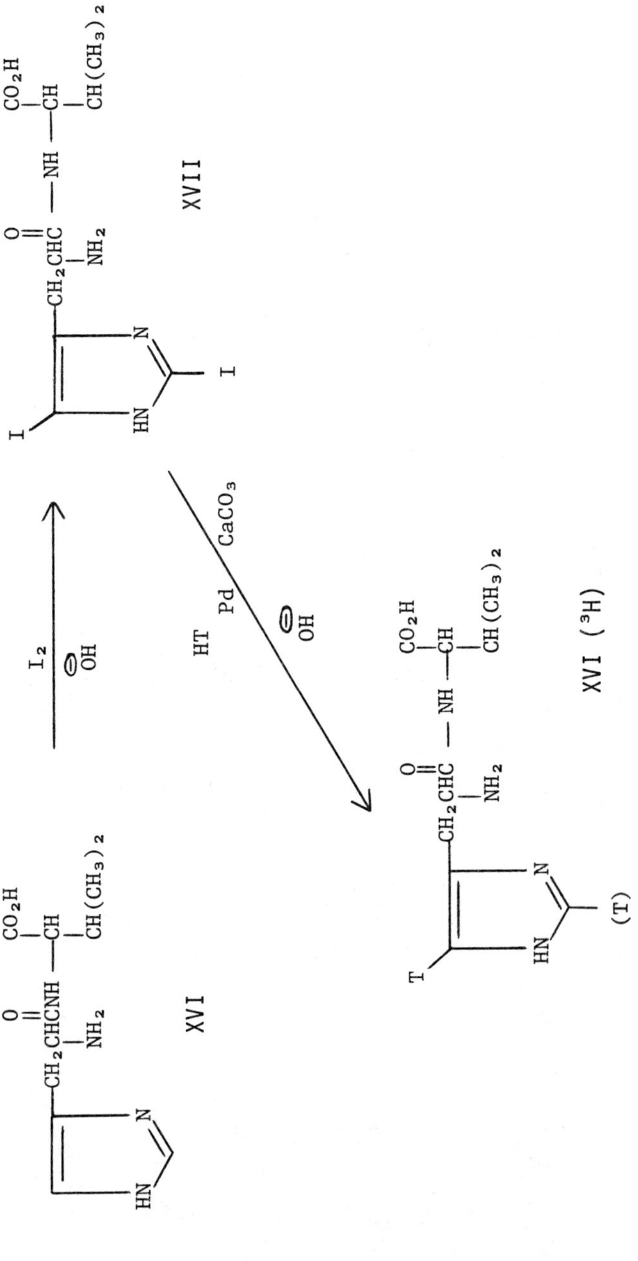

XVI
XVII
XVI (³H)
I_2
OH
HT
Pd
$CaCO_3$

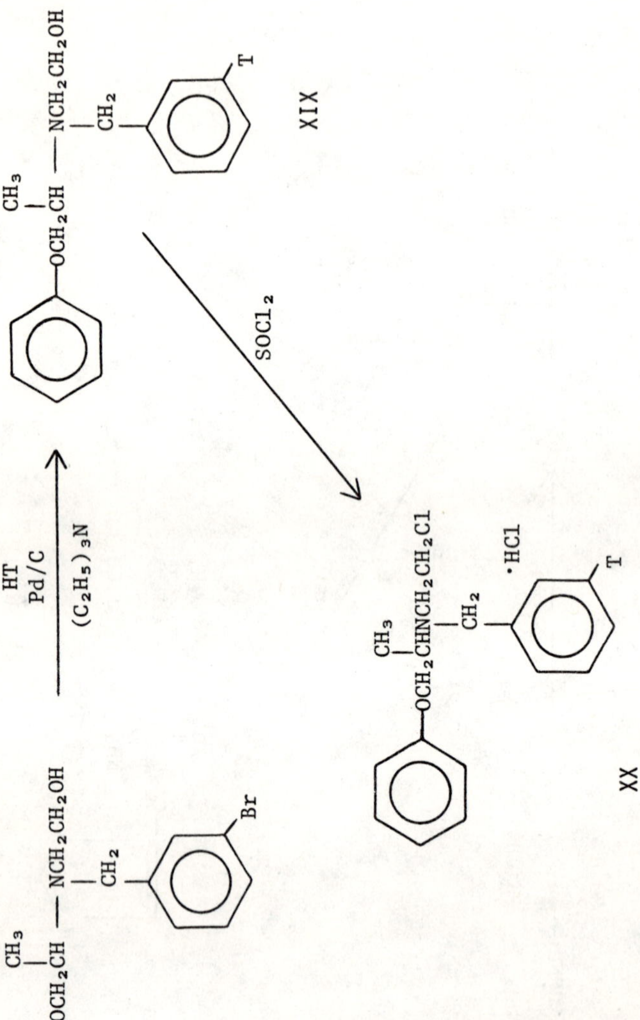
XVIII
XIX
XX
HT
Pd/C
(C2H5)3N
SOCl2
·HCl

In collaboration with an investigation of α-receptor sites of adrenergic blocking agents, we were asked to prepare high specific activity ^{3}H-phenoxybenzamine. We directed our attention to the specific labeling by placing the tritium in the meta position of the benzyl group (Mendelson, et al, 1966). The bromine-substituted precursor XVIII was dehalogenated with tritium by an atmospheric reduction in triethylamine with 5% palladium on carbon. The N-(phenoxyisopropyl)-N-(benzyl-3-^{3}H) ethanolamine, XIX, was converted to ^{3}H-phenoxybenzamine, XX, by treatment with thionyl chloride.

Radiolytic decomposition of all ^{3}H-phenoxybenzamine samples (0.59-6.23 Ci/mM) was evident after storage for a period of 2-3 months.

RADIOCHEMICAL STABILITY

Radiation Stability

The decomposition may be due to a direct radiation effect or an effect produced from a reactive species produced in the solvent (secondary). The lability of a specific chemical compound is described by its G(-M) value: the number of molecules changed per 100 ev adsorbed. The theoretical value is three and the average decomposition rate for a tritiated compound is 0.03% per mci/mM/yr. Bayly and Evans (Bayly and Evans, 1966, 1967) have carried out extensive studies on the stability and storage of radiochemicals. The following precautions can lengthen the shelf-life of labeled compounds:

Dilute out the compound in an inert matrix (<1 mci/ml of solvent)

Addition of a free radical scavenger, e.g. less than 5% ethanol

Low storage temperature; however, aggregation of compound due to freezing of solvent may speed up decomposition.

A vivid example of the radiation induced breakdown of solvent was the accident I was involved in when a sealed thick-wall ampule containing 1.3 Ci of carrier-free elemental sulfur-35 in 15 ml of benzene suddenly exploded on scratching the neck of the vial (Maass, et al, 1962). Needless to say one should use screw-top containers for shipment or open ampules as soon as possible after receipt.

Chemical Stability

Because of the very small amounts of material involved, small chemical or microbiological contamination can cause untoward effects not seen with larger amounts of chemical.

A sudden loss of the identity of sodium acetate-^{14}C given to a biochemist was ascribed to this effect.

Bio-or Metabolic Stability

A best possible guess on the fate of the chemical compound due to metabolism must be made before initiating the radiosynthesis. Suffice it to say the Drug Metabolism department should be consulted before tagging objectives are set. In labeling the diuretic, Triamterene (Blackburn and Burghard, 1966), XXII, with tritium for cytological and double labeling studies we decided to label both the meta and para positions of the 6-phenyl group by treating the dry Grignard of the mixed bromotoluenes, XXI, with tritium oxide on a vacuum line.

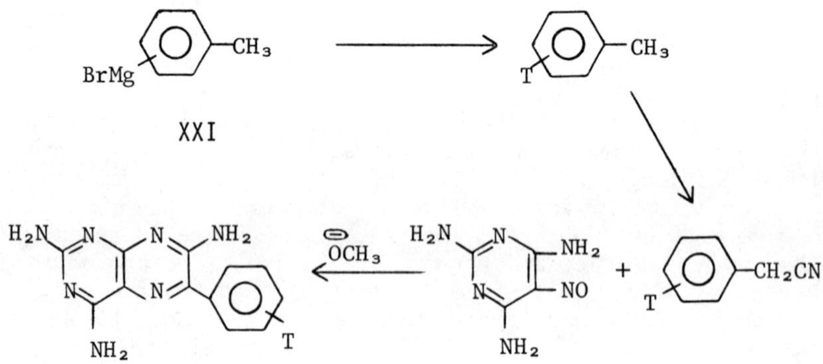

XXII

Preliminary dog experiments showed that there was tritium loss as tritium oxide and there was evidence of the p-hydroxylation of the 6-phenyl group. The preparation was repeated using pure m-bromotoluene for specific labeling in the meta position.

In addition to the normal metabolites, more complete biological degradation may yield fragments containing the labeling isotope. These fragments may be incorporated into normal tissue constituents. This is called nonmetabolic residue and the radioactivity may not be related to the normal metabolites (Rosenblum, et al, 1971). Of course this phenomenon must be minimized if one is determining animal organ and tissue residues.

NIH Shift

Udenfriend, Witkop and coworkers at the National Institutes of Health discovered that tritium, deuterium and halides on aromatic rings may undergo intramolecular migration to an adjacent position upon enzymatic hydroxylation of certain substrates, Scheme A (Chem. & Eng. News, 45 (8), 42, 1951; 45, (17) 57, 1967).

The label retention pathway may actually predominate. A cationoid intermediate has been postulated to explain the results.

Biological loss can be determined by comparing the specific activity of the major metabolite with the parent compound, detection of free tritiated water in the urine, and measurement of the $^{14}C/^{3}H$ ratios of double labeled compounds.

SPECIFICITY OF TRITIUM LABEL

In reviewing the literature, one becomes aware of the increased use of direct synthesis to insure that the major position of the label is known with some degree of certainty (N, nominal label). Critical studies may require that the position be known with higher certainty. For example, we may want to use tritiated metal hydrides for reducing carbonyl or a homogeneous catalyst for addition across a double bond to insure higher specificity. We may also determine tritium distribution by degradation. Otto and Juppe (Otto and Juppe, 1966) prepared biphenyl-2-^{3}H, -3-^{3}H and -4-^{3}H by tritium dehalogenation of the corresponding bromobiphenyls with palladium on barium sulfate. By the degradation scheme depicted below (Scheme B) the label specificity was determined to be 96.1, 99.3 and 98.7%, respectively, for the above tritiated biphenyls. Simon and Floss (Simon and Floss, 1967) have published a book devoted to determination of isotope distribution in labeled compounds.

Stereochemical or confirmation specificity has been shown particularly in the tritiation of steroids. The classic example of 1α or β label obtained from $\Delta^{1,4}$-androstandiene-3,17-dione; Osinski employed palladium on carbon in dioxane to give principally the 1β, 2β label in

STEROID STEREOSELECTIVITY

SCHEME A

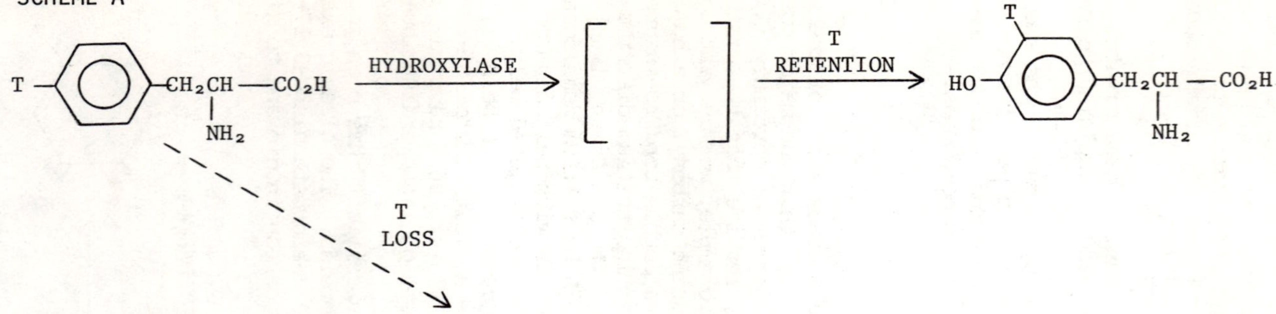

SCHEME B

ISOTOPE DISTRIBUTION BY CHEMICAL DEGRADATION

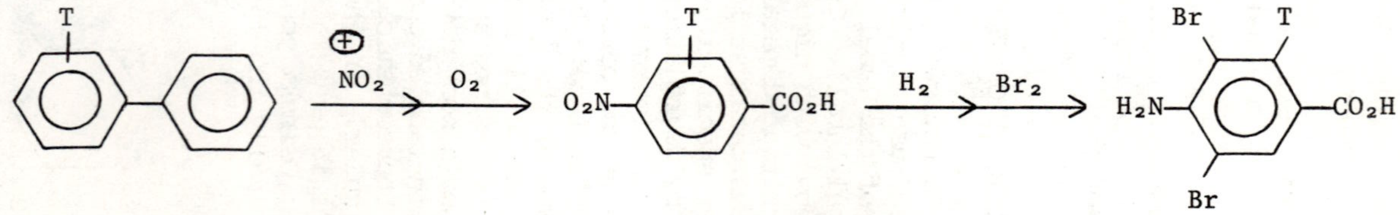

low yields (poor selectivity for $\Delta^{1(2)}$). Djerassi and Gutzwiller (Djerassi and Gutzwiller, 1966) employed chlorotris (triphenylphosphine)rhodium(I) in obtaining the 1α, 2α label with much higher selectivity. The catalyst is known to differentiate different olefinic bonds. There are numerous papers which have shown stereoselective steroid processes occurring in dehydration, dehydrobromination and enzymatic conversions. We look forward to greater applications using tritium as a sensitive and selective probe into the stereoselective reactions of other structural classes of compounds.

HANDLING TRITIUM COMPOUNDS IN THE LABORATORY

The laboratory facilities required to handle tritium are similar to those required for all radioisotopes. An excellent description of the facilities and safety is in the manual, SAFE HANDLING OF RADIONUCLIDES, published by the International Atomic Energy Agency (1973). We have employed the laboratory classification system described since 1965 with excellent results. The laboratory facility (A, B, or C) is gauged to the relative radiotoxicity, activity and physical form of the isotope in question. The classification system has promoted the wider use of radioisotopes in our laboratories; small amounts of tritium labeled compounds can be handled in an ordinary chemical laboratory (Class C area). Such quantities are often sufficient for biochemical tests or studies of chemical mechanisms. The high pressure hydrogenation of a labeled chemical compound is occasionally required. We will then cover the floor and walls of a cubicle holding a 2 9/16" rocking autoclave with adsorbent paper (Figure 1). A wood box with a filtered air intake and exhaust is placed over the autoclave as shown in Figure 2. The cubicle exhaust trunk is attached to the exhaust filter. At the end of the reduction the vessel is vented then resealed and sent to the radiosynthesis laboratory for product recovery (Figure 3).

The manipulation of tritium compounds should be carried out in a segregated area within a Class A laboratory. The high activity and difficulty in controlling contamination requires shoe covers, frequent smears of facilities to prevent spread of contamination and urine assay of chemists to check possible ingestion. We prefer to have the initial multicurie catalytic tritiation carried out by a radiochemical supplier† who specializes in the technique and is set-up for handling

†The New England Nuclear Corp., Boston, Mass., has routinely carried out our initial tritium incorporation.

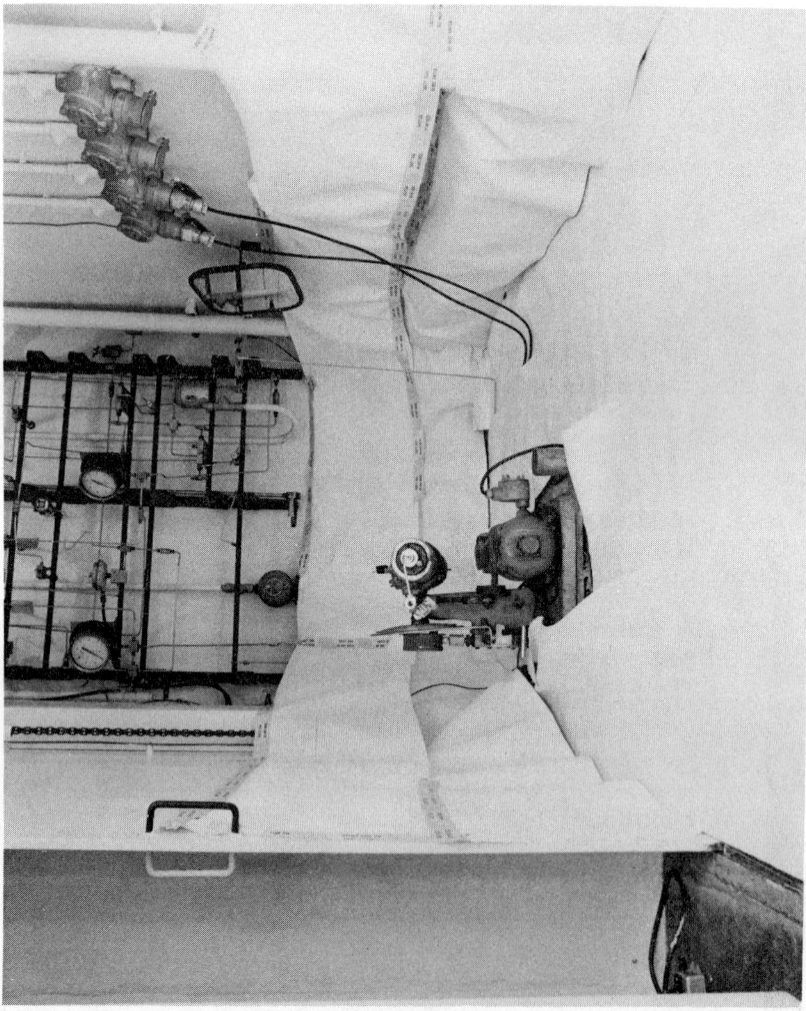

Figure 1

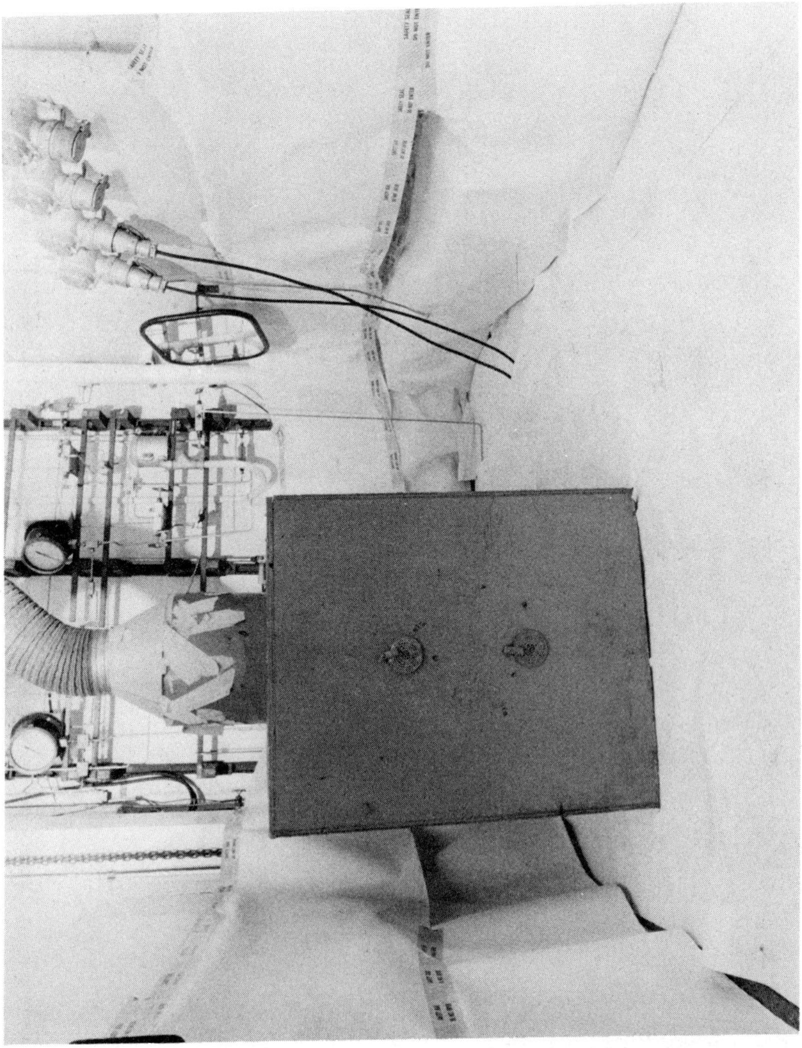

Figure 2

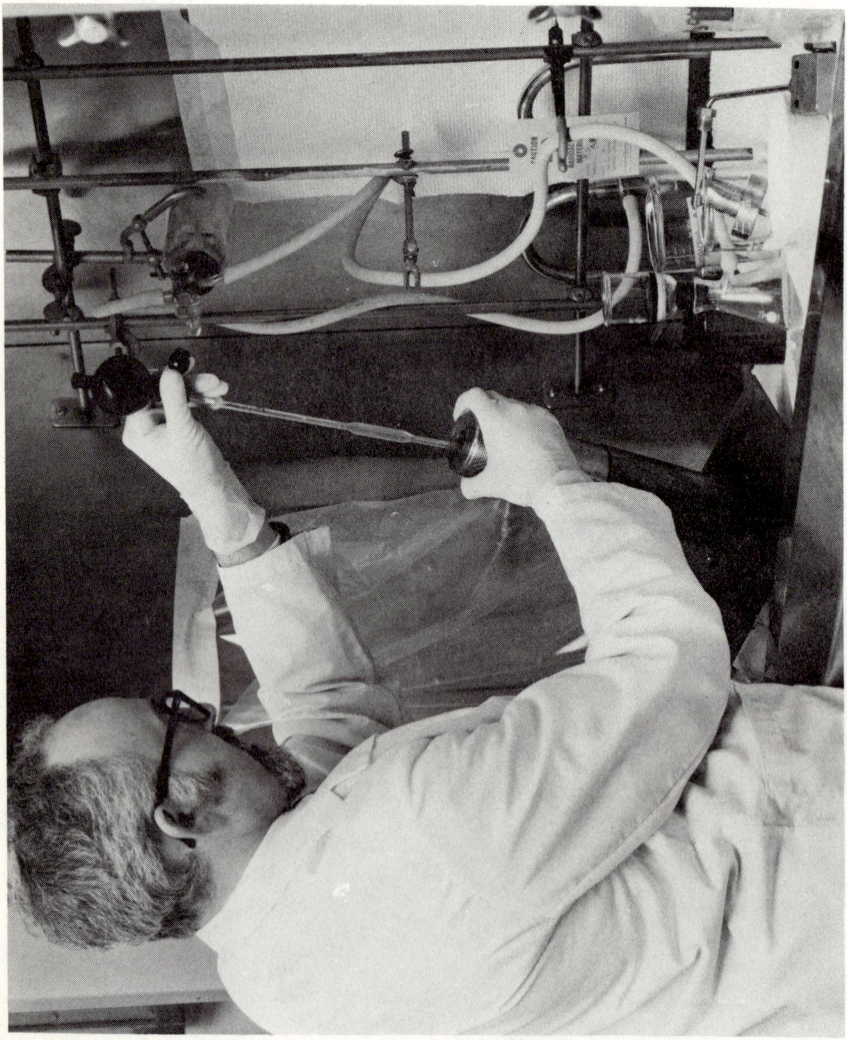

Figure 3

multicurie tritium gas on a routine basis. We, however, explore thoroughly the methodology with a series of cold runs with an atmospheric microhydrogenation apparatus (Harrison and Harrison, 1964) (Figure 4). We send the procedure and the essential raw materials to the radiochemical company for the custom tritiation. They filter the catalyst and remove most of the labile tritium by evaporation of a polar solvent.

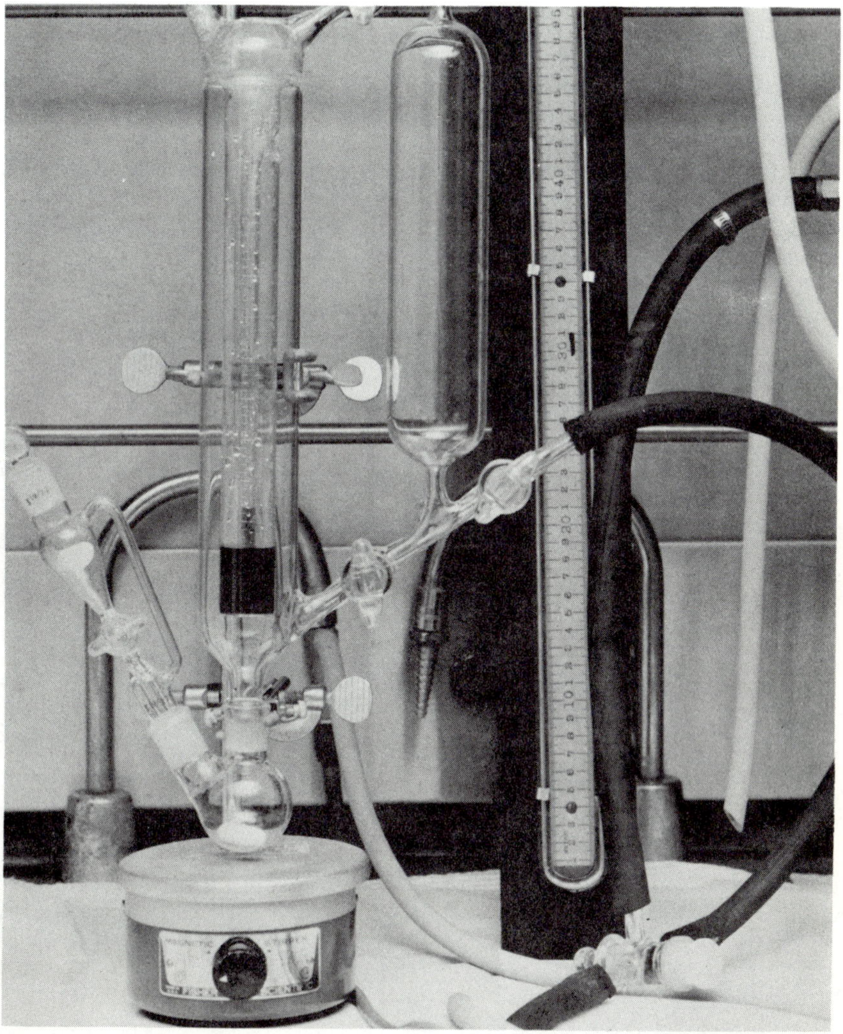

Figure 4

Figure 5

To handle such volatile tritium intermediates as tritium oxide, we constructed a general vacuum manifold similar to that reported by Isbell (1959) at the National Bureau of Standards. This unit was recently discarded in favor of several small manifolds designed for the specific chemical operations. An example of a vacuum manifold for manipulating volatile compounds is shown in Figure 5.

Operations with solid tritium compounds of high specific activity should be carried out in a closed safety enclosure like the glove box depicted in the next picture (Figure 6) to

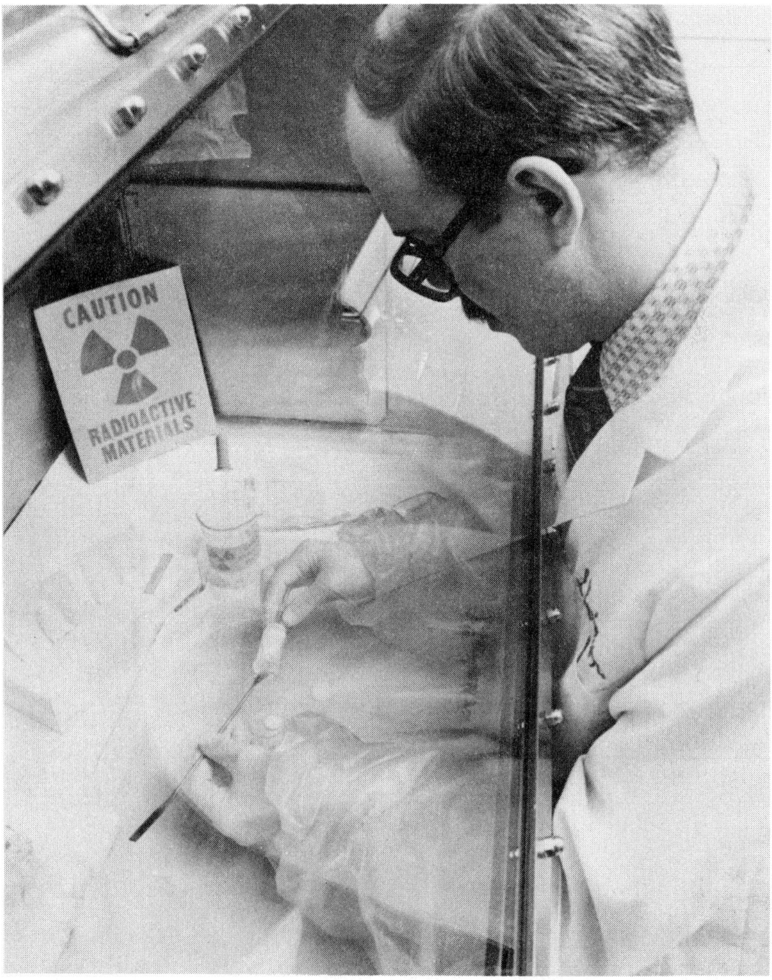

Figure 6

prevent spread of contamination. The gloves should be washed with solvent before removing the hands from the safety enclosure. Amersham uses tritium process boxes (Evans, 1974) equipped with foot-operated hand washers.

LOOK INTO THE FUTURE

Jacobson and coworkers (1964) prepared estradiol-6, 7-^{3}H with 56-93% theoretical tritium content. Three milligrams of the unsaturated precursor was catalytically tritiated with a microhydrogenator. They proceeded to carry out a four-step ultramicro synthesis with 10-100 micrograms. The intermediates and products were separated by preparative paper chromatography.

We feel the use of microchemical techniques and high speed liquid chromatography can revolutionize high specific activity radiosynthesis. The reduced synthetic scale can increase the speed and safety of multistep sequences. Carrier dilution can be applied to the final product to increase shelf life.

ACKNOWLEDGMENTS

We acknowledge the contribution of the following radiochemists to our growing knowledge and expertise with tritium: Mr. M. Emas, Mr. G. Burghard, Mr. A. Post, Dr. W. Mendelson, Dr. V. Spaziano, Mr. L. Weaner, Mr. A. Villani.

REFERENCES

1. Bayly, R. and Evans, E., J. Labelled Compds., 2, 1 (1966).
2. Bayly, R. and Evans, E., J. Labelled Compds., 3, (supplement 1), 349 (1967).
3. Blackburn, D. and Burghard, G., J. Labelled Compds., 2, 62 (1966).
4. Blackburn, D., Burghard, G. and Post, A., Euratom, Proceedings of the Conference on Methods of Preparing and Storing Marked Molecules, Eur 3746d,f,e, Brussels, Nov. 28, 1966, p 1135.
5. Brundish, D. E. and Wade, R., J. Chem. Soc. Perkin I, 2875 (1973).
6. Brunings, K. J., J. Am. Chem. Soc., 69, 205 (1947).
7. Burghard, G., unpublished results, (1967).
8. Djerassi, C. and Gutzwiller, J., J. Am. Chem. Soc., 88, 4537 (1966).

9. Evans, E. A., Tritium and its Compounds, 2nd. edition, Halsted Press, John Wiley and sons, N. Y., 1974.
10. Gill, E. W. and Jones, G., J. Labelled Compds, 8, 237, (1972).
11. Harrison, I. T. and Harrison, S., Chem. and Ind. (London), May 16, 834 (1964).
12. Isbell, H. S., J. Res. of Natl. Bur. Stnds., 63A, 177 (1959).
13. Jacobson, H. I., Sahn, N. N., Newmann, H. G., Calucci, V. and Jensen, E. V. (University of Chicago), Synthesis, Storage and Manipulation of Tritiated Steroid Estrogens of High Specific Activity, paper presented at the 1964 Oak Ridge Radioisotope Conference.
14. Kaiser, C., Swagzdis, J., Flanagan, T., Lester, B., Burghard, G., Green, H. and Zirkle, C., J. Med. Chem., 15, 1146 (1972).
15. Kalberer, F. and Rutschmann, J., Helv. Chim. Acta, 46, 586 (1963).
16. Koch, G. K., J. Labelled Compds., 5, 110 (1969).
17. Maass, A. R., Flanagan, T. L., Blackburn, D. and Smyth, M., Health Physics, 9, 731 (1962).
18. Mendelson, W., Blackburn, D. and Spaziano, V., Euratom, Proceedings of the Conference on Methods of Preparing and Storing Marked Molecules, Eur 3746d,f,e, Brussels, Nov. 28, 1966, p 815.
19. Otto, P. Ph. H. L., and Juppe, G., J. Labelled Compds., 2, 349 (1966).
20. Rosenblum C., Trenner, N. R. and Wolf, D. E., J. Labelled Compds., 7, 225 (1971).
21. Rothchild, Seymour, Advances in Tracer Methodology, Vol. I (1963), Vol. II (1965), and Vol. III (1966), Plenum Press, N. Y.
22. Simon, H. and Floss, H. G., Anwendung von Isotopen in der Organischen Chemie und Biochemie, Band I. Bestimmung der Isotopenverteilung in Markierten Verbindingen, 1967, Springer-Verlag, N. Y.
23. van Kordelaar, J., Favier, J. and Kitcher, J., J. Labelled Compds., 9, 635 (1973).
24. Yavorski, P. M. and Gorin, E., J. Am. Chem. Soc., 84, 1071 (1962).

ENZYMATIC EPOXIDATION AND OXYGEN ACTIVATION

SHELDON W. MAY

School of Chemistry,
Georgia Institute of Technology,
Atlanta, Georgia 30332

An enzyme system consisting of three protein components has been isolated from *Pseudomonas oleovorans* by Coon and co-workers (Peterson *et al.*, 1966, 1967; Peterson and Coon, 1968; McKenna and Coon, 1970; Lode and Coon, 1971; Ueda *et al.*, 1972). In the presence of NADH and molecular oxygen, this system catalyzes the ω-hydroxylation of fatty acids and the terminal hydroxylation of alkanes. The three protein components have been identified as rubredoxin, reductase, and ω-hydroxylase.

Rubredoxin is an iron-sulfur protein of molecular weight 19,000 containing one or two iron atoms per molecule. This enzyme, which apparently functions as an electron carrier in the system, has been purified to homogeneity (Lode and Coon, 1971) and its amino acid sequence has been determined (Benson *et al.*, 1971). It has also been shown that in the presence of reductase and NADH, rubredoxin catalyzes the reduction of octylhydroperoxide to 1-octanol (Boyer *et al.*, 1971). However, hydroperoxides have not been identified as free intermediates in the hydroxylation reaction.

The reductase component has also been purified to homogeneity. It is flavoprotein of molecular weight 55,000 which transfers electrons from NADH to rubredoxin (Ueda *et al.*, 1972; Ueda and Coon, 1972). In contrast, the hydroxylase is very unstable and attempts to purify this enzyme to homogeneity have so far been unsuccessful. Recent work suggests that this protein contains non-heme iron and requires the presence of phospholipid for hydroxylation activity (Ruettinger *et al.*, 1974).

Recently, we reported that this enzyme system also catalyzes the conversion of terminal olefins to the corresponding 1,2-oxides (May and Abbott, 1972, 1973 a,b; May *et al.*, 1973, 1974; May and Schwartz, 1974). The epoxidation reaction was found to require the presence of all three protein components as well as NADH and molecular oxygen. The substrate 1-octene, which contains both a terminal methyl group and a terminal double bond, is converted to both 7-octene-1-ol and 1,2-epoxyoctane. On the other hand, 1,7-octadiene, which does not have a terminal methyl group, is converted exclusively to 7,8-epoxy-1-octene, and this product, in turn, is further oxidized to 1,2-7,8-diepoxyoctane. On the basis of cofactor oxidation rates, 1,7-octadiene is more reactive than either 1-octene or octane, and this result indicates that the epoxidation reaction can proceed even more readily than the hydroxylation reaction. A 1:1 stoichiometry was obtained between the amount of NADH oxidized and the amount of product formed, which strongly suggests that the ωHS is acting as a mixed-function oxidase in catalyzing the epoxidation reaction.

The exact relationship between the epoxidation and hydroxylation reactions catalyzed by the ωHS is presently unclear. Both reactions proceed at comparable rates, are inhibited by cyanide, affected similarly by pH, and show the same cofactor selectivities. These facts suggest that epoxidation and hydroxylation proceed via similar mechanisms. However, since the hydroxylase used in our studies was not homogeneous, it is conceivable that the two reactions are actually associated with two distinct, though similar, enzymes present in our hydroxylase preparations. Furthermore, even if the same "oxygenase" is involved, it is not known whether the reactions occur at the same active site or whether the functional role of each protein component is the same in both reactions. On the other hand, it appears that the two reactions are not completely independent, since we find that they are mutually competitive under conditions where the concentrations of rubredoxin, reductase, O_2 and NADH are high enough to support both reactions simultaneously at their maximal rates. Accordingly, detailed substrate specificity studies were carried out in order to define the relationship between these two reactions and to compare the epoxidation reaction of this enzyme system with those catalyzed by other microbial and mamalian systems (Van der Linden, 1963; Bloom and Schull, 1955, Coronelli _et al._, 1964; Maynert _et al._, 1970; Jerina _et al._ 1968, 1970; Udenfriend, 1971; Grover _et al._, 1973; Yamamoto and Bloch, 1970).

Table I shows the effect of varying the carbon chain length on both the epoxidation of a series of dienes and on the hydroxylation of a series of alkanes. It is evident that for both epoxidation and hydroxylation maximal activity is observed at a carbon chain length of C_8. In the case of epoxidation, as the chain length is increased, the relative rate of reaction decreases somewhat, but dodecadiene still has 85% of the reactivity of octadiene. On the other hand,

TABLE I

EFFECT OF CARBON CHAIN LENGTH ON EPOXIDATION[(a)]			EFFECT OF CHAIN LENGTH ON HYDROXYLATION (data from Peterson and Coon, 1968)	
Substrate	Relative Rate of NADH Oxidation (340nm)[(b)]	Product Detected by Gas Chromatography	Substrate	Relative Rate of Cofactor Oxidation
1,5-Hexadiene	40	(c)		
1,6-Heptadiene	67	6,7-Epoxy-1-heptene	Hexane	75
1,7-Octadiene	100	7,8-Epoxy-1-octene	Heptane	87
1,8-Nonadiene	87	8,9-Epoxy-1-nonene	Octane	100
1,9-Decadiene	92	9,10-Epoxy-1-decene	Nonane	99
1,10-Undecadiene	80	10,11-Epoxy-1-undecene	Decane	80
1,11-Dodecadiene	85	11,12-Epoxy-1-dodecene	Dodecane	47
			Hexadecane	3

[(a)]Reaction conditions as for TABLE IV, May and Abbott, 1973a. [(b)]A relative rate of 100 corresponds to the net oxidation of 6.6 nmole of NADH per minute. [(c)]At low concentrations, the peak corresponding to 5,6-Epoxy-1-hexene is obscured by the large peak from the hexane used in the extraction procedure (see May and Abbott, 1973a).

when the carbon chain length is decreased by only 2 carbon atoms, to hexadiene, the relative rate of reaction falls off much more rapidly. In contrast, for the hydroxylation reaction, increasing the carbon chain length has a much more drastic effect on the relative rate of reaction but decreasing the carbon chain to hexane results in only a 25% decrease in reactivity.

If this trend were continued to even shorter substrates, one would expect to reach a point where only the hydroxylation reaction proceeds at a significant rate. As shown in Table II, this is indeed the case, and both propylene and 1-butene are hydroxylated to the corresponding unsaturated alcohols, but not epoxidated. McKenna and Coon (1970) have reported that whereas internal methylene groups of straight-chain substrates are inert to hydroxylation, cyclohexane stimulates the rate of cofactor oxidation by the ω-hydroxylation system. As shown in Table II, we find that internal double bonds of both straight-chain and cyclic compounds are inert to epoxidation. Cyclohexene is converted to both isomeric cyclohexene-ols, which are formed in comparable amounts, but not to epoxycyclohexane. Similarly, neither cis-5-decene nor 2-octene are epoxidated. Styrene is not epoxidated at either the exocyclic double bond or the aromatic ring, and it has been reported that several other aromatic compounds are not hydroxylated (McKenna and Coon, 1970). As expected, gas chromatographic analysis showed that cyclohexanol is formed from cyclohexane.

TABLE II. *Substrate Specificity*[a]

Substrate	Product Detected by Gas Chromatography
Propylene	Allyl Alcohol
1-Butene	3-Buten-1-ol
Cyclohexene	2-Cyclohexene-1-ol 3-Cyclohexene-1-ol
Cyclohexane	Cyclohexanol
cis-5-Decene	cis-5-Decen-1-ol
Styrene	---

[a]Standard reaction conditions were used except that gaseous substrates were introduced via a modified procedure (May et al., unpublished results).

The fact that on the basis of cofactor oxidation rates octadiene is more reactive than either octane or octene, suggests that with straight-chain substrates of moderate carbon chain length, epoxidation proceeds more readily than hydroxylation. However, since it is conceivable that octane and octadiene bind to the enzyme with vastly different affinities, the relative rates of the two reactions were compared using a single substrate capable of undergoing either reaction, e.g., 1-decene. The results are shown in Table III. Two separate experiments were carried out under different reaction conditions and in both cases substantially more 1,2-epoxydecane than 9-decene-1-ol was formed. In both experiments, an excellent agreement was obtained between the total amount of NADH oxidized and the total amount of product formed. This indicates that under the reaction conditions, no significant enzymatic oxidation of 1,2-epoxydecane or 9-decene-1-ol occurs, and, therefore, the product distribution is a reliable indication of the relative rates of the epoxidation and hydroxylation reactions.

Taken together, these results surely do not fit into any straight forward chemical reactivity pattern. In peracid epoxidation reactions, for example, cyclohexene and 2-hexene are more than twenty times as reactive as 1-hexene, reflecting the stimulatory effect of alkyl substitution on electrophilic addition to the double bond. Similarly, no significant effect of carbon chain length on epoxidation of a terminal double bond is observed, and even propylene is 80% as reactive as 1-octene (Swern, 1947; House, 1972). In the enzymatic reaction, our results with the moderately long straight chain alkene, 1-decene, reveal the expected predominance of epoxidation over hydroxylation, presumably reflecting the greater inherent reactivity of the π electron system of the double bond. Yet, the reactivity patterns observed with internal olefins, short chain substrates and the homologous series of terminal olefins are inexplicable in these terms. Cyclic compounds clearly bind to the enzyme and undergo hydroxylation even though the methylene carbons of straight chain alkanes are inert to hydroxylation. Yet, the internal double bonds of cyclic compounds are just as unreactive toward epoxidation as those of straight chain compounds. It is obvious that factors other than simple chemical reactivity differences are being reflected in our results with the enzymatic system.

We have previously suggested that the configurations of the methylene chains of hydrocarbon substrates critically affect the mode of binding of such compounds to the active sites of oxygenases (May and Abbott, 1973a). The notion that such factors are indeed among those being reflected in the substrate reactivity patterns we observe is supported by recent results with a homologous series of imidoester inhibitors (Schwartz, R.D., unpublished observations).

TABLE III. *Relative Rates of Epoxidation and Hydroxylation for 1-Decene*

Experiment	Reaction Time (sec)	nMoles NADH Oxidized[a]	nMoles Epoxide Formed	nMoles Alcohol Formed	Total nMoles Product
A[b]	470	22	13	7	21
B[c]	1450	52	39	12	51

[a]These values are corrected for the endogeneous oxidation of NADH (May and Abbott, 1973a). [b]Reaction conditions as for Table I, except that 1 μmole substrate in 20 μl-acetone was added to the reaction mixture. [c]Reaction mixture contained 500 μg "hydroxylase" and 7 μg reductase.

Addition of the methyl imidoesters IE6, IE8 or IE12 to the functioning enzymatic epoxidation system causes rapid inhibition of the reaction. Table IV lists results obtained with imidoesters of varying carbon chain length. The pattern of relative potency of inhibitors parallels that of substrate reactivity, as would be expected if changes in the methylene chain primarily affect the binding step in the catalytic process. In view of the simple chemical structures of the substrates and inhibitors, hydrophobic interactions must play a key role in the binding process.

There is currently considerable speculation as to the nature of the "activated oxygen" species in reactions catalyzed by mixed-function oxidases. Hamilton first suggested that an "oxenoid" species is generated in these reactions by transfer of two electrons to oxygen prior to, or concurrent with, transfer of an oxygen atom to the substrate (Hamilton, 1964). This mechanism has found particular favor in accounting for the reactions of P-450 - containing oxygenases — particularly the "NIH shift" which occurs during aromatic hydroxylation reactions (Jerina, 1973). A number of organic model systems which undergo oxenoid-type reactions have also been designed and investigated (Hamilton, 1969, 1971, 1973; Jerina, 1973).

TABLE IV. *Imidoester Inhibition of Epoxidation by Crude Enxyme Extract*

Imidoester	% Inhibition[a]
IE2	3
IE3	7
IE4	47
IE5	47
IE6	85
IE8	94
IE10	83
IE12	91
IE13	89
IE14	93
IE15	36
IE18	22

[a]Percent inhibition is defined as $\frac{C-I}{C} \times 100$, where C and I are the total amount of epoxide synthesized after 60 min in the absence and presence of imidoester, respectively. The standard assay conditions used and the details of the experimental procedure are described elsewhere (May et al., in preparation).

In the case of the P. oleovorans epoxidation system, the data in Table II and the high reactivity of octadiene relative to octane or octene are in accord with the expected reactivity of an electrophilic "oxenoid" species. However, from much of the other data reported here we conclude that if such a species is indeed involved in reactions catalyzed by this enzyme system, its reactivity is moderated with unusual severity by factors related to the mode of substrate binding. Our recent discovery of the high stereoselectivity of the epoxidation reaction is in line with this conclusion. We have found that when octadiene is incubated with whole cells of P. oleovorans, more than 90% of the 7,8-epoxy-1-octene formed is the R-(+) isomer (May and Schwartz, 1974). This example points up the critical role of substrate binding, since production of optically active epoxides requires binding of octadiene in such a way as to totally preclude oxygen attack from one face of the octadiene molecule (May and Schwartz, 1974).

In contrast to several other oxygenase systems, it has been suggested that free superoxide is not involved in reactions catalyzed by the *P. oleovorans* enzyme system. We have reported (May and Abbott, 1973a) that the hydroxylation and epoxidation reactions are not supported by a xanthine oxidase-based superoxide generating system and are not inhibited by superoxide dismutase. Similarly, Ueda and Coon (1972) have shown that the rubredoxin-dependent reduction of cytochrome c is not inhibited by superoxide dismutase. On the other hand, we have recently reported that the rubredoxin and reductase components of the ωHS jointly catalyze the oxidation of epinephrine in a process which does appear to involve superoxide (May *et al.*, 1973). Yet, despite this apparent difference in reaction mechanisms, epinephrine oxidation and epoxidation are mutually competitive processes, with the extent of inhibition of epoxidation by epinephrine diminishing with epinephrine concentration.

That the enzymes of the ω-hydroxylation system do not produce significant amounts of free superoxide, by some unknown mechanism, as a side reaction during epoxidation is clear from the oxygen stoichiometry data in Table V. On the other hand, the NADH/adrenochrome stoichiometry data we have reported (May *et al.*, 1973) indicate that the enzymes do not simply generate a small amount of initiator which oxidizes epinephrine in a self-perpetuating process.

TABLE V. *Oxygen Stoichiometry for the Epoxidation of Octadiene*

Product	Reaction Time (sec)	nMoles of O_2 Used[a]	nMoles of Product Detected	Ratio of Product/O_2
7,8-Epoxy-1-octene	270	100	110	1.1

[a]This value is corrected for the endogenous consumption of oxygen which occurs in the absence of substrate.

A mechanism consistent with our data involves the reaction of reduced rubredoxin with oxygen to form a rubredoxin-superoxide complex, which is insensitive to superoxide dismutase. This complex could either participate directly in epoxidation (with the "hydroxylase" and substrate) or initiate an epinephrine oxidizing sequence similar to that proposed by Misra and Fridovich (1972), which would be expected to be sensitive to superoxide dismutase. Extensive dissociation of this complex in the presence of hydroxylase and substrate (under which conditions epinephrine oxidation still proceeds) is ruled out by the data of Table V and by the insensitivity of epoxidation to direct inhibition by superoxide dismutase or stimulation by xanthine oxidase. However, it is likely that in the absence of compounds capable of reacting directly with the complex, dissociation to give free superoxide does occur.

It is recognized that several aspects of the proposed mechanism remain to be clarified. For example, the data presented here do not preclude the possibility that epinephrine binds to reduced rubredoxin before oxygen, and thereby influences the oxygen activation mechanism. If this were true, a superoxide species could be involved in adrenochrome formation but not in the usual oxygenase reactions of the *P. oleovorans* system. In addition, the usual caution must be exercised in interpreting data of the sort presented here since evidence exists that other species may be derived from superoxide (Khan, 1970; Beauchamp and Fridovich, 1970) and that SDM may not be totally specific for superoxide (Finazzi Agro *et al.*, 1972). Nevertheless, we regard the proposed mechanism as a working hypothesis, on which current investigations are being based.

REFERENCES

1. Beauchamp, C., and Fridovich, I., J. Biol. Chem., 245, 4641, (1970).
2. Benson, A., Tomoda, K., Chang, J., Matsueda, G., Lode, E.T., Coon, M.J., and Yasundobu, K.T., Biochem. Biophys. Res. Comm., 42, 640, (1971).
3. Bloom, A.C., and Schull, G.M., J. Am. Chem. Soc., 77, 5767, (1955).
4. Boyer, R.F., Lode, E.T., and Coon, M.J., Biochem. Biophys. Res. Comm., 44, 925, (1971).
5. Coronelli, C., Kluepfel, D., and Sensi, P., Experentia, 20, 208, (1964).
6. Finazzi Agro, A., Giovagnoli, C., DeSole, P., Calabrese, L., Totilio, G., and Mondovi, B., FEBS Lett., 21, 183, (1972).

7. Grover, P.L., Hewer, A., and Sims, P., FEBS Lett., 34, 63, (1973).
8. Hamilton, G.A., J. Am. Chem. Soc., 86, 3391, (1964).
9. Hamilton, G.A., Adv. Enzymol., 32, 55-96, (1969).
10. Hamilton, G.A., Progr. Bioorg. Chem., (E.T. Kaiser and F.J. Kezdy, eds.), Vol. 1, pp. 83, Wiley, N.Y., (1971).
11. Hamilton, G.A., Ann. N.Y. Acad. Sci., 212, 4, (1973).
12. House, H.O., Modern Synthetic Reactions, 2nd ed., pp. 296-307, W.A. Benjamin, Inc., Menlo Park, Ca., (1972).
13. Jerina, D.M., Daly, J.W., Witkop, B., Zaltzman-Nirenberg, P., and Udenfriend, S., Arch. Biochem. Biophys., 128, 176, (1968).
14. Jerina, D.M., Daly, J.W., Witkop, B., Zaltzman-Nirenberg, P., and Udenfriend, S., Biochemistry, 9, 147, (1970).
15. Jerina, D.M., Chem. Technol, 4, 120, (1973).
16. Khan, A.V., Science, 168, 476, (1970).
17. Lode, E.T., and Coon, M.J., J. Biol. Chem., 246, 791, (1971).
18. May, S.W., and Abbott, B.J., Biochem. Biophys. Res. Comm., 48, 1230, (1972).
19. May, S.W., and Abbott, B.J., J. Biol. Chem., 248, 1725, (1973a).
20. May, S.W., and Abbott, B.J., Abstracts, 166th National Meeting of the American Chemical Society, Chicago, Ill., Biol. No. 218, (1973b).
21. May, S.W., Abbott, B.J., and Felix, A.F., Biochem. Biophys. Res. Comm., 54, 1540, (1973).
22. May, S.W., and Schwartz, R.D., J. Am. Chem. Soc., 96, 4031, (1974).
23. May, S.W., Abbott, B.J., and Schwartz, R.D., Abstracts, 168th National Meeting of the American Chemical Society, Atlantic City, N.J., Sept. 1974, Ind. Eng. Chem. No. 26.
24. Maynert, E.W., Foreman, R.L., and Watabe, T., J. Biol. Chem., 245, 5234, (1970).
25. McKenna, E.J., and Coon, M.J., J. Biol. Chem., 245, 3882, (1970).
26. Misra, H.P., and Fridovich, I., J. Biol. Chem., 247, 3170, (1972).
27. Peterson, J.A., Basu, D., and Coon, M.J., J. Biol. Chem., 241, 5162, (1966).
28. Peterson, J.A., Kusnose, M., Kusnose, E., and Coon, J.J., J. Biol. Chem., 242, 4334, (1967).
29. Peterson, J.A., and Coon, J.J., J. Biol. Chem., 243, 329, (1968).
30. Ruettinger, R.T., Olson, S.T., Boyer, R.F., and Coon, M.J., Biochem. Biophys. Res. Comm., 57, 1011, (1974).
31. Swern, D., J. Am. Chem. Soc., 69, 1969, (1947).

32. Udenfriend, S., Ann. N.Y. Acad. Sci., 179, 295, (1971).
33. Ueda, T., Lode, E.T., and Coon, M.J., J. Biol. Chem., 247, 2109, (1972).
34. Ueda, T., and Coon, M.J., J. Biol. Chem. 247, 5010, (1972).
35. Van der Linden, A.C., Biochim. Biophys. Acta, 77, 157, (1963).
36. Yamamoto, S., and Bloch, K., J. Biol. Chem., 245, 1670, (1970).

Abbreviations Used: IE2, methyl acetimidate; IE3, methyl propanoimidate; IE4, methyl butanoimidate; IE5, methyl pentanoimidate; IE6, methyl hexanoimidate; IE8, methyl octanoimidate; IE10, methyl decanoimidate; IE12, methyl dodecanoimidate; IE13, methyl tridecanoimidate; IE14, methyl tetradecanoimidate; IE15, methyl pentadecanoimidate; IE18, methyl octadecanoimidate; ωHS, ω-hydroxylation system.

The financial support of Research Corporation and the National Science Foundation (BMS74-20830) is gratefully acknowledged.

EXAMPLES OF THE USE OF DTA FOR THE STUDY OF CATALYTIC ACTIVITY AND RELATED PHENOMENA

PATRICK K. GALLAGHER, DAVID W. JOHNSON, JR.
and EVA M. VOGEL

Bell Laboratories, Murray Hill, New Jersey 07974

ABSTRACT

Advantages of the use of differential thermal analysis (DTA) for the rapid screening of potential catalysts are given. Evidence is presented which indicates that there is a close correlation between DTA results and more conventional reactor studies. The effects of heating rate and gaseous flow rate are also presented.

The method is used to study the catalytic activity of a variety of substituted lanthanum manganates(III) for the oxidation of CO. Finally, the effect of Pt additions on this activity, particularly in the presence of SO_2, was determined by DTA.

INTRODUCTION

Locke and Rase (1960) pointed out the value of differential thermal analysis (DTA) as a useful technique for rapidly screening potential catalysts. They followed the adsorption of H_2 on a supported Ni catalyst used for hydrogenation. This adsorption leads to an endothermic peak in the DTA experiment which they studied as a function of catalyst pretreatment and the presence of poisons. As Locke and Rase (1960) indicate, DTA is both a rapid and inexpensive method for screening potential catalysts, for evaluating the effects of various pretreatments, and for determining the poisoning tendencies of agents added to either the gas stream or the catalyst. This speed and simplicity provide opportunities for extensive surveys which are otherwise inconvenient.

Recently there has been a renaissance of the technique for this purpose induced by the search for suitable catalysts to control automotive emissions. Johnson and Gallagher (1973) and Johnson et al (1974) investigated the oxidation of CO, C_6H_{14}, and C_2H_6 in the presence of various oxide catalysts having the perovskite structure. The effects of Pt additions to the catalysts and poisoning by SO_2 in the gas stream were described. Similarly, Wedding and Farrauto (1974) and Farrauto and Wedding (1973) performed similar studies using copper chromite catalysts.

The preceding studies utilized commercial apparatus in the conventional dynamic mode. Papadatos and Shelstad (1973) demonstrated that valuable information could also be obtained when using such instruments in an isothermal fashion. Ishii et al, (1974) described a simple gas-flow DTA method which they used to study the chlorination of some ores of magnesium. This technique was also extended to studies of the thermal decomposition of $KClO_4$ catalysed by various preparations of Fe_2O_3 by Furuichi et al, (1974), and to a variety of other catalytic reactions by Ishii (1974).

In many DTA experiments the equipment is not designed to flow the gas directly through the sample bed but rather through the sample compartment and simply over the sample bed. Under these conditions Wedding and Farrauto (1974) have shown that the DTA signal is not directly proportional to the amount of sample used, and Johnson and Gallagher (1973) have further established this point by demonstrating that diluting the catalyst with an inert material does not produce a proportional decrease in apparent activity based upon the DTA pattern.

In this work the effects of some of the other experimental variables, i.e., flow rate, heating rate, and temperature regime, will be described. The conventional DTA presentation is compared with analytical measurements of the extent of conversion to substantiate the validity of this technique. A conventional catalytic reactor is also adapted to provide a simulated DTA output, and the degree of conversion is simultaneously determined by chemical analysis for comparison.

Having established the viability of the DTA method for catalytic studies, it is then used to briefly study the effects of bulk composition, mode of catalyst preparation, and Pt content upon the catalytic activity of a variety of oxides having the perovskite crystal structure for the oxidation of CO. The tendency of SO_2 to poison such catalysts is also investigated.

EXPERIMENTAL PROCEDURES

Apparatus

A catalytic reactor employed in earlier studies of supported catalysts by Gallagher et al (1974 and 1975 a,b) is illustrated in Fig. 1. The percent conversion of CO to CO_2 is measured as a function of both the inlet and outlet temperatures

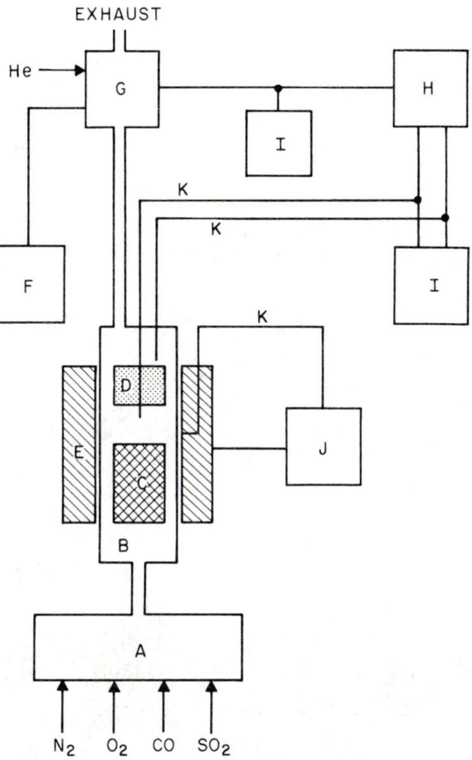

A) FLOW METERS
B) SILICA TUBE
C) HEAT EXCHANGER
D) SAMPLE
E) KANTHAL FURNACE
F) TIMER FOR SAMPLING CYCLE
G) GAS CHROMATOGRAPH
H) DATA ACQUISITION
I) RECORDERS
J) PROGRAMMER CONTROLLER
K) THERMOCOUPLES

Fig. 1. Block Diagram of Catalytic Reactor Using Monolithic Supported Samples.

of the gas stream as indicated by thermocouples immediately before and after the supported sample. These same thermocouples are used to generate a differential signal which can then be plotted as a function of inlet temperature to provide a comparison with DTA plots and simultaneously with the results of chemical analysis.

In another series of experiments a DuPont Model 900 DTA unit was used in place of the furnace assembly in Fig. 1. The DSC cell was modified as shown in Fig. 2 to provide for the chemical analysis of the gas stream. The exit gases from the

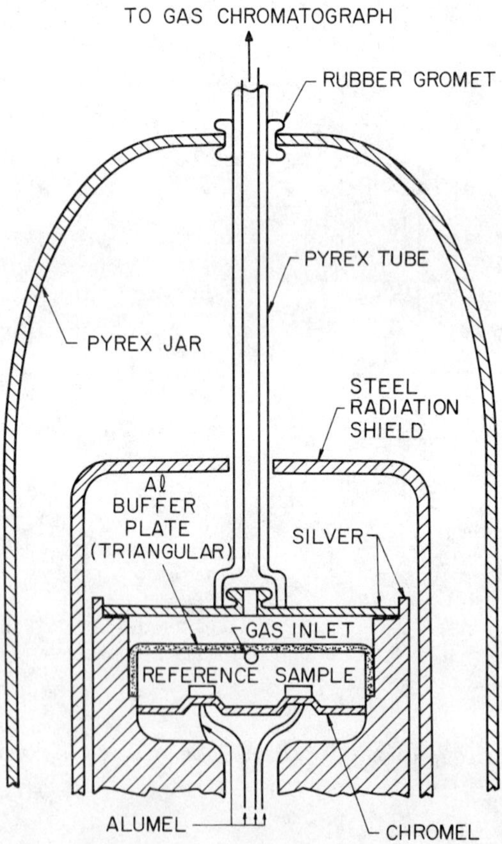

Fig. 2. Dupont DSC Cell Modified for Simultaneous Chemical Analysis of the Gaseous Products. Not to scale.

sample compartment were led directly to the input of the gas chromatograph in Fig. 1. Efforts were made to make the volume of this path as small as possible in order to maximize the resolution and minimize the time lag until analysis. A flow of 15 cm^3 min^{-1} through the chromatograph was maintained by applying gentle controlled suction at the exhaust port of the chromatograph. The aluminum buffer plate shown in Fig. 2 was necessary to minimize the effects of the brief pulse of back pressure resulting from the sampling valve of the chromatograph.

For those DTA studies in which the gas stream was not analyzed, the DSC cell was used as supplied without the special exit tube or buffer plate shown in Fig. 2. The initial concentrations of CP grade gases were 2% CO, 2% O_2, 96% N_2, and when desired 150 ppm of SO_2. Nominally 15 mg samples were used with an empty Al pan as reference.

Catalyst Preparation

Copper chromite was Baker grade catalyst which had been heated to 500°C for 2 hrs in O_2. Preparation and characterization of the other catalysts is described in detail in the publications referred to at the time of the discussion of the results herein and summarized in the appropriate figure caption. Surface areas were determined by N_2 adsorption (BET method). Pt analysis were performed by emission spectrographic techniques.

RESULTS AND DISCUSSION

Some DTA plots based upon the inlet and outlet temperatures of the reactor are compared with the results of the simultaneous chemical analysis in Fig. 3. The three samples were selected to cover a wide range of operation. The ΔT signal has been corrected to remove the strong baseline slope normally present and only fortuitously corresponds to the % conversion scale in Fig. 3. The magnitude of ΔT is, of course, dependent upon the space velocity since the actual amount of heat produced is proportional to the space velocity at constant % conversion and input concentrations. It is evident that even in these crude DTA experiments that the differential signal is a sensitive technique for determining the onset of the reaction. It closely parallels the course of the reaction up to 40% conversion over the whole temperature range and does much better at lower temperatures, i.e., more active catalysts. The more active catalysts accomplish the oxidation in a narrower range of temperature. Both the narrow and lower range of temperature minimize the error due to base line compensation. The close correspondence in Fig. 3 indicates that the simplistic DTA

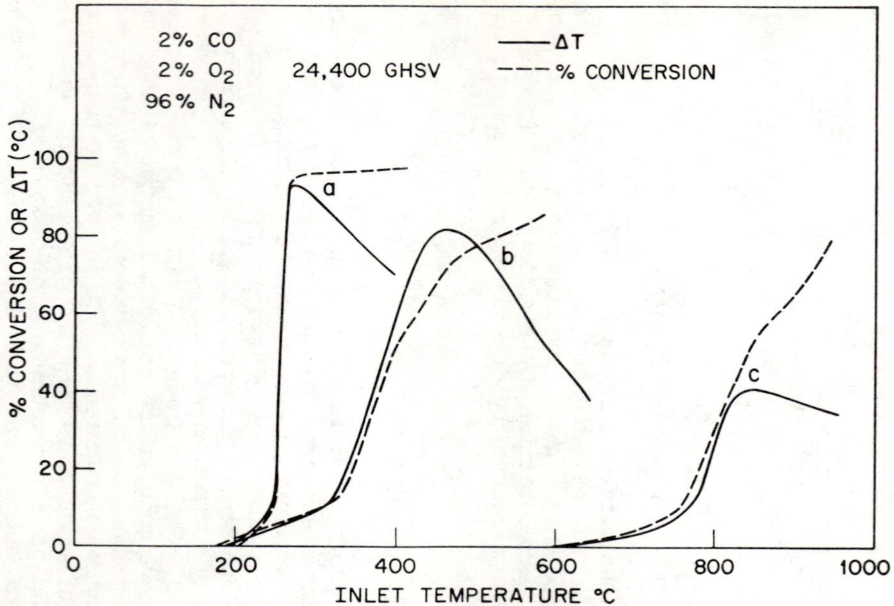

Fig. 3. Difference in Temperature (ΔT) of the Gas Stream Before and After the Sample in the Catalytic Reactor and the Degree of CO Oxidation as a Function of the Gas Temperature Just Prior to the Supported Catalyst. a) Pt on Corderite (Engelhard Ind. PTX); b) $La_{0.5}Sr_{0.5}MnO_3$ (Derived from coppt. carbonate decomposed at 900°C for 2 hr. in air and fired on corderite support at 900°C for 4 hr. in air); c) Same as above with 50 ppm of SO_2 in the gas stream.

approach should be quite suitable for screening such highly exothermic reactions.

By using conventional DTA apparatus the need for base line compensation can be eliminated, smaller samples can be used, and the whole experimental process can be simplified. However, the gas stream does not pass through the catalyst bed but around it as can be seen in Fig. 2. Hence, it is desirable to establish to what extent the active component is reacted in order to correlate the DTA curve with the degree of conversion. Clearly the heating rate and gaseous flow rate will affect this correlation. Similarly the temperature at which the reaction occurs is also important because the DTA signal is dependent upon thermal properties of the cell compartment and the extent to which the reaction occurs homogeneously, both of which are dependent upon the temperature.

The effects of heating rate are considered first, and Fig. 4 shows DTA curves for copper chromite catalysts at heating rates of 2°, 5°, and 10°C min^{-1} and a constant flow rate through the sample compartment of 100 cm^3 min^{-1}. Points are also given for the equilibrium isothermal values at various temperatures. Only the curves during the heating portion of the cycle are presented to minimize clutter. In general, the cooling curve tends to be displaced about 3-10°C towards lower temperatures during rapid change in ΔT provided there is no change in the catalyst.

It is evident from Fig. 4 that the effects of changes in heating rate are very small. Consequently there is little reason not to use the more rapid rates for general studies. However, for some of the preliminary studies to correlate DTA curves with % conversion it is more appropriate to work at slower heating rates in order to obtain more analytical points during the experiment. One other point to be made from Fig. 4 is the constancy of base line (dashed line) which will be left off subsequent figures.

The results of the analytical data, associated with the experiments in Fig. 4, are summarized in Fig. 5. Also shown is the extent of the reaction in the absence of an added catalyst. The effect of this blank has been subtracted from the curves for copper chromite. This correction will be made without further comment for subsequent curves involving the extent of conversion. This correction becomes more significant at temperatures above 400°C and, no doubt, is at least partially responsible for the increased scatter in corrected data above 400°C. The conclusion from Fig. 5 is that the extent of the reaction occurring in the DTA cell is largely independent of heating rate.

This fact combined with the similarity of the DTA curves as a function of heating rate (Fig. 4) means that the ΔT signal

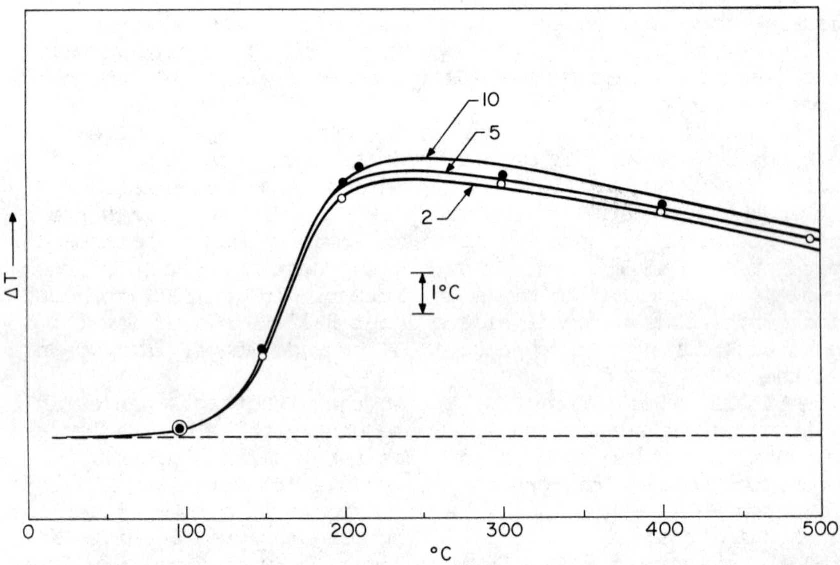

Fig. 4. Ascending DTA curves of copper chromite (15.3 mg, 18.0 m^2g^{-1}) at various heating rates in a flow of 2% CO, 2% O_2, 96% N_2 at 100 cm^3min^{-1}. Isothermal values shown as points (solid upon cooling--open upon heating). The dashed line represents the base line in the absence of a sample.

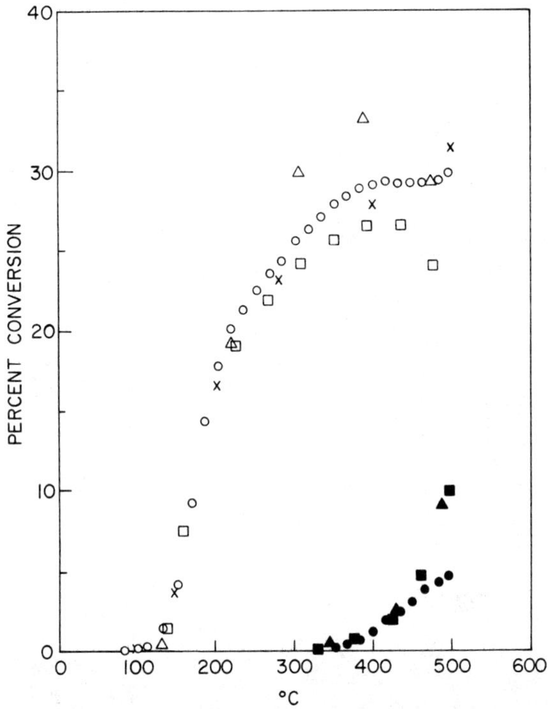

Fig. 5. *Oxidation of CO over copper chromite (15.3 mg, 18.0 m^2g^{-1}) in the DTA cell as a function of temperature. The flow was 100 cm^3 min^{-1} of 2% CO, 2% O_2, 96% N_2. Closed symbols represent the extent of the homogeneous reaction in the absence of a catalyst*

O *2°C* min^{-1} Δ *10°C* min^{-1}

□ *5°C* min^{-1} X *isothermal*

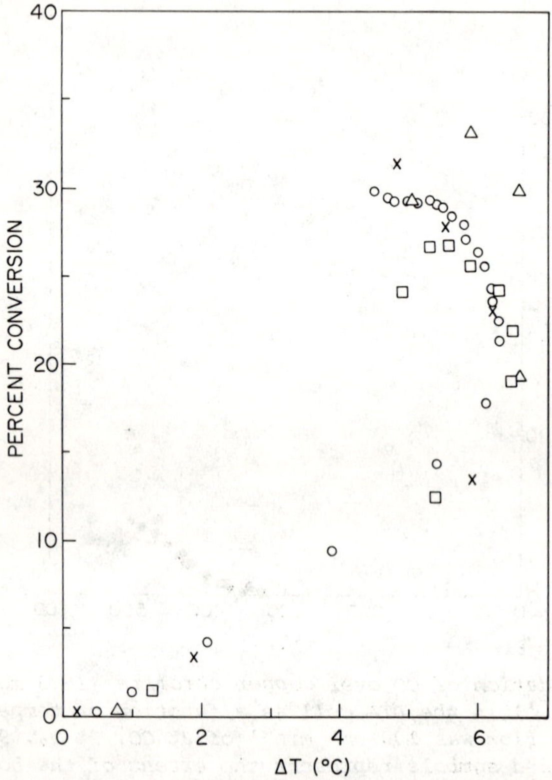

Fig. 6. Comparison of the oxidation of CO over copper chromite (15.3 mg, 18.0 m^2g^{-1}) in the DTA cell with the ΔT measured. The flow was 100 cm^3 min^{-1} of 2% CO, 2% O_2, 96% N_2

O *2°C* min^{-1} Δ *10°C* min^{-1}

□ *5°C* min^{-1} X *isothermal*

should be a reasonably good indication of the extent of the reaction regardless of the heating rate in this region of temperature. Figure 6 is a plot of these factors. This conclusion is consistent with the approximately straigth line relationship between % conversion and ΔT during the initial portion of the DTA curve. However, this relationship breaks down above ∿15% conversion for copper chromite.

The departure from linearity in Fig. 6 is attributed to three factors. First, the reaction must be limited by mass transport at some point because the gases are simply passed through a relatively large chamber containing a small pile of catalyst. From Fig. 6 it appears that about a third of the gas has the opportunity to react with the catalyst at this flow rate since reactor studies indicate at least 90% conversion by 500°C. The dependence of this upper limit of conversion upon flow rate (vide infra) is also indicative of this limitation. Second, the blank will tend to reduce the ΔT signal compared to the % conversion. That portion which occurs homogeneously or on surfaces of the cell rather than on the catalyst will of course not induce a ΔT signal. In one experiment which was inadvertently carried to a high temperature, the ΔT signal dropped to 0 around 700°C. As described earlier in reference to Fig. 5, however, at least an attempt was made to subtract out this effect. Third, there is a general trend in DTA that the magnitude of the ΔT signal will decrease with both increasing temperature and ΔT simply due the thermal properties of the cell, particularly the thermal diffusivity between the sample and reference thermocouples. Fortunately these limitations in the correlation between ΔT and % conversion at higher conversions are not important to the use of the technique for determining whether or not a particular material is or is not catalytically active.

The % conversion is plotted versus ΔT for several flow rates at constant heating rate in Fig. 7. Clearly, working at higher flow rates leads to a better DTA signal because more CO molecules, albeit a lower fraction, are being oxidized. Figure 8, however, shows that there is more involved than simply the oxidation of more CO molecules. The number of calories evolved was calculated from the flow rate, concentration of CO, % conversion, and the heat of combustion of CO. The points corresponding to the two slower flow rates fall together; however, the higher flow gives a significantly larger value of ΔT per calorie.

Because the % conversion is small for the higher flow rates, the error in the determination of the calories evolved at that flow is largest, but the difference is much too great to be explained by this factor alone. This apparent increased sensi-

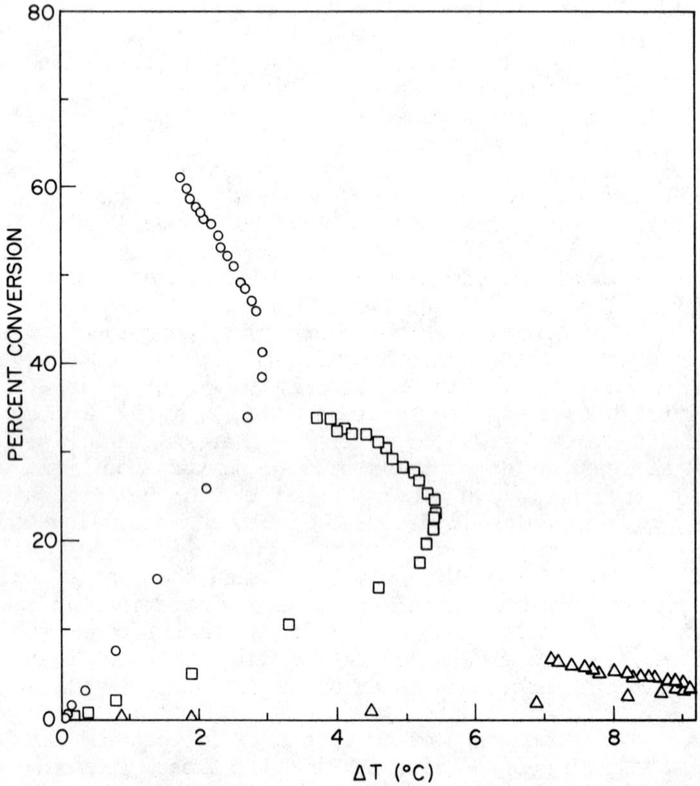

Fig. 7. *Comparison of the oxidation of CO over copper chromite (16.0 mg, 18.0 m^2g^{-1}) in the DTA cell with the ΔT measured. The heating rate was 2°C min^{-1} in a gas stream of 2% CO, 2% O_2, 96% N_2*

O *20 cm^3 min^{-1}*

□ *100 cm^3 min^{-1}*

Δ *500 cm^3 min^{-1}*

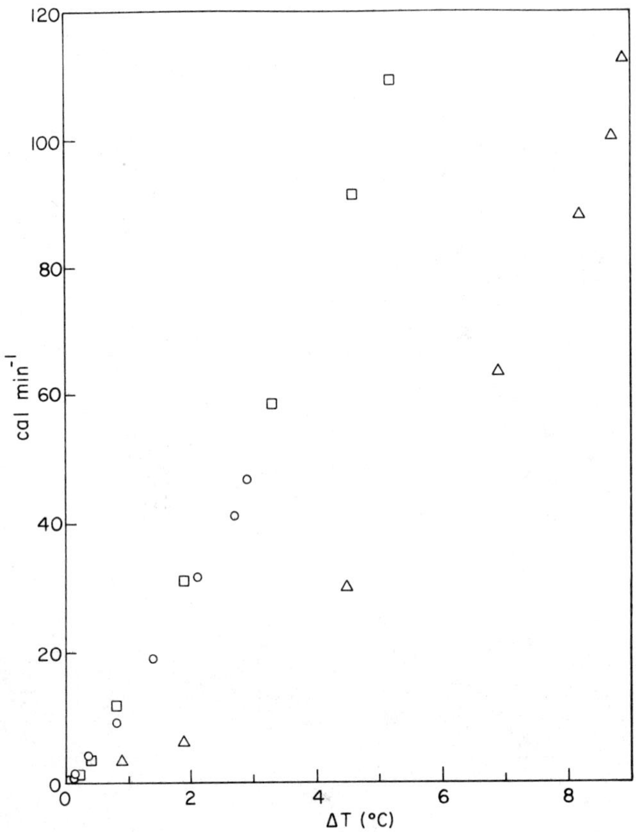

Fig. 8. *Comparison of the calories evolved by the oxidation of CO over copper chromite (16.0 mg, 18.0 m^2g^{-1}) in the DTA cell with the ΔT measured. The heating rate was 2°C min^{-1} in a gas stream of 2% CO, 2% O_2, 96% N_2.*

O	*20 cm^3 min^{-1}*
□	*100 cm^3 min^{-1}*
Δ	*500 cm^3 min^{-1}*

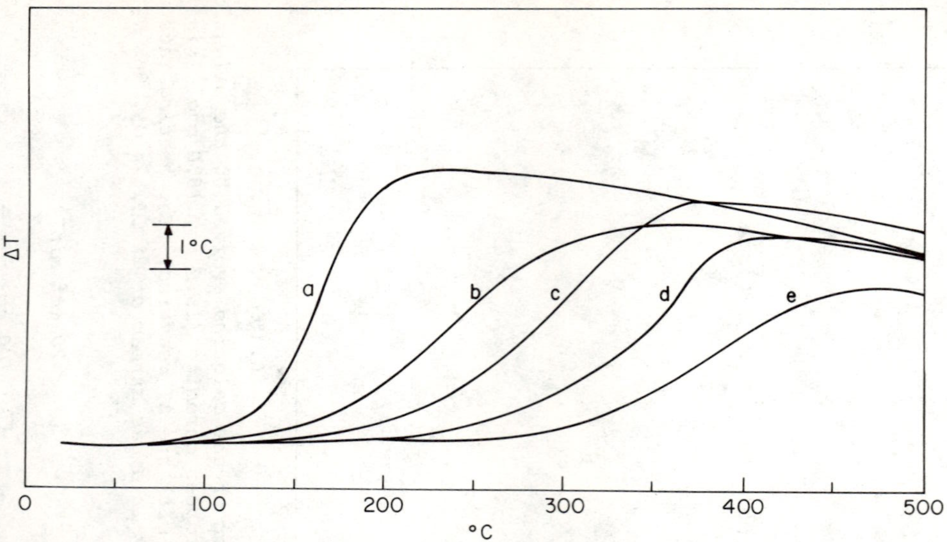

Fig. 9. *DTA curves for oxidation of CO using catalysts with a wide variety of activities. The heating rate is 2°C min^{-1} in a gas stream of 2% CO, 2% O_2, 96% N_2 at 100 cm^3 min^{-1}.*

a *copper chromite, 16.0 mg, 18.0 m^2g^{-1}*

b *$La_{0.5}Sr_{0.5}MnO_3$, 16.0 mg, 31.5 m^2g^{-1}*

c *$La_{0.75}K_{0.25}MnO_3$, 15.3 mg, 18.7 m^2g^{-1}*

d *$La_{0.7}Pb_{0.3}MnO_3$, crystals, 3000 ppm Pt, 16.4 mg, 3.1 m^2g^{-1}*

e *$La_{0.7}Pb_{0.3}MnO_3$, crystals, 1600 ppm Pt, 16.3 mg, 2.0 m^2g^{-1}*

tivity of higher flow rates is attributed to the breakdown of the mixing or flow patterns within the DTA cell caused by the buffer plate (Fig. 1). At high flows the entering gas stream tends to split above and below this plate and the reduced residence time prevents good mixing and equilibration with the catalyst. Consequently the exit gas which is drawn off the top to the chromatograph probably contains a lower fraction of CO_2 than the overall average. If this is the entire explanation, then the values of % conversion in Fig. 7 are low by about a factor of 2 for the highest flow rate. Regardless of the explanation of this phenomena it would seem advisable to perform DTA experiments at the higher flow rate, 500 $cm^3\ min^{-1}$, in order to take advantage of the increased sensitivity.

The remaining factor to be considered prior to utilizing DTA is the temperature regime in which the reaction occurs. The previous results reported herein were for copper chromite which begins to be active around 100°C (Fig. 4). Figure 9 presents DTA results for a number of catalysts selected to exhibit activities at higher temperatures but within the temperature spectrum of the instrument. The relative order of the catalytic activities is obvious and agrees qualitatively with earlier work on supported materials in the reactor described earlier. The crushed crystals of $La_{0.7}Pb_{0.3}MnO_3$ have lower surface areas of 2-3 m^2g^{-1} as compared to the others of 18-19 m^2g^{-1}.

The data were replotted for the early portion of the DTA curve as % conversion and cal min^{-1} versus ΔT in Fig. 10. They seem to form a line having a slope of approximately 1 cal min^{-1} per °C of ΔT. This slope is similar to those observed in Fig. 8 for copper chromite in various flows. The points representing crushed crystals of $La_{0.7}Pb_{0.3}MnO_3$ with 1600 ppm of Pt are subject to considerable error because approximately half of the % conversion measured is due to the blank. Clearly there are differences in thermal conductivity and other factors which preclude exact quantitative data on a range of materials without individual calibration. Nevertheless the techniques are obviously suitable for determining 1) if the material is active in a particular atmosphere, 2) if so, in what temperature range, and 3) approximately how active.

Having established the value and limitations of the DTA technique for catalytic studies, several examples will be discussed to demonstrate its applicability. These examples will also involve the catalytic oxidation of CO. The general aspects covered are 1) catalyst composition and preparative technique, e.g., substituted $LaMnO_3$; 2) the effects of trace active components, e.g., Pt; and 3) studies of poisoning, e.g., by SO_2.

In the extensive study of Johnson et al (1975) on the effects of substitution in $LaMnO_3$ upon the catalytic properties, they prepared numerous compositions by several methods. Figure

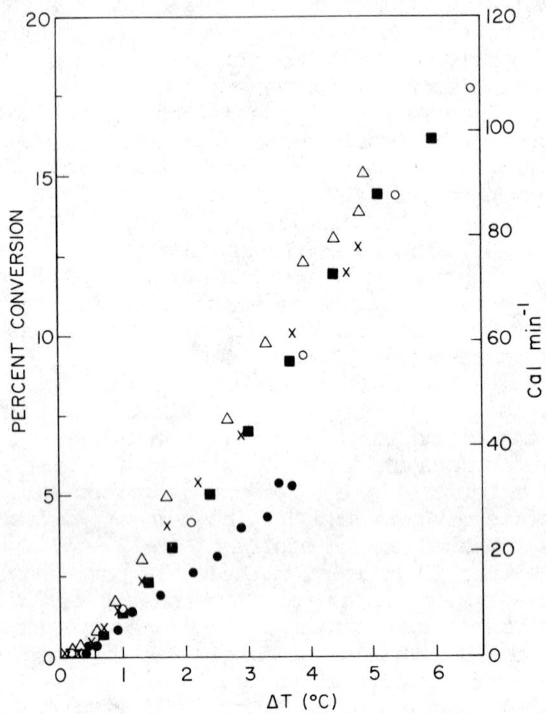

Fig. 10. *Comparison of the oxidation of CO over various catalysts in the DTA cell with the ΔT measured. The heating rate was 2°C min^{-1}in a gas stream of 2% CO, 2% O_2, 96% N_2 at 100 cm^3 min^{-1}.*

O copper chromite, 16.0 mg, 18.0 m^{2}g^{-1}
Δ $La_{0.5}Sr_{0.5}MnO_3$, 16.0 mg, 31.5 m^{2}g^{-1}
□ $La_{0.75}K_{0.25}MnO_3$, 15.3 mg, 18.7 m^{2}g^{-1}
X $La_{0.7}Pb_{0.3}MnO_3$, crystals, 3000 ppm Pt, 16.4 mg, m^{2}g^{-1}
• $La_{0.7}Pb_{0.3}MnO_3$, crystals, 1600 ppm Pt, 16.3 mg, m^{2}g^{-1}

11 shows DTA curves for several compositions prepared from freeze dried solutions. Precise comparisons are complicated by variations in surface areas. This is demonstrated in Fig. 12 where a DTA curve is given for $La_{0.5}Sr_{0.5}MnO_3$ prepared by different methods and consequently having different surface areas. It can be seen that this effect, assuming that it is solely due to differences in surface areas, can be very critical in deciding the relative order of the catalytic activity in Fig. 11. The materials shown in Fig. 11 were all prepared by freeze drying techniques and have high surface areas. The order of these activities is consistent with that observed for the supported studies in the catalytic reactor, Johnson et al (1975).

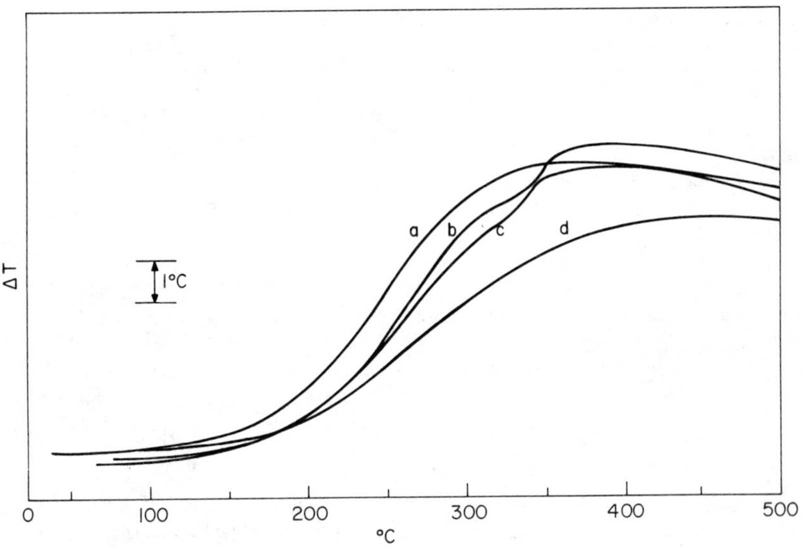

Fig. 11. *DTA curves for the catalytic oxidation of CO using some substituted perovskites prepared by freeze drying and fired at 700°C in O_2 for 2 hrs. Heating rate was 10°C min^{-1} in 2% CO, 2% O_2, 96% N_2 at 500 $cm^3 min^{-1}$*

a *$La_{0.5}Sr_{0.5}MnO_3$, 15.8 mg, 31.5 m^2g^{-1}*

b *$LaMn_{0.5}Ni_{0.5}O_3$, 15.5 mg, 24.0 m^2g^{-1}*

c *$LaMn_{0.67}Mg_{0.33}O_3$, 15.8 mg, 17.0 m^2g^{-1}*

d *$La_{0.5}Pb_{0.5}MnO_3$, 15.8 mg, 17.5 m^2g^{-1}*

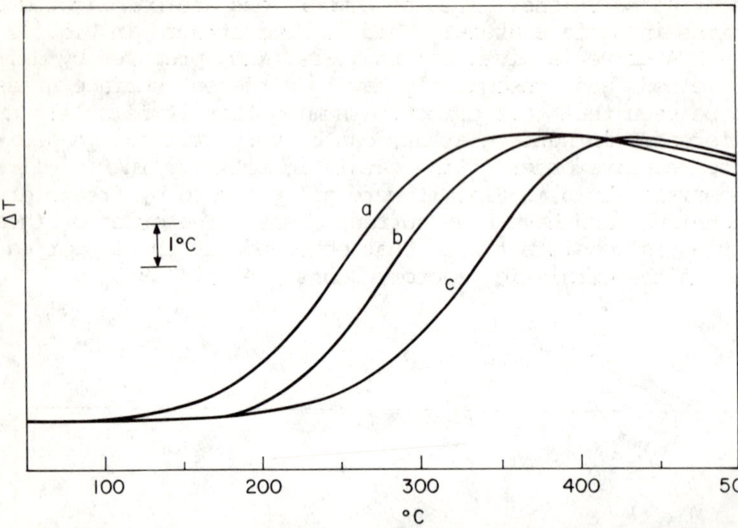

Fig. 12. DTA curves for the catalytic oxidation of CO using $La_{0.5}Sr_{0.5}MnO_3$ prepared by several methods. Heating rate was 10°C min^{-1} in 2% CO, 2% O_2, 96% N_2 at 500 cm^3 min^{-1}.

a *Freeze dried, 500°C-2 hr-Vac, 700°C-2 hr-O_2, 15.1mg, 31.5 m^2g^{-1}*

b *Spray dried, 850°C-2 hr-Vac, 700°C-2 hr-O_2, 15.2mg 16.0 m^2g^{-1}*

c *Coprecipitated, 800°C- 2 hr-Vac, 700°C-2 hr-O_2, 15.4 mg, 8.0 m^2g^{-1}*

Gallagher et al (1975 a,b) have discussed the effects of Pt on the catalytic reactivity of both supported and unsupported powders of $La_{0.7}Pb_{0.3}MnO_3$. DTA curves for many of these same powders are shown in Figs. 13 and 14 for the crushed single crystals and polycrystalline materials respectively. Obviously DTA is highly sensitive to the Pt content of these materials. Lead substituted perovskites have substantial activity as indicated by Pt free material having a high surface area in Fig. 11 and a lesser extent by the Pt free material having a lower surface area in Fig. 14. As the Pt content is increased to a level which depends on many factors, its catalytic activity predominates. This along with its ramifications and implications is discussed in detail elsewhere, Gallagher et al (1975a). The

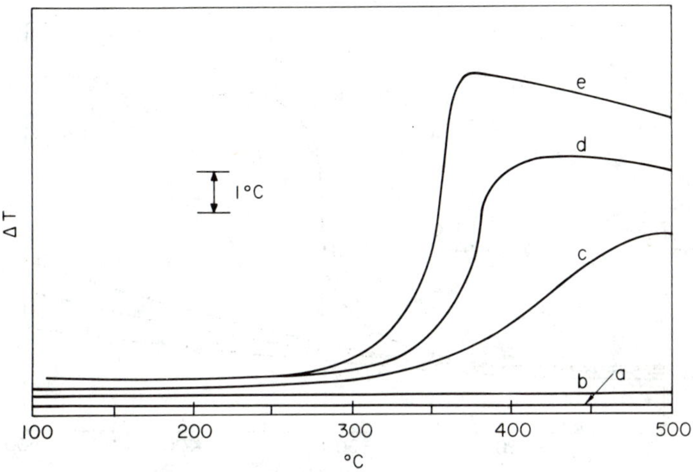

Fig. 13. *DTA curves for the catalytic oxidation of CO using crushed single crystals of $La_{0.7}Pb_{0.3}MnO_3$ with varying Pt contents. Heating rate was 10°C min^{-1} in 2% CO, 2% O_2, 96% N_2 at 500 cm^3 min^{-1}. Samples were equilibrated in above atmosphere at 700°C for 16 hrs, etched for 5 min in 5% HNO_3 at 65°C, and refired as before but for only 30 min.*

a *<10 ppm Pt, 15.8 mg, 0.6 m^2g^{-1}*
b *90 ppm Pt, 15.1 mg, 1.6 m^2g^{-1}*
c *570 ppm Pt, 15.6 mg, 1.6 m^2g^{-1}*
d *2200 ppm Pt, 15.1 mg, 0.8 m^2g^{-1}*
e *5200 ppm Pt, 15.9 mg, 0.6 m^2g^{-1}*

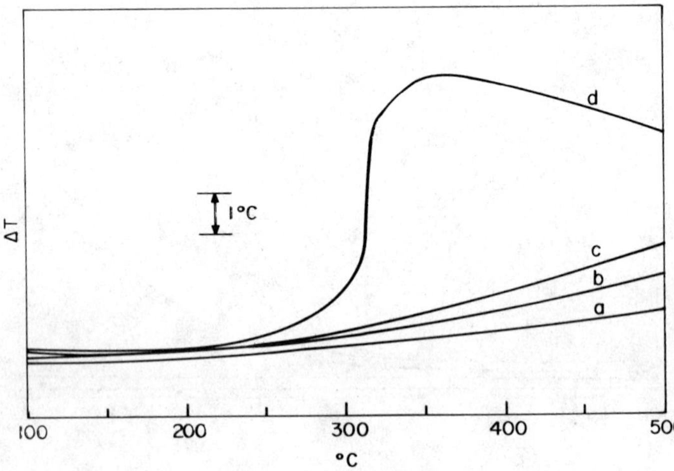

Fig. 14. DTA curves for the catalytic oxidation of CO using polycrystalline samples of $La_{0.7}Pb_{0.3}MnO_3$ with varying Pt contents. Heating rate was 10°C min^{-1} in 2% CO, 2% O_2, 96% N_2 at 500 cm^3 min^{-1}.

a *Hot pressed, <10 ppm Pt, 15.4 mg, 2.3 m^2g^{-1}*
b *Freeze dried, <10 ppm Pt, 15.3 mg, 4.2 m^2g^{-1}*
c *Freeze dried, 210 ppm Pt, 15.8 mg, 4.5 m^2g^{-1}*
d *Freeze dried, 2400 ppm Pt, 15.6 mg, 2.7 m^2g^{-1}*

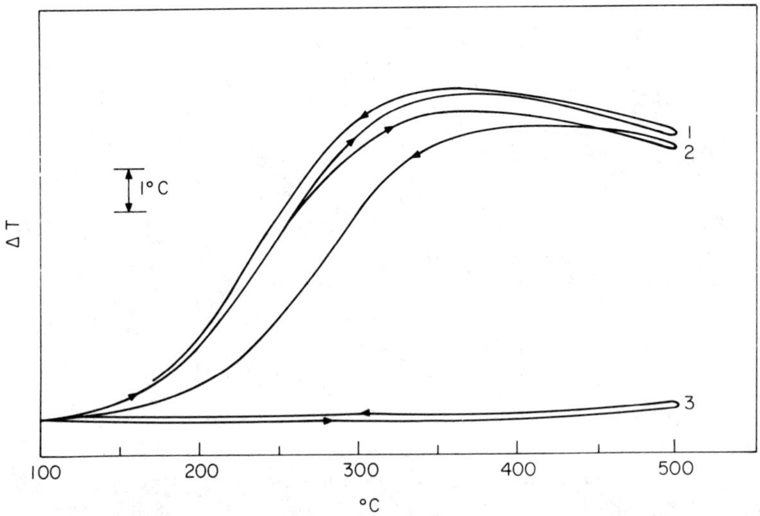

Fig. 15. *DTA curves for the catalytic oxidation of CO using $La_{0.5}Sr_{0.5}MnO_3$ (15.7 mg, 28.6 m^2g^{-1}). Heating rate was 10°C min^{-1} in 2% CO, 2% O_2, 96% N_2 at 500 cm^3 min^{-1}*

1. *1st cycle, no SO_2*
2. *2nd cycle, 150 ppm SO_2 in gas stream*
3. *3rd cycle, 150 ppm SO_2 in gas stream*

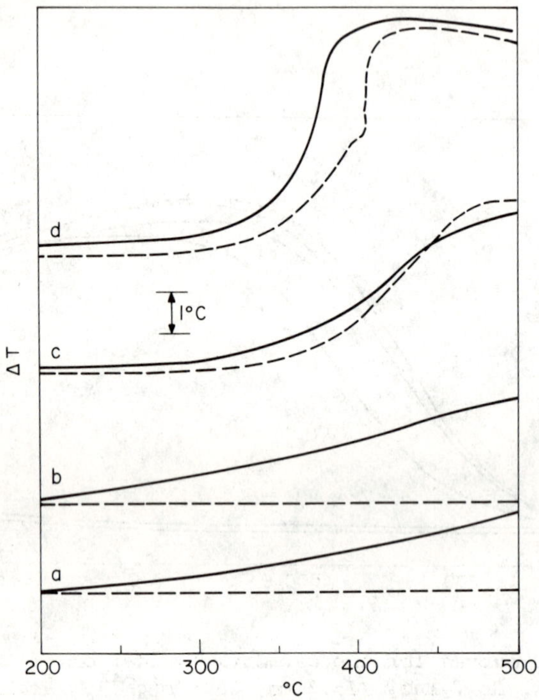

Fig. 16. DTA curves for the catalytic oxidation of CO using $La_{0.7}Pb_{0.3}MnO_3$ with varying Pt contents. Heating rate was 10°C min^{-1} in 2% CO, 2% O_2, 96% N_2 at 500 cm^3 min^{-1}. Dashed curves have 150 ppm SO_2 added to the gas stream. Crystals were treated as in Fig. 13.

a *Freeze dried, <10 ppm Pt, 15.3 mg, 4.2 m^2g^{-1}*
b *Freeze dried, 210 ppm Pt, 15.8 mg, 4.5 m^2g^{-1}*
c *Crushed crystals, 570 ppm Pt, 15.6 mg, 1.6 m^2g^{-1}*
d *Crushed crystals, 2200 ppm Pt, 15.1 mg, 0.8 m^2g^{-1}*

point to be made here is that DTA can be successfully used to study trace amounts of catalytically active materials.

Poisoning is another important aspect of catalysis and Fig. 15 shows the usefulness of DTA for such investigations. The curves are for $La_{0.5}Sr_{0.5}MnO_3$. Curve 1 is the first cycle which had no SO_2 in the gas stream. At the start of the second cycle 150 ppm of SO_2 was added to the gas stream. There was some poisoning during the heating portion but at high temperatures the role of poisoning was increased and was more pronounced during the cooling portion. By the third cycle the catalyst was essentially inactive. Earlier DTA work, Johnson et al (1974), has shown how similar perovskite catalysts can be rejuvenated by heating at 900°C.

Since Pt catalysts are only marginally poisoned by SO_2 under these conditions, Johnson et al (1974) have shown that small Pt additions are effective in combating this poisoning in perovskite catalysts, and Wedding and Farrauto (1974) and Farrauto and Wedding (1973) have demonstrated that unspecified Pt additions will also provide copper chromite with resistance. Figure 16 shows DTA curves for selected samples from Figs. 13 and 14 both in the absence and presence of SO_2. Those with $\leq$200 ppm of Pt were totally deactivated while those with greater amounts were quite resistant to poisoning by SO_2. This fact is described and discussed in greater detail for supported samples by Gallagher et al (1975b) and for unsupported powders by Johnson et al (1974).

CONCLUSIONS

1. DTA is a quick and convenient technique for many preliminary catalytic studies.
2. If only DTA results are being obtained, then relatively rapid heating rates (10-20°C min^{-1}) will reduce the time required and relatively high gaseous flow rates will increase the sensitivity of the technique.
3. The blank due to the reaction on the surfaces of the cell and the extent of the homogeneous reaction will become important at higher temperatures depending upon your materials of construction. This and other inherent problems associated with DTA will diminish the sensitivity at higher temperatures.
4. In a brief survey of substituted perovskites $La_{0.7}Sr_{0.5}MnO_3$ proved quite active.
5. In a series of samples of $La_{0.7}Pb_{0.3}MnO_3$ with varying small amounts of Pt, DTA proved sensitive to the Pt activity which added to that of the perovskite itself.
6. The poisoning effect of SO_2 upon oxide catalysts was readily detected by DTA. Small additions of Pt were shown to

be beneficial in resisting poisoning by SO_2.

7. Results of DTA experiments and the conclusions therefrom are in agreement with other more detailed work using supported samples in a catalytic reactor.

ACKNOWLEDGMENTS

The authors are grateful to Messrs. F. Schrey for preparation of the spray dried and coprecipitated materials, measurements of many of the surface areas, and the Pt analysis; to D. L. Nash for assistance in the Pt analysis; to J. P. Remeika for providing most of the single crystal samples; to J. W. Fleming for help in the preparation of Pt free single crystals; and to W. W. Rhodes for the preparation of the freeze dried samples.

REFERENCES

1. Farrauto, R. J., Wedding, B., J. Catal., 33, 249 (1973).
2. Furuichi, R., Ishii, T., Kobayashi, K., J. Therm. Anal., 6, 305 (1974).
3. Gallagher, P. K., Johnson, D. W. Jr., Remeika, J. P., Schrey, F., Vogel, E. M., Voorhoeve, R. J. H., Mater. Res. Bull, 10, 529 (1975a).
4. Gallagher, P. K., Johnson, D. W. Jr., Vogel, E. M., Schrey, F., Mater. Res. Bull., 10, 623 (1975b).
5. Ishii, T., "Thermal Analysis: Comparative Studies of Materials," pp 195-212, Kodansha Ltd., Tokyo, 1974.
6. Ishii, T., Furuichi, R., Kobayashi, Y., Thermochimica Acta 9, 39 (1974).
7. Johnson, D. W. Jr., Gallagher, P. K., Thermochimica Acta, 7, 303 (1973).
8. Johnson, D. W. Jr., Gallagher, P. K., Vogel, E. M., Schrey, F., Proceedings of the 4th International Conference on Thermal Analysis, Budapest, Hungary, July 1974, to be published.
9. Johnson, D. W. Jr., Gallagher, P. K., Rhodes, W., Schrey, F., Presented at the 77th Annual Meeting of the American Ceramic Society, Washington D. C., May 1975. To be published J. Am. Cer. Soc.
10. Locke, C. E. Rase, H. F., Ind Eng. Chem., 60, 515 (1960).
11. Papadatos, K., Shelstad, K. A., J. Catal. 28, 116 (1973).
12. Wedding, B., Farrauto, R. J., Ind. Eng. Chem., Process Des. Develop. 13, 45 (1974).

CATALYST ACTIVITY VIA THERMAL ANALYSIS

JOHN R. KOSAK, Ph.D.

E. I. duPont de Nemours & Co., Inc.
Deepwater, N. J. 08023

INTRODUCTION

The greatest problem associated with the manufacture of a hydrogenation catalyst is the ability to consistently duplicate the performance of that catalyst. Second only to this problem is the ability to measure the performance of the catalyst system. A catalyst may be compared with respect to many of its physical parameters (e.g. metal content, metal surface area, substrate surface area, pore size, pore volume, etc.), but the most useful criteria of catalyst reproducibility are related to catalyst activity and/or selectivity. One general procedure for determining catalyst activity is to duplicate the reaction in a laboratory scale autoclave and measure either the rate of hydrogen consumption and/or product yield per unit of time. Where selectivity is also critical the hydrogenation must be completed and the product subjected to extensive analyses.

This approach is tedious, time consuming, expensive and subject to variability in the quality of the starting materials used.

This work was undertaken to evaluate the applicability of high pressure differential scanning calorimetry (DSC) to the evaluation of catalyst activities for powdered hydrogenation catalysts. DSC records the differential heat flow between a sample and reference material as a function of temperature or time. The integrated area, therefore, is directly proportional to the total heat evolved. Thus, in the case of a hydrogenation catalyst contacted with hydrogen under pressure, the total heat evolved is proportional to the quantity of gas consumed (i.e. the amount required to activate and/or chemisorb) by the catalyst, and should be a measure of catalyst activity.

APPARATUS

The instrument used in this study consists of a standard

Du Pont DSC cell, described by Baxter, enclosed in a specially designed pressure housing and a Model 900 Du Pont Thermal Analyzer. The pressure module is shown in Figure 1.

Both the top of the cell and the cylinder are removable by simply unscrewing the three bolts with the fingers. Normally only the top is removed for loading and unloading samples. "O"-ring seals, set in grooves in the top and bottom portions of the enclosure, seat against the side wall of the cylinder and increase their contact as either pressure or vacuum is applied.

The assembly is designed to operate at a maximum working pressure of 1000 psig and to attain a vacuum of under 10 microns. The cell, valving and connections are diagrammed in Figure 2.

One or more supply gases (connected to tank regulators set at 1000 psig or lower) enter the cell through the input connector and are controlled by the input control valve. The incoming gas enters the cell through a port in the baseplate and

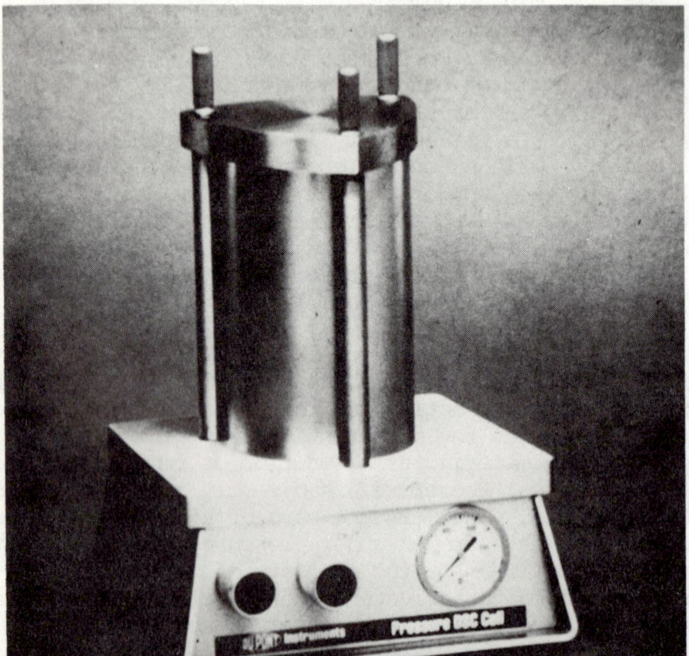

Figure 1.

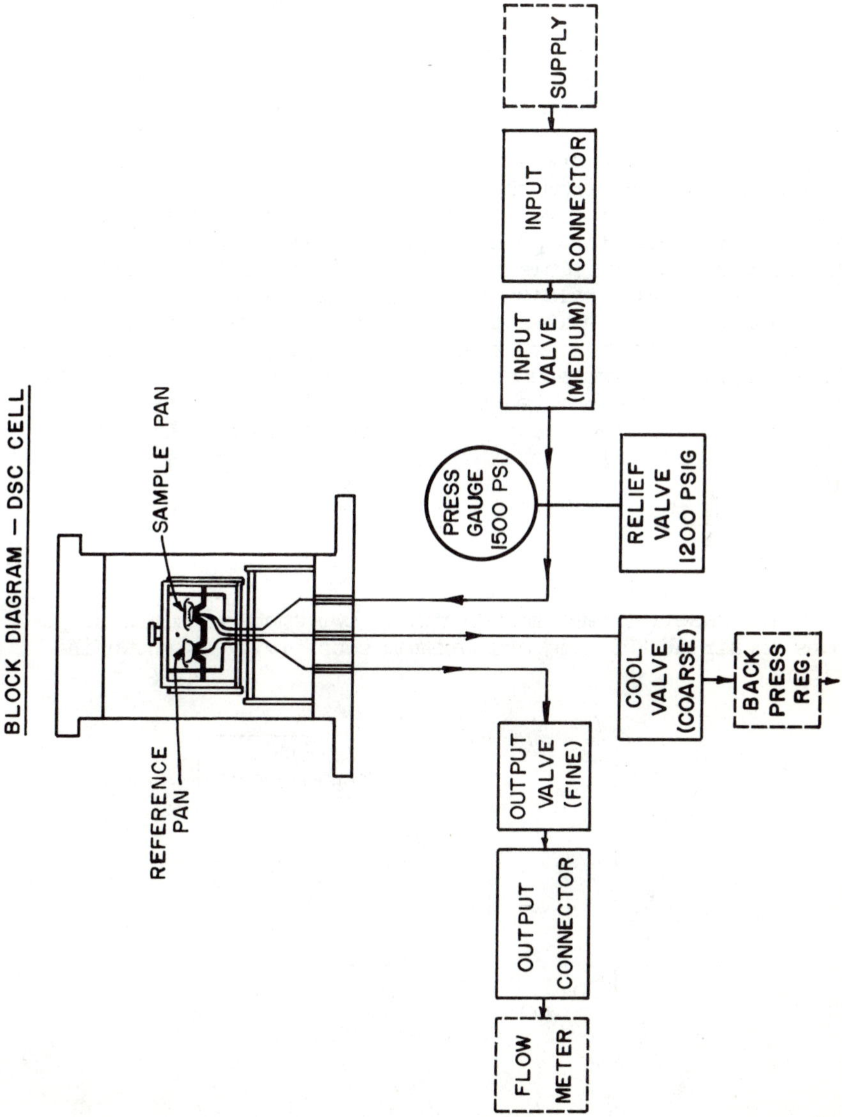
BLOCK DIAGRAM – DSC CELL
SAMPLE PAN
REFERENCE PAN
SUPPLY
INPUT CONNECTOR
INPUT VALVE (MEDIUM)
PRESS GAUGE 1500 PSI
RELIEF VALVE 1200 PSIG
COOL VALVE (COARSE)
BACK PRESS REG.
OUTPUT VALVE (FINE)
OUTPUT CONNECTOR
FLOW METER

Figure 2.

floods the entire volume of the cylinder. A spring loaded safety valve in the supply line between the input valve and the baseplate is set to open at 1200 psig. A pressure gauge with a maximum range of 1500 psig (red lined above 1000 psig) is also connected at this point. Gas exiting the cell during purging enters the sample chamber through a hole in the silver lid, passes over the sample, through an orifice in the side wall of the sample chamber and out through a tube connected to the baseplate. Gas flow rate during purging is controlled by a fine (14 turn) high pressure needle valve. A flow meter may be connected to a line from the exhaust valve. A third valve, labeled "cool control valve" in Figure 2, is connected internally in the cell to a point under the heating block outside of the sample chamber. This is a coarse (3 turn) stop valve which is used to rapidly release pressure from the cell after a run is completed. Since it has the largest orifice it is the point at which a vacuum system is normally connected to the unit. If true constant pressure conditions are required, a back pressure regulator may be connected to the "cool control valve".

A more detailed description of the Pressure DSC cell including a survey of general applications is given by Levy et al, (1970).

CALIBRATION

The baseline performance run on two empty aluminum sample pans in air at 800 psig and maximum sensitivity is shown in Figure 3.

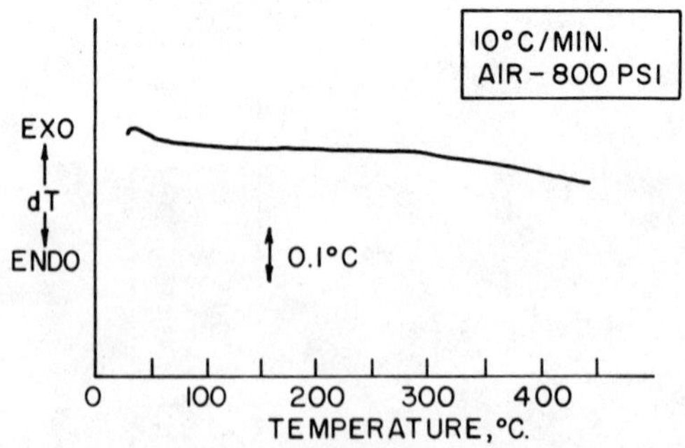

FIGURE 3

BASELINE - PRESSURE DSC

Figure 3.

At ambient pressure, performance of the Pressure DSC cell is identical to that of a standard Du Pont DSC cell. At elevated pressures, there is some change (less than 10% over the entire temperature and pressure range) in the calibration coefficient (E) used to calculate the energies involved in a thermal transition. For this work, the cell was calibrated using an indium metal standard over the pressure range of interest in both helium and hydrogen atmospheres. Figure 4 shows the melting endotherm of indium under H_2 at 200 psig.

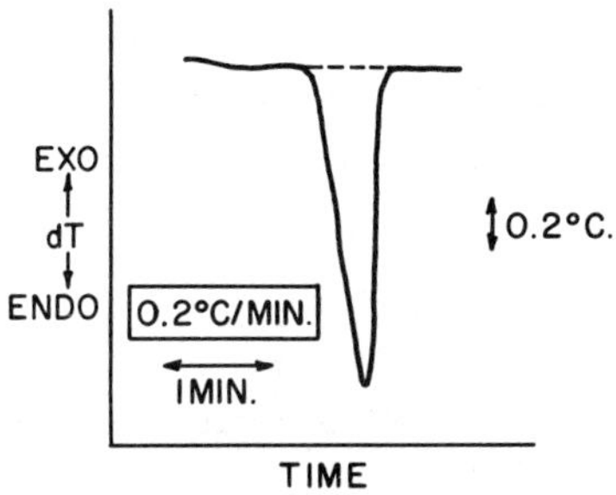

FIGURE 4

PRESSURE DSC CELL CALIBRATION
INDIUM MELT

H_2 AT 200 psig

$$E = \frac{dH \times M \times C}{A \times dT_s}$$

E = 280 mcal/min/°C

Figure 4.

dT vs. temperature and dT vs. time scans were obtained simultaneously by attaching a Honeywell Electronic 194 strip chart recorder in parallel with the X-Y recorder in the Du Pont Thermal Analyzer. The time-base scan allows expansion of peak areas for ease of measurement. The calibration coefficient (E) is calculated from the dT-time scan using the following equation.

$$E = \frac{dH \times M \times C}{A \times dTs} = 280 \text{ mcal/min/deg C}$$

where: dH = 6.79 cal/gm (from Handbook of Chem. & Physics) = heat of fusion for Indium.
M = 14.85 mg (indium sample weight).
C = 3 in/min (chart speed).
A = 4.32 sq. in. (Measured from dT-time curve).
dTs = 0.25 deg/in. (Y axis or dT sensitivity).

EXPERIMENTAL

Materials

Catalyst samples used, with the exception of two, are commercial products. Cylinders of Matheson Spectro Grade helium and hydrogen were used throughout the investigation.

General Test Procedure

Catalyst samples are dried in a vacuum oven below 60°C and with a slight nitrogen bleed on the oven. A portion of the dry sample is weighed directly into an aluminum pan (the weight generally ranges from 5-20 mg.). The aluminum pan containing the sample and an empty reference pan are placed on their respective platforms in the DSC sample chamber.

> Weighed portions of catalyst substrate (e.g. carbon, alumina, etc.) may be used as a reference material to distinguish between the heat of physical adsorption on the substrate and that of chemisorption on the metal. This step is not necessary where the only object is to compare close weights of metals on the same substrate. Heat capacity differences using an empty reference pan versus a reference pan containing an inert reference material were not significant in relation to the total heat being measured. An empty pan was used for convenience.

The cell is closed, evacuated to about 5 mm Hg pressure, then heated to a predetermined temperature for a given catalyst and held a minimum of 5 min. The vacuum pump is in operation throughout this time and normally a total of 15-30 minutes. The vacuum is released slowly and helium gas is introduced into the cell until a minimum pressure of 200 psig is established. Also, the steady state flow of gas through the cell is adjusted to 1-2 cfh. A stable base line is established on the time base recorder. The gas flow through the cell is then changed from helium to hydrogen producing an almost instantaneous exotherm which is recorded on the time base recorder. The H_2/He pressure ratio must be greater than 1 and generally is in the range of 1.1 to 1.5. The thermogram is complete and scanning is discontinued when the recorder pen returns to a stable base line. The hydrogen is displaced from the cell by venting to atmospheric pressure and sweeping the cell with Helium.

Figure 5. Calculation for Heat of Chemisorption (dH_C)

Using external time base recorder.

$$dH_c = E\,\frac{A \cdot dT_s}{M \cdot C \cdot X} \qquad = \text{cal./g. precious metal (PM)}$$

where,

E = mcal./°C-min. (experimentally determined, calibration coefficient).
A = Peak area, sq. in.
dT_s = Y-axis sensitivity, °C/in.
M = Sample mass, mg.
C = Chart speed, in./min.
X = Precious metal content of a specific catalyst (Wgt. fraction).

RESULTS

Catalyst Activities

A typical dT-time exotherm curve for a 5% Pd/carbon catalyst is shown in Figure 6. No attempt was made to separate the heat of chemisorption from that of physical adsorption on the carbon. Six replicates of this sample gave an average heat of chemisorption (dH_C) of 390 ± 5 cal/gm. palladium.

The heat of chemisorption for various catalyst samples is given in Table I.

DISCUSSION

The relative activity of one catalyst system vs. another catalyst system for a specific hydrogenation process cannot be predicted from the heat of chemisorption values alone because of differences in the type of substrates - carbon vs. alumina, or carbon A vs. carbon B. Different substrates reflect differ-

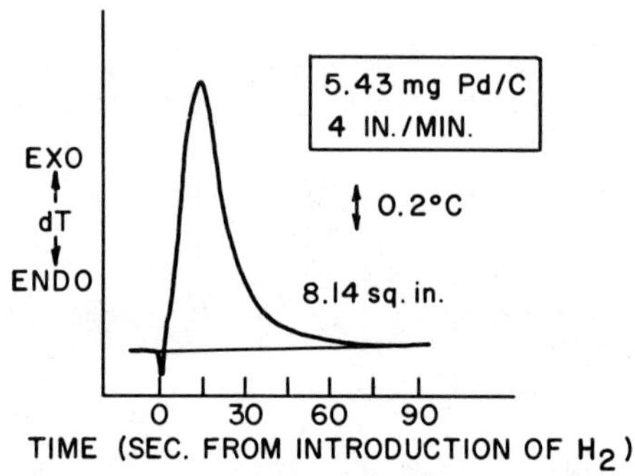

FIGURE 6

5% Pd/CARBON AT
200 psig H_2, 75°C

$$dH = \frac{8.14 \times 0.2 \times 263}{5.43 \times 4 \times .05}$$
$$= 393 \text{ cal/gm}$$

TABLE I. *Evaluation of Catalyst Activities by DSC at 200 Psig H_2, 75°C*

Catalyst	Cal/gm PM
5% Pd/Carbon	390*
5% Pd/Calcium Carbonate	212
5% Pd/Alumina	188
5% Pt/Alumina	142
5% Pt/Carbon	280
59% Ruthenium Oxide	298

PM - Precious Metal

*(Average of 6 runs: 390 ± 5 cal/gm PM)

ences in structure, surface area, pore size, pore volume, etc. which, in turn, affect catalyst metal surface area plus catalyst metal distribution on the substrate. However, it is interesting to note that the order of activity observed for the hydrogenation of nitrobenzene using these catalysts agrees with the DSC results. For example, Pd is more active than Pt and carbon is superior to alumina as a substrate. To compare the validity of the DSC method for the determination of catalyst activity a commercial catalyst was analyzed by the DSC method and the data plotted against the activity number given by the supplier. The results are shown in Figure 7. All of the catalyst was of satisfactory activity. In case a the DSC activity test was found to be a more accurate measure of catalyst performance than the activity number. In cases b and c, the catalyst was found to be of better than minimum activity but not in agreement with the supplier's high activity number.

dHc - Pt/C Area Ratio Correlation

A number of samples of a 5% platinum on carbon catalyst were evaluated for activity by the pressure DSC method and also examined by ESCA (Electron Spectroscopy for Chemical Analysis) to get a relative measure of the platinum surface area. As can be seen from Table II, the relative heat of chemisorption values obtained by pressure DSC and the relative Pt/C ratios obtained by ESCA parallel each other and are in perfect agreement.

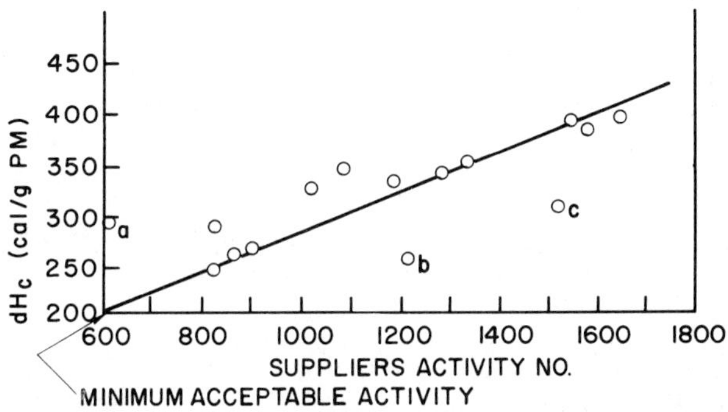

FIGURE 7

ACTIVITY TEST CORRELATION

5% Pd/C CATALYST

TABLE II. *5% Pt/C Catalyst, Relative Ranking of Catalyst Activity and Elemental Ratios of ESCA Measurements*

Pt/C Surface		DSC	Activity
Ratio	Rel. Ranking	dH_C	Rel. Ranking
.033	1	10	1
.080	2	18	2
.117	3	45	3
.121	4	85	4
.165	5	192	5

SAMPLE VARIABILITY

Variability in catalyst activity for different samples of a given catalyst system can be seen in Table III.

Deterioration in activity of a catalyst is sometimes attributed to aging of the catalyst. Based on the dH_C values obtained for the different catalysts in Table III, there appears to be essentially no deterioration in activity with aging. This was confirmed for the platinum and palladium catalysts by hydrogenating o-nitrotoluene and finding hydrogenation rate and product quality to be unchanged.

Effect of Pressure

Reproducible results are obtained in the pressure range of 200-500 psig. Higher pressures were not investigated. Pressures below 200 psig give erratic results. The cause of erratic results below 200 psig is probably incomplete diffusion of hydrogen gas to the catalyst surface. The ratio of the hydrogen to helium pressures must be greater than 1.0 to provide positive displacement of the helium by hydrogen.

Effect of Temperature

The effect of temperature on the dH_C value depends on a second variable - mode of deposition of the catalyst metal on the substrate. If the catalyst metal is dispersed, principally, at the surface of the substrate, there is no temperature effect. This is shown for two catalyst systems in Table IV.

However, if the catalyst metal is dispersed throughout the substrate, then there will be a substantial increase in the exotherm with increasing temperature. This is shown in

TABLE III.

SAMPLE VARIABILITY

dH_C VS. CATALYST SAMPLE

CATALYST TYPE	SAMPLE NO.	dH_C (CAL/G PM)
5% Pt/C	1-1	248
	1-2	262
	1-3	228
	1-4	250 (DAY 1)
		241 (DAY 100)
		262 (AVG. - 9 RUNS)
5% Pd/C	2-1	336
	2-2	376
	2-3	393
	2-4	292 (DAY 1)
		247 (DAY 52)
		271 (AVG. - 8 RUNS)
5% Ru/C	3-1	393
	3-2	388
	3-3	271
	3-4	421 (DAY 1)
		407 (DAY 40)
		414 (AVG. - 2 RUNS)

TABLE IV. *Catalyst Metal Dispersed on Surface of Carbon Support*

Catalyst	Sample No.	Temp. (°C)	dH_C (cal./g. PM)
4.5% Pd, 0.5% Pt/C	1	100	193
		150	220
	2	100	257
		150	241
	3	100	44
		150	34
5% Pt/C	1	100	195
		150	188
	2	100	205
		150	190

Table V for a 1% Pt/C catalyst.

TABLE V. *Catalyst Metal Dispersed Throughout the Carbon Support, 1% Pt on Carbon*

Run No.	Temp. (°C)	dH_C (cal./g. Pt)
1	100	87
2	150	213
3	150	253
4	150	215
5	190	265
6	190	269
7	190	289

It should be noted that consistent results are obtained at a given temperature.

Selectivity Applications

The synergistic effect obtained by the use of bimetallic catalyst systems has been discussed by Webster and Bond. This synergism was detected readily by the pressure DSC method when comparing a 5% palladium on carbon catalyst vs. a mixed 4.5% palladium plus 0.5% platinum on the same carbon support (Table VI). Preparation of the catalysts was by the same procedure with the exception of the composition of the bimetallic catalyst.

The magnitude of the rate improvement for the reduction of o-nitrotoluene corresponds well with the heat of chemisorption values observed by the thermal analysis method. Both methods indicate a large improvement in activity with the bimetallic catalyst.

Prior to routine use of the DSC activity test procedure, a batch of bimetallic catalyst was giving very slow reduction rates. Comparison of "good" catalyst with the "poor" catalyst in the pressure DSC test showed the following difference (Figure 8).

The double peaks in the "good" sample (i.e. Run 1) are believed to represent chemisorption by both metals in the catalyst. The missing peak in the "poor" catalyst sample is attributed to poisoning of the more susceptible platinum metal specie. Furthermore, the decreased reactivity correlates with the loss of synergism by the secondary metal as discussed by Bond and Webster. On the basis of this data, some of the se-

TABLE VI. *Bimetallic Synergism*

Catalyst Type	Sample No.	dH_c*	dH_c Ratio	Normalized Redn. Rate**
5% Pd/C	1	182	1.0	1.0
	2	204	1.12	1.02
	3	247	1.36	1.25
4.5%Pd+0.5%Pt/C	1	394	2.16	2.08
	2	420	2.31	2.34
	3	506	2.78	2.58

*Calories per gram of precious metal. **Based on moles of hydrogen consumed per gram of precious metal per minute. Hydrogenation system is the reduction of o-nitrotoluene at 120° ± 5°C and 500 psig hydrogen pressure.

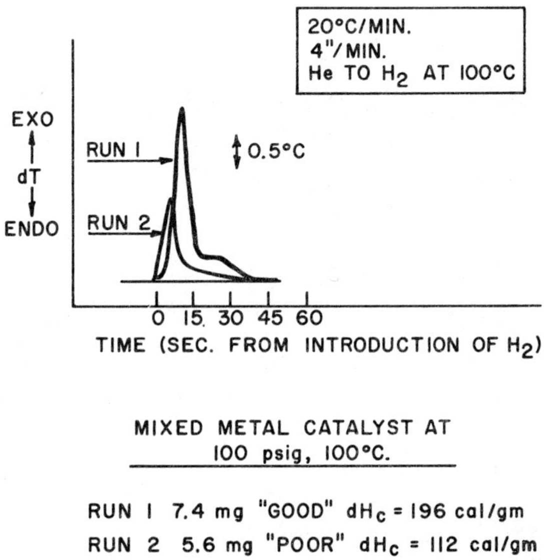

Figure 8.

condary metal as Pt/C catalyst was physically admixed to the batch of "poor" catalyst and activity was increased to a usable (but less than normal) level.

An example where the pressure DSC method was useful in defining a catalyst selectivity as well as activity problem involved the use of a 5% ruthenium on alumina catalyst (Figure 9). This particular catalyst was used to affect ring hydrogenation of a specific organic compound without concommitant cleavage of the ether linkage also present in the structure. Measuring dH_c of this catalyst system at 100°C showed that "good" catalyst has a charactersitic thermogram pattern in addition to having a high dH_c value. "Poor" samples, on the other hand, do not have the characteristic thermogram pattern and always showed a lower dH_c value. In this particular case, washing the catalyst with 10% caustic did not improve the catalyst and in fact, made it worse. The cause of "poor" catalyst was never resolved. However, it was possible to monitor new lots of catalyst as they were received and predict, accurately, "good" and "bad" lots so that plant manufacture could be maintained with a minimum of upsets.

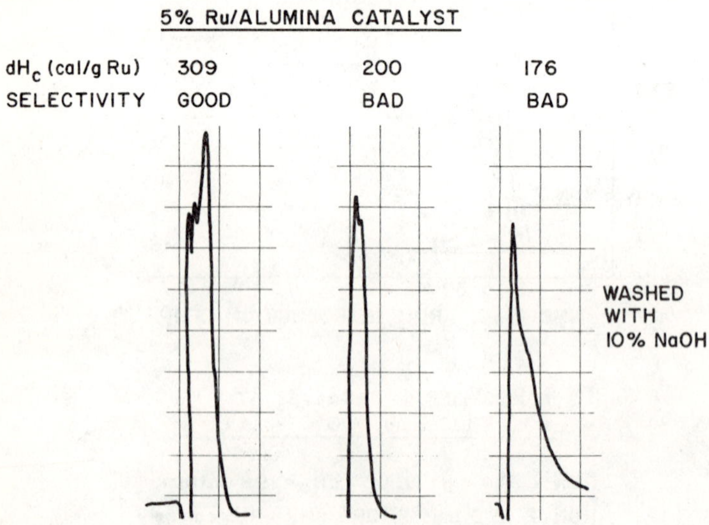

Figure 9.

SUMMARY

A reliable procedure based on pressure DSC has been developed for determining the relative activity of catalyst samples within a given powdered catalyst system.

The pressure DSC method is a practical tool for monitoring catalyst quality.

Unique thermogram patterns were found for two catalyst systems (the 5% Pd-Pt bimetallic on carbon and the 5% Ru on alumina) which defined both activity and selectivity characteristics of these catalyst systems.

The pressure DSC method is a useful means for preliminary screening of candidate catalyst systems.

ACKNOWLEDGMENT

The author wishes to thank Mr. W. E. Collins (DuPont, Photo Products Department, Instr. Prod. Div.) for his assistance in developing the initial working parameters for the catalyst activity procedure.

REFERENCES

1. Baxter, R. in Schwenker, R. and Garn, P. (Eds.) Thermal Analysis, Academic Press, N. Y. pp 65-84 (1969).
2. Bond, G. C. and Webster, D., Ann, N. Y. Acad. Sci., 158, pp 541 (1969).
3. Levy, P., Nieuweboer, G., Semanski, L., Submitted to Thermochimica Acta (Feb., 1970). Presented - American Chemical Society Meeting, Houston, Texas, Feb., 1970.

POLYMER ATTACHED CATALYSTS

ROBERT H. GRUBBS, E. M. SWEET and SAWIT PHISANBUT

Department of Chemistry, Michigan State University
East Lansing, Michigan 48824

Polymer attached reagents, particularly transition-metal catalysts, have been demonstrated to have many of the advantages of both heterogeneous and homogeneous analogs (Grubbs and Kroll, 1971; Grubbs, et al, 1973; Pittman and Evans, 1973; Bailar, Jr., 1974; Collman, 1972; Haag and Whitehurst, 1973). The studies to date have demonstrated that the attachment of a known homogeneous reagent to a cross-linked polystyrene copolymer produces an insoluble catalyst which retains catalytic activity and is easily separated from reactants and products. These reagents in some cases show considerable selectivity on the basis of overall molecular size and polarity (Grubbs and Kroll, 1971). For example, cyclohexene was reduced 32 times faster than Δ^2-cholestene by a rhodium(I) catalyst attached to 2% cross-linked polystyrene through a phosphine link. Similar catalysts suspended in a polar solvent reduce non-polar in preference to polar olefins (Grubbs, et al, 1973).

The effect of catalyst loading levels on the rate and selectivity of the attached catalyst was investigated. A sample of Rh(I) attached catalyst was prepared by the following method (Pittman and Evans, 1973; Michalska and Webster, 1973; Bailar, Jr., 1974).

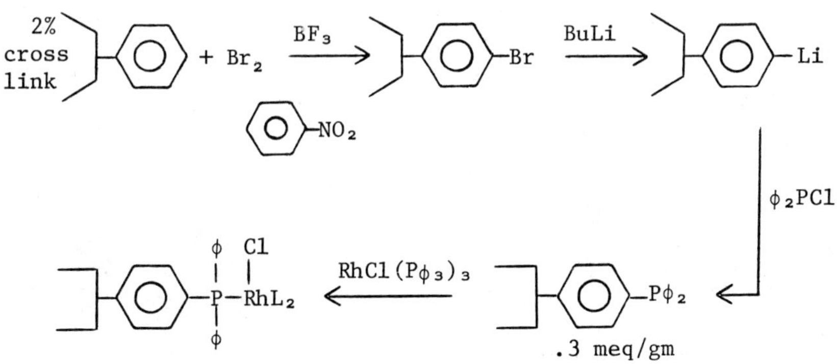

From earlier work, it is known that the rhodium is bound to the polymer by more than one phosphine link (Pittman and Evans, 1973; Michalska and Webster, 1973; Bailar, Jr., 1974; Kroll, 1974). The phosphines, although they are on the average separated by about 30 styrene units, are mobile enough on this low cross-linked polymer to chelate with a metal center. Rhodium was loaded on the polymer at two levels. The polymer was equilibrated with a deficiency of catalyst to produce a phosphine/rhodium ratio of 10. A large excess of rhodium complex produced a metal loading with a phosphorous/rhodium ratio of 5.

The saturated beads (P/Rh ≃ 5) reduced cyclohexene (1M) at a rate of 2.3×10^{-2} ml/sec/gm of catalyst. Under similar conditions, the deficient beads (P/Rh ≃ 10) reduced cyclohexene at a rate of 0.48×10^{-2} mlH_2/sec/gm. As was expected, the rate of the deficient beads was lower. However, the 4.8 fold decrease is greater than would be expected on the basis of the decreased metal loading. This is apparently due to the difference in the relative amounts of free phosphine in the lightly loaded catalysts. The rate of hydrogenation with homogeneous $RhCl(P\phi_3)$ is decreased significantly by added phosphine (O'Conner and Wilkinson, 1968). Since the phosphines attached to the low cross-linked polymer are mobile, the rate of reaction in the polymer will depend upon the concentration of the free phosphine that is accessible to the metal center in the polymer. Consequently, the lower the loading, the lower the activity per equivalent of metal catalyst present.

The relative rates of reduction of a series of olefins of different molar volumes are listed in Table I. As can be seen from these data, the selectivity of the catalyst attached to polystyrene beads which are the most active are also the most selective on the basis of molecular volumes. For example, the saturated beads reduce β-pinene 8×10^{-3} times as fast as cyclohexene, whereas, the deficient catalyst reduces β-pinene ca. 3.5×10^{-2} times as fast as cyclohexene.

Since the polymer surrounding the catalyst imposes a diffusion barrier between the bulk solution and the catalytic center, a concentration gradient of reactant will be developed between the bulk solution and the interior of the polymer bead. The magnitude of the gradient will be greatest for the most active catalyst. When comparing olefins of different molar volumes (different diffusion rates), the olefin with the largest molar volume should result in the greatest concentration gradient in the polymer. Consequently, the most active catalyst will place the greater demand on the bulk solution to supply substrates which will result in an enhancement of the differences in the diffusion rates of two olefins. The

TABLE I. *Rate of Reduction of Cyclohexene(1M) With Poly-Rh(I) in Toluene*

	Saturated Beads	Deficient Beads
Cyclopentene	1.75±.03	1.80±.1
Cyclohexene	1.00±.05	1.00±.04
Cycloheptene	0.805±.05	0.97±.06
Cyclooctene	0.43±.08	0.64±.05
Beta Pinene	0.08±.003	0.35±.02
Cyclohexene	2.276×10^{-2} ml/sec/gm	0.478×10^{-2} ml/sec/gm

results suggest that the selectivity of polymer attached catalyst toward substrates of different sizes can be controlled by the loading of the catalyst on the polymer support.

The rates of hydrogenation of cyclohexene with the Rh(I) catalysts were strongly dependent on the solvent used to swell and suspend the polymer support. The relative rates of reduction of cyclohexene in benzene, toluene and cyclohexane are presented in Table II. Although the relationship between rate

TABLE II. *Poly-Rh(I) Catalyst Reduction of Cyclohexene(1M)*

Solvent	Rate/g($\times 10^2$)	Rel. Rate	Swelling Ratio
Benzene	5.80±.39	1.000	3.9
Toluene	0.478±.069	0.082	3.45
Cyclohexane	0.144±.018	0.025	1.9

and swelling ability of the solvent is complex, the rate decreases with decreasing swelling ratio† of the polymer in the solvent. Cyclohexane is a poor swelling solvent for the polymer and requires about 24 hours to reach maximum swelling. The better the solvent swells the polymer, the lower the diffusion barrier to the substrate.

The selectivity (Grubbs and Kroll, 1971) of the catalysts toward olefins of different molar volumes (size) was also dependent on the solvent used. The results of the reduction of a series of olefins of different sizes (calculated from the molar volumes) are given in Table III. Small changes in swelling ratios of the solvent produce small changes in the selectivity for the range of olefins chosen. The major differences are between toluene (swelling ratio ≃ 3.45)† and cyclohexane (swelling ratio ≃ 1.9)†. These differences should be enhanced by the uses of the catalyst saturated polymer. In general, the lower the swelling ratio of the solvent used, the greater the selectivity on the basis of size.

TABLE III. *Relative Rate of Reduction of Olefins With Deficient Rh(I)-Poly Catalyst*

Substrate(1M)	Diameter	Benzene	Toluene	Cyclohexene
Cyclopentene	6.542	2.10±.20	1.80±.15	1.73±.38
Cyclohexene	6.698	1.00±.07	1.00±.04	1.00±.15
Norbornene	7.030		1.31±.04	
Cycloheptene	7.184	0.95±.07	0.97±.06	0.537±.039
Cyclooctene	7.444	0.68±.04	0.64±.05	0.105±.08
Beta Pinene	7.932	0.14±.04	0.35±.02	
Camphene	7.996		0.29±.01	

†Measured as ratio of volume of solvent equilibrated polymer to dry polymer.

The results presented above† and the data presently being gathered should allow the design of catalysts of controlled activity and selectivity.

Since these catalysts show a selectivity for substrates of different sizes, experiments were designed to determine the change in regioselectivity of a homogeneous catalyst on polymer attachment. A reagent which could distinguish between similar functionalities in the side chain and the ring nucleus of a steroid could prove useful in a number of reactions. A model system was designed to test if the attachment of a reagent to a polymer would impart the desired regioselectivity.

The general structure of the model compounds is;

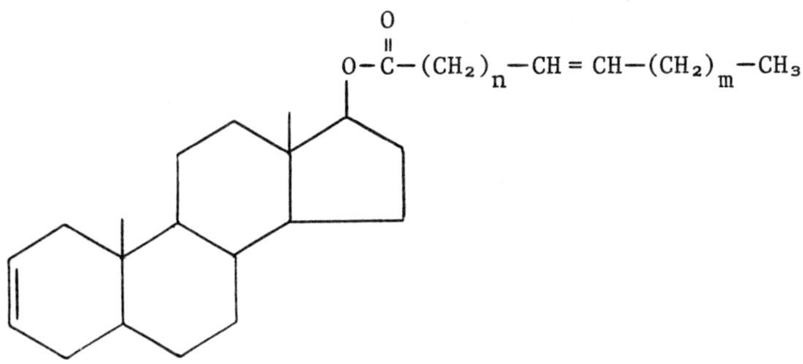

These compounds were chosen due to the ease of preparation of a series of related compounds differing only in the separation of the functionality of the side chain functionality from the steroid nucleus. Also, the ester bond is easily cleaved for independent analysis of the extent of reaction of the two functionalities in the reaction products. 5α-Androst-2-ene, 17β-ol prepared by the method of Fetizon (Fetizon and Jurian, 1969) was esterified with a number of unsaturated acid chlorides. The compounds that were prepared are listed in Table IV.

Each compound was hydrogenated using tris(triphenylphosphine)chlororhodium in solution and attached to phosphinated 2% cross-linked polystyrene beads. Each reaction was terminated when 1 eq. of hydrogen had been absorbed.

At the end of the reaction period, the esters were hydrolyzed in aqueous base and the steroid and acid were separated by extraction. The percent reduction of the steroid and the acid was determined by n.m.r. and mass spectrometry. The re-

†All rates were measured at 1 atm. of H_2 using a digital gas burette to monitor the uptake of gas with time.

TABLE IV. *Reduction of Steroids in Benzene*

	ESTER N	ESTER M	% ACID REDUCED[a] / % Δ^2 STEROID REDUCED BEADS	% ACID REDUCED[a] / % Δ^2 STEROID REDUCED HOMOGENEOUS	SELECTIVITY FACTOR
I.	1	0	84/15(5.6)	64/37(1.73)	3.3
II.	2	0	87/12(7.2)	65/34(1.9)	3.8
III.	3	0	86/13(6.7)	66/35(1.87)	3.6
IV.	8	0	90/11(8.2)	71/29(2.4)	3.4
V.	7	8 trans	69/32(2.2)	52/49(1.05)	2.1
VI.	11	8 cis	74/27(2.75)	60/39(1.5)	1.9

[a] ±2%

sults are presented in Table IV. As expected from earlier work (Pittman and Evans, 1973; Michalska and Webster, 1973; Bailar, Jr., 1974), the reductions using the attached catalysts were slower than the corresponding homogeneous reaction. In all cases, the regioselectivity of the catalyst was increased by a factor of 2 to 4.

The corrected increased selectivity of the terminal double bonded compounds were essentially identical. This result suggests that attaching the catalyst to the polymer results in an increase in the steric restrictions around the catalytic center.

The selectivity in a more polar solvent was dependent on the separation of functionality. Earlier work had indicated that a polar protic solvent produced two counter balancing effects. These solvents produce a polar gradient which drives non-polar molecules into the polymer matrix while decreasing the swelling ratio of the polymer (Grubbs, et al, 1973). The first effect increases rates and the second produces a retardation.

These effects appeared to balance for moderately sized non-polar olefins in 1:1 benzene:ethanol solvent. Table V demonstrates the effect of solvent change on a short(n=2) and a long chained ester(n=8). In both of these esters, the selectivity factor in benzene was of the order of 3.5. There was little change in the selectivity of the pentenyl ester on increasing the solvent polarity; however, the dodecenyl ester gave a factor greater than 2.5 on changing the solvent to 1:1 benzene:ethanol from pure benzene. It appears as though the selectivity in benzene is mostly due to an increase in bulk of the catalyst ligands (O'Conner and Wilkinson, 1968), but in the more polar solvents, there can be a selective absorption of portions of the molecule into the polymer. Changes due to ligand steric effects should be rather insensitive to the separation of the two functional groups, whereas, selective absorption should increase as the separation of reactant sites on the substrate increases.

TABLE V. *Reduction of Steroids in 1:1 Ethanol:Benzene*

	ESTER	RATIO REDUCED ACID / REDUCED STEROID	RATIO IN EtOH:BENZENE / BENZENE
	N		
II.	2	5.0	0.7
IV.	8	≥20.0[a]	≥2.4

[a]No unsaturated acid was detectable by n.m.r.

These results suggest that polymer attached reagents may show a new and useful type of regioselectivity.

We wish to acknowledge the support of the Army Research Office at Durham, the Eli Lilly Company and Chevron for this work.

REFERENCES

1. Bailar, Jr., J. C., Cat. Rev. Sci. Eng., 17 (1974).
2. Collman, J. P., Hegedus, L. S., Cooke, M. P., Norton, J. R., Dolcetti, G., and Marquardt, D. N., J. Am. Chem. Soc., 94, 1789 (1972).
3. Fetizon, M., Jurian, M. and Anh, N. T., Chem. Commun., 112 (1969).
4. Grubbs, R. H. and Kroll, L. C., J. Amer. Chem. Soc., 93, 3062 (1971).
5. Grubbs, R. H., Kroll, L. C. and Sweet, E. M., J. Macromol Sic. Chem., A7(5), 1047 (1973).
6. Haag, W. and Whitehurst, D. D., Catalysis, Vol. 1; North Holland, New York, 1973, p 29.
7. Kroll, L. C., 1974, Thesis, Michigan State University.
8. Michalska, F. M. and Webster, D. E., Platinium Metal Reviews, 18, 65 (1973).
9. O'Conner, C. and Wilkinson, G., J. Chem. Soc., A, 2665 (1968).
10. Pittman, C. U. and Evans, G. O., Chem. Tech., 560 (1973).

POLYMER-ANCHORED HOMOGENEOUS HYDROGENATION CATALYSTS AND THEIR USE IN MULTISTEP SYNTHETIC REACTIONS

CHARLES U. PITTMAN, JR.*, STEPHEN JACOBSON, LARRY R. SMITH, WILLIE CLEMENTS, AND HARLIN HIRAMOTO

Department of Chemistry, The University of Alabama, University, Alabama 35486

The discovery and utilization of homogeneous hydrogenation catalysts has expanded in importance recently (James, 1973). Commercial processes based on homogeneously catalyzed routes are becoming increasingly important (Szonyi, 1968) suggesting that commercial homogeneous hydrogenation processes may not be far off. Since homogeneous hydrogenations can exhibit great selectivity, they do offer certain real advantages unmatched by heterogeneous processes. For example, conjugated dienes are specifically 1,4-hydrogenated to cis-olefins in the presence of η^6-arenetricarbonyl-chromium (Frankel and Butterfield, 1969) and trienes, such as cyclododecatriene, can be reduced with 97% selectivity to cyclododecene (Fahey, 1973 a,b) using $Ru(PPh_3)_2(CO)_2(Cl)_2$.

(1)

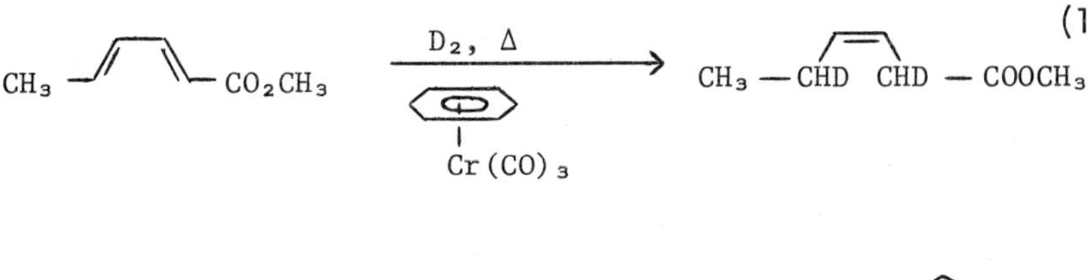

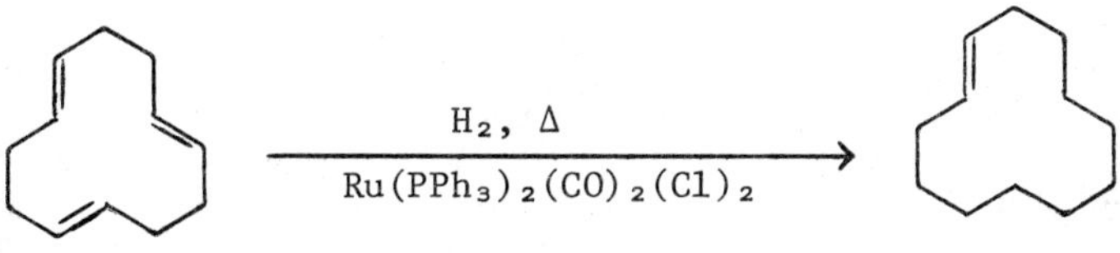

In fact, polymer-anchored homogeneous catalysts promoting hydrogenation (Grubbs et al, 1971, 1973, 1974; Pittman, 1975a) hydroformylation (Pittman et al, 1974, Haag and Whitehurst 1972) and oligomerization (Evans, Pittman et al, 1974; Pittman and Smith, 1975b) have been prepared, and reviews of such heterogenized homogeneous catalysts have appeared (Bailar, 1974; Pittman, 1973). Grubbs and coworkers have demonstrated that the polymer matrix offers advantages besides simple immobilization. By controlling the degree of resin swelling, the diffusion of large and small olefins into the resin could be tailored so that smaller olefins could be hydrogenated faster than large ones using resin-anchored $Rh(PPh_3)_3Cl$ catalysts. Thus, a size selectivity was conferred on the catalyst system by the polymer. Similarly, greater positional selectivity was observed in the reduction of nonequivalent olefinic sites within the same molecule when using anchored $Rh(PPh_3)_3Cl$. Finally, by using the polymer to "matrix isolate" one reactive catalytic site from another, polymer anchored titanocene was generated from anchored biscyclopentadienyltitaniumdichloride. Pittman and Smith (1975c) showed that two catalysts, anchored to the same polymer, could be used to conduct sequential multistep organic synthesis such as cyclooligomerization - hydroformylation sequences.

EXPERIMENTAL SECTION

Hydrogenations were carried out in stainless steel autoclaves. Solvent, resin-anchored catalyst, and substrate were added and thoroughly degassed by freeze thaw methods or by purging with hydrogen. Where the substrate was a gas, it was directly added from a tank through nitrogen or hydrogen purged lines. Time was then allowed for the polymers to thoroughly swell. The autoclave was heated to the reaction temperature and hydrogen was added with vigorous agitation. The product distribution during or following the reaction was followed by GLC.

Benzene, THF, and toluene were dried over CaH_2 for at least 24 hr. and then distilled under nitrogen. Similar care was taken to dry all solvents. Nitrogen, hydrogen, and carbon monoxide were obtained commercially (99+%) and used as received. Organometallic complexes were obtained from Strem or Pressure Chemical Co. GLC separations were done on a Varian Model 90-P gas chromatograph using Carbowax 20-M or SE-30 (15% on 100-120 N chromasorb P, 6 ft. x ¼ in.) or OV-17 (5% on glass packings). The ir, ^{1}H NMR, uv, and mass spectra were obtained on a Beckman Ir-33, Perkin Elmer R20B or Varian Ha-100, Cary 14, and Perkin-Elmer Hitachi RMU-6M instruments respectively.

Anchoring Catalysts to Diphenylphosphinated Polystyrene Resins

The diphenylphosphinated resins (styrene-divinylbenzene resins, Biobeads SX-1 and SX-2, 200-400 mesh, were brominated under Friedel Crafts conditions and then treated with excess $NaPPh_2$ in THF) were swollen in a stirred toluene solution at 100° under nitrogen. A toluene solution of the homogeneous catalyst to be anchored was then added, and the reaction was conducted at 100° for 24 hr. The solvent was decanted, and the polymer beads were extracted (Soxhlet) for 24 hr. with benzene under nitrogen to remove any unbound catalyst and free PPh_3 from the resin. The total phosphine and metal contents were controlled as described in detail elsewhere (Smith, 1974).

Reaction of 1,3-Butadiene with Polymer-Bound Bis(triphenylphosphine)nickel Dicarbonyl

Into a dry, nitrogen purged stainless steel bomb were placed polymer-bound bis(triphenylphosphine)-nickel dicarbonyl (0.50g, 0.179 mmol of Ni), and dry benzene (10 ml, nitrogen saturated). The system was degassed by several freeze-thaw cycles, 1,3-butadiene (10.4g, 193 mmol) was added at -78°, and the bomb was warmed to 110-115°, and shaken for 24 hr. After cooling to 25° and venting in a -78° trap, 1.03g (9.9%) of the 1,3-butadiene was recovered. GLC of the reaction mixture gave cis-1,2-divinylcyclobutane (15.2%), 4-vinylcyclohexene (31.7%), (Z,Z)-1,5-cyclooctadiene (41.2%), and (E,E,E)-1,5,9-cyclododecatriene (11.9%). The products were collected by preparative GLC, and their ir, NMR, and mass spectra were found to be identical with commercially available samples.

Reactions which had reached completion gave no divinylcyclobutane. The polymer-bound catalysts were recycled after filtration under nitrogen, washed with dry, nitrogen saturated benzene, vacuum dried, and then carried through an identical reaction procedure again.

Reaction of Butadiene's Cyclooligomers with Homogeneous and Polymer-Bound $(Ph_3P)_3RhCl$

Into a dry, nitrogen purged stainless steel bomb were placed tris(triphenylphosphine)rhodium chloride (0.0925g, 0.1 mmol) 4-vinylcyclohexene (2.55g, 23.6 mmol), and dry benzene (10 ml. nitrogen saturated). Dissolved gases were removed by several freeze-thaw cycles; the bomb was flushed with hydrogen, pressurized to 350 psig with hydrogen, and heated at 50°, with shaking for 4 hr. After cooling to 25°, a pressure of 225 psig was noted. Analytical GLC showed only ethylcyclohexane. Samples were collected by preparative GLC. The ir spectrum showed no =C—H or C=C bands, but now exhibited a band at 1380 cm^{-1}

(terminal methyl). The mass spectrum was identical with an authentic sample of ethylcyclohexane as was the NMR spectrum. Similar reactions with 1,5-cyclooctadiene and 1,5,9-cyclododecatriene also resulted in complete hydrogenation to cyclooctane and cyclododecane.

Analogous reactions using polymer-bound tris(triphenylphosphine)rhodium chloride (0.50g, 0.083 mmol of Rh) also resulted in complete hydrogenation. The polymeric catalyst was recycled by filtering under nitrogen, washing with dry benzene, and being carried through the reaction procedure again. The resin catalyst could be stored dry in air for extended periods.

Hydrogenation of Butadiene's Cyclooligomers by Homogeneous and Polymer-Bound $(Ph_3P)_2RuCl_2(CO)_2$

Into a dry, nitrogen purged stainless steel bomb were placed dichlorodicarbonylbis(triphenylphosphine)ruthenium (0.070g, 0.09 mmol), triphenylphosphine (0.262g, 1 mmol), cyclooctadiene (1.76g, 16.3 mmol), and dry benzene (20 ml, nitrogen saturated). Dissolved gases were removed by several freeze-thaw cycles, the bomb was flushed with hydrogen, pressurized to 150 psig with hydrogen, and both were heated and shaken at 145° for 5 hr. After cooling to 25° , a pressure of 100 psig was noted. Analytical GLC showed three peaks. These were collected by preparative GLC, giving unreacted 1,5-cyclooctadiene (2.5%), (Z)-cyclooctene (93.7%) (parent ion m/e 110; ir (KBr) 3000 (s), 2910 (vs), 2840 (vs), 1655 (m), 1475 (s), 1455 (s), 900 (s), 755 (vs), 705 (s) cm^{-1}; NMR ($CDCl_3$) δ5.61 (m, 2H, olefinic protons), and 1.54 (s, 4H, saturated ring protons) and 1.5 (m, 4H, saturated ring protons)) and cyclooctane (3.94%) (parent ion m/e 112; ir no $C{\equiv}C$ or $=C-H$; NMR ($CDCl_3$) δ1.54 singlet).

When 4-vinylcyclohexene was the substrate, GLC of the reaction mixture showed unreacted 4-vinylcyclohexene (2.9%), 4-ethylcyclohexene (85.3%) (parent ion m/e 110; ir 3020 (m), 2940-2820 (s), 1660 (w), 1450 (s), 1380 (m), 1140 (m), 1005 (m), and 810 (s) cm^{-1}; NMR ($CDCl_3$) δ4.76-5.38 (complex, 2H, olefinic protons), 1.97-1.14 (complex, 9H, methylene protons), and 0.72 (t, 3H, J=7.2 Hz-CH_3)) and ethylcyclohexane (11.8%) (parent ion m/e 112; ir and NMR spectra were identical to an authentic sample).

Analogous reactions using polymer-bound $(Ph_3P)_2RuCl_2(CO)_2$ (0.5g, 0.053 mmol of Ru) also gave the same selective hydrogenation reactions when conducted at 165-170°. The polymeric catalyst was recycled by filtration under nitrogen from a reaction solution, washed with benzene, vacuum dried (25° , 0.05 Torr), and carried through the reaction procedure again.

Reactions using the catalysts with excess bound phosphine

group but no added PPh_3 (1.0g, .10 mmol Ru) and 1.5-cyclooctadiene (1.76g, 16.3 mmol) gave cyclooctane 45.9%, cyclooctene 54.1%, and no 1,5-cyclooctadiene.

Butadiene Cyclooligomerization and Sequential Hydrogenation with Homogeneous and Polymer-Bound $(PPh_3)_2Ni(CO)_2$ and $(PPh_3)_3RhCl$

Into a dry, nitrogen purged stainless steel bomb were placed $(PPh_3)_2Ni(CO)_2$(0.10g, 0.156 mmol), $(PPh_3)_3RhCl$(0.10g, 0.108 mmol), and dry benzene (10 ml, nitrogen saturated). After degassing, 1,3-butadiene (10.0g, 187 mmol) was added at -78 °, and the reaction was maintained at 100° with shaking for 24 hr. After cooling, no unreacted 1,3-butadiene was recovered. Aliquots of solution were removed for analyses. A nitrogen purge was used in some cases to remove traces of unreacted 1,3-butadiene. The reactor was flushed with hydrogen, pressurized to 500 psig, heated to 60°, and shaken for 24 hr. Analytical GLC of the 1 ml of solution from the first step showed 4-vinylcyclohexene (17.3%), (Z,Z)-1,5-cyclooctadiene (66.7%), and (E,E,E)-1.5,9-cyclododecatriene (15.9%), identified as reported. Only traces (less than 1%) of unreacted 1,3-butadiene were found. After cooling to 25°, a pressure of 95 psig was noted, and the excess hydrogen was vented. Analytical GLC showed three products which were collected by preparative GLC. Using ir, Ms, and NMR, the products were identified as ethylcyclohexane, cyclooctane, and cyclododecane.

Similar reactions using polymer-bound bis(triphenylphosphine)nickel dicarbonyl (0.5g, 0.179 mmol of Ni) and polymer-bound tris(triphenylphosphine)rhodium chloride (0.5g, 0.083 mmol of Rh) gave the same products but in somewhat lower yields. Oligomerization of butadiene resulted in recovery of unreacted butadiene (17.3%) with the products (82.7%) being 4-vinylcyclohexene (20.3%), (Z,Z)-1,5-cyclooctadiene (51.9%), and (E,E,E)-1,5,9-cyclododecatriene (26.5%). The hydrogenation yield was approximately 85%, giving a complex mixture of ethylcyclohexane (18.1%), cyclooctane (48%), and cyclododecane (21.6%), along with small amounts of various mono-, di-, and trienes.

Analogous reactions using resin (1.0g) containing both catalysts on the same beads, bis(triphenylphosphine)nickel dicarbonyl (0.424 mmol of Ni g^{-1}) and tris(triphenylphosphine) rhodium chloride (0.30 mmol of Rh g^{-1}), gave yields almost identical with the catalysts on separate resin beads.

Butadiene Cyclooligomerization and Sequential Selective Hydrogenation to Monoenes with Homogeneous and Polymer-Bound $(Ph_3P)_2Ni(CO)_2$ and $(Ph_3P)_2RuCl_2(CO)_2$

Into a dry, nitrogen purged stainless steel bomb were placed $(PPh_3)_2Ni(CO)_2$(0.10g, 0.157 mmol), $(PPh_3)_2RuCl_2(CO)_2$ (0.10g, 0.13 mmol), triphenylphosphine (0.262g, 1 mmol), and dry benzene (20 ml, nitrogen saturated). After degassing, 1,3-butadiene (6.6g, 122 mmol) was added at -78° and the reaction was heated 24 hr. at 100° with constant shaking. After cooling, unreacted butadiene (1.02g, 15.5%) was recovered by venting into a cold trap. A 1-ml aliquot was removed and a nitrogen purge was employed to remove any dissolved, unreacted butadiene. The reactor was flushed with hydrogen, pressurized to 150 psig with hydrogen, and heated with shaking at 140° for 14 hr. Analytical GLC of the 1-ml sample from the first step found 4-vinylcyclohexene (21.8%), (Z,Z)-1,5-cyclooctadiene (68.8%), and (E,E,E)-1,5,9-cyclododecatriene (9.4%). These were identified as described previously.

After cooling to 25°, the hydrogen pressure was 81 psig. Analytical GLC of the reaction mixture now found ethylcyclohexane (2.6%), 4-ethylcyclohexene (18.5%), 4-vinylcyclohexene (0.6%), cyclooctane (3.2%), (Z)-cyclooctene (61.4%), 1,5-cyclooctadiene (4.2%), cyclododecene (9.3%), and only traced of cyclododecane, cyclododecadiene, and 1,5,9-cyclododecatriene.

The system was recycled by distilling off solvent and products under vacuum, reintroducing dry benzene (20 ml, nitrogen saturated) condensing in 1,3-butadiene at -78°, and beginning the reaction procedure over again. A temperature of 145° must not be exceeded or the $(Ph_3P)_2Ni(CO)_2$ will decompose, giving black, finely divided nickel.

Analogous reactions using the catalysts, bis(triphenylphosphine)nickel dicarbonyl (0.5g, 0.179 mmol of Ni) and dichlorodicarbonylbis(triphenylphosphine)ruthenium (0.5g, 0.053 mmol of Ru), bound to separate batches of resin gave overall results only slightly lower than the homogeneous reaction. Recovery of the polymer was accomplished by filtration; however, the $(Ph_3P)_2Ni(CO)_2$ had decomposed during the period of hydrogenation when a temperature of 165° was employed.

Hydrogenation of 1,5-Cyclooctadiene and 4-Vinylcyclohexene with Homogeneous and Polymer-Bound $Ir(CO)(Cl)(PPh_3)_2$

The title compounds, benzene and catalyst were charged to a bomb, as described previously, and heated to 170° under 150 psig of hydrogen. Other temperatures and pressures were used and these, plus the stoichiometry used is described in Tables I-IV. The products were analyzed by GLC as described above.

RESULTS AND DISCUSSION

Hydrogenation of 1,5-Cyclooctadiene Using Anchored Ir(CO)(Cl)$(PPh_3)_2$

In order to study the effect of excess anchored phosphine ligand on hydrogenations catalyzed by Ir$(PPh_3)_2$(CO)(Cl), a large number of diphenylphosphinated, microporous, styrene-divinylbenzene resins were prepared. First, a group of resins containing the following precentages of P were prepared: 1.25% (corresponds to 12% of the rings phosphinated), 1.91% (18%), 2.02% (19%), 4.02% (38%), and 10.28% (96%). Each of these resins was reacted with Ir(CO)(Cl)$(PPh_3)_2$ to give a series of resin-anchored catalysts where P/Ir ratios of 3, 4, 7, 11, 12, and 22 were achieved. Thus, hydrogenation rates could be studied as a function of the P/Ir ratio over a range of total phosphine levels.

1,5-Cycloooctadiene, 1, was selectively hydrogenated over Ir(CO)(Cl)$(PPh_3)_2$ and its styrene-divinylbenzene-anchored analog. Successive isomerization to 1,4-COD, 2, and 1,3-COD, 3, occurs followed by hydrogenation to cyclooctene, 4, and, finally, cyclooctane, 5.

(3)

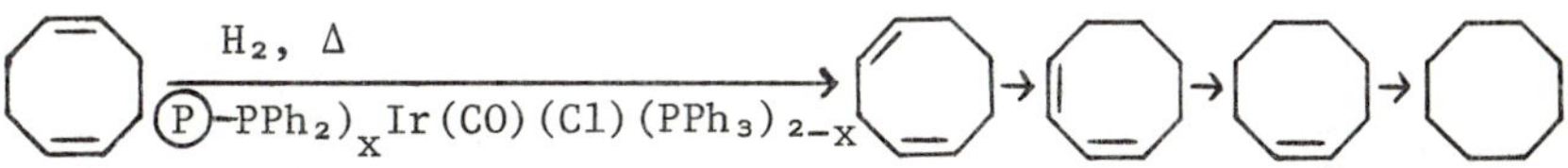

1 2 3 4 5

The rates of homogeneously and heterogeneously catalyzed reactions were compared using the same solvent volume and equal amounts of catalyst at equal P/Ir ratios. Excess PPh_3 was added to Ir$(PPh_3)_2$(CO)(Cl), when employed homogeneously, to achieve P/Ir=2. When compared at equal P/Ir ratios, the rate of hydrogenation was significantly faster using the anchored catalyst whenever the P/Ir ratio was <5. This occurred even at a high total % P on the polymer. At high % P, the dissociation equilibria Ⓟ-$PPh_2)_2$Ir(CO)(olefin)-(Cl) $\rightleftharpoons$ Ⓟ-PPh_2 + Ⓟ-PPh_2)Ir(CO)-(olefin)(Cl), which operates in the hydrogenation (Burnett et al, 1973, Strohmeier and Onada, 1969), would be repressed. The magnitude of this anchored/homogeneous rate ratio is quite large at low P/Ir ratios as can be seen by comparing reactions 1 with 2 and 3 with 4 in Table I. All these reactions were carried out in benzene at 170° and 150 psig H_2.

The anchored/homogeneous rate ratio drops as the P/Ir ratio increases. For example, at P/Ir = 7, the hydrogenation rate, using the anchored catalyst, was slower at 4.02% P (38% of rings phosphinated) but faster at 1.25% P (12% of rings phosphinated) than the corresponding homogeneous rates. Increasing P/Ir to 12 or 20-22, it was found the homogeneous hydrogenation (and 1,5 ⟶ 1,4 and 1,4 ⟶ 1,3-COD isomerization) rates were greater. This behavior suggests that anchored phosphine moieties are not sufficiently mobile (i.e. on the time scale of key steps in the mechanism) to intercept anchored coordinatively unsaturated Ir intermediates at a rate equivalent to dissolved PPh_3 when P/Ir < 5. This is true even when the anchored phosphine concentration is high (i.e. 4.02 or 10.28%).

TABLE I. *Comparative 1,5-COD Hydrogenation Rates, Catalyzed by Resin-Anchored or Homogeneous Ir(CO)(Cl)(PPh₃)₂ in Benzene at 170° and 150 psig H_2.*[a]

Reaction Number	Mode	P/Ir	Total % P on Polymer	Time Hr	Conversion %4	%5
1	homogeneous	3	—	33	37	3
2	anchored	3	4.02	0.25	57	4
3	homogeneous	4	—	48	44	3
4	anchored	4	10.28	0.25	69	5
5	homogeneous	6	—	72	68	11
6	anchored	7	4.02	93	5	0
7	anchored	7	1.25	0.20	68	6
8	homogeneous	14	—	72	50	1
9	anchored	12	2.53	94	19	2
10	anchored	12	4.02	96	6	0
11	homogeneous	22	—	72	15	0
12	anchored	22	4.02	93	18	2

[a]All reactions employed 0.064 mmol Ir, 27.8 mmol 1,5-COD in 15 ml of benzene.

A highly unusual temperature effect was also observed using the anchored catalyst system. The hydrogenation rate was observed to increase as the temperature decreased from 170° to 80-65° at P/Ir = 4. As shown in Table II, this occurred at total % P levels of 1.91% and 10.28%. Assuming the same mechanism operates over this temperature range, the observation can be explained in terms of the mobility of anchored $-PPh_2$ groups. This mobility should increase significantly as the temperature is raised from 80° to 170°. Thus, excess anchored $-PPh_2$ can more efficiently repress the dissociation equilibrium at higher temperatures, thereby retarding the rate. If the net activation energy of the isomerization-hydrogenation sequence is fairly low, this increased Ⓟ $-PPh_2$ mobility could retard the rate more than the temperature rise would increase it.

Holding the P/Ir ratio constant (i.e. at 7, 11, 12, 22), the hydrogenation rate (using the anchored system) decreased as the total % P on the polymer was increased, whenever P/Ir < 5. This is in accord with greater chance of anchored $-PPh_2$ encountering a coordinatively unsaturated metal site as the total P density increases. In the range P/Ir = 3-4, the rates are very sensitive to changes in total % P and P/Ir. The role of diffusion was also briefly examined by comparing the rates using 1% and 2% divinylbenzene (DVB) resins where P/Ir = both 4 and 12 and total % P = 2.0. The rates of isomerization and hydrogenation at 170°, using 1% and 2% DVB, were almost identical. Since the bead size and both P/Ir and % P were the same, this shows that diffusion was not limiting at 170°.

TABLE II. *Rate Variation With Temperature in 1,5-COD Hydrogenations Catalyzed by Resin-Anchored Ir(CO)(Cl)(PPh_3)$_2$.*[a]

Reaction Number	% P on Polymer	Time Hr	Temperature, °C	Conversion %4	%5
13	1.91	1.1	170	46	4
14	1.91	6	65	61	25
15	10.28	0.20	170	52	4
16	10.28	1.5	80	55	14

[a]Each reaction run with P/Ir = 4; 0.064 mmol Ir, 27.8 mmol 1,5-COD in 15 ml benzene.

Selective Hydrogenation of 4-Vinylcyclohexene

The series of resin-anchored Ir(CO)(Cl)$(PPh_3)_2$ catalysts, described above, was also employed in hydrogenations of 4-vinylcyclohexene. They promoted the selective hydrogenation of 6 to ethylcyclohexene, 7, in benzene. Both homogeneous and anchored Ir(CO)(Cl)$(PPh_3)_2$ catalyzed the formation of 7 in 93-99% selectivity over ethylcyclohexane, 8 (see Table III). The selectivity varied at different % P loadings and P/Ir ratios using the anchored catalyst (see runs 8-13, Table III). In

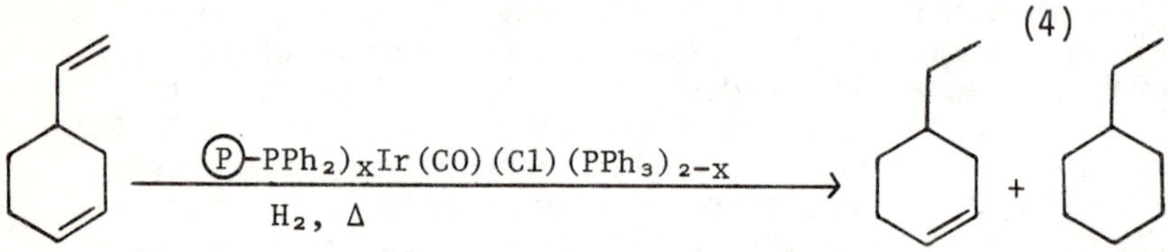

(4)

6 7 8

general, the selectivity dropped as the % P in the resin increased but was very high at low values of % P. This high selectivity was similar to that achieved by Lyons (1973) who selectively hydrogenated 1,3- and 1,4-cyclohexadiene to cyclohexene (97 to 99% selectivity at 62-69% yield) in DMA at 83° and 30 psig H_2 using Vaska's complex homogeneously.

The anchored catalyst could be conveniently separated from the product solution and recycled. The resin could be exposed to air without loss of activity. The rate was slower during the first cycle than on succeeding cycles using the anchored catalyst (Table IV). This was due, in part, to an induction period associated with the formation of an iridium hydride intermediate. IR analyses of the resin, after the first cycle, exhibited new Ir-H absorptions at 2210 and 2090 cm^{-1} suggesting trapped Ⓟ—PPh_2)Ir(CO)(olefin)(Cl)H_2 or Ⓟ—$PPh_2)_2$Ir(CO)(Cl)H_2 is present.

The rates increased upon successive recycling. In many cases, the rates were faster using the anchored catalyst than using the catalyst homogeneously, when compared at equal P/Ir ratios. For example, at P/Ir = 4 and 80°, the rate using the anchored catalyst (1.91% P) was twice that of the homogeneous rate. At P/Ir = 7 and 80°, the anchored catalyst (1.25% P) promoted hydrogenation three times faster on its third cycle than the homogeneous rate. This can be explained in terms of dissociation equilibrium, where excess phosphine reduces the concentration of the 4-coordinate Ir intermediate thought to

$$\text{Ⓟ}\!-\!PPh_2)_2Ir(CO)(olefin)Cl \rightleftharpoons \text{Ⓟ}\!-\!PPh_2)Ir(CO)(olefin)Cl + \text{Ⓟ}\!-\!PPh_2$$

be the active catalyst which oxidatively adds H_2 in the rate determining step. Upon dissociation of an anchored $-PPh_2$ group, various bond rotations could occur retarding the rate of reverse reaction. The entropy of Ⓟ—PPh_2 recombination might be unfavorable relative to the equivalent process in the use of the catalyst homogeneously. In agreement with this potential explanation is the observation that when compared at high P/Ir ratios (i.e. 12 or greater) and high P loadings on the polymer (4.02% or 10.28%), the homogeneous rates were always faster than the rates using the anchored systems. For example see run 6 versus 13 in Table II.

TABLE III. *Example Selective Hydrogenations of 4-Vinylcyclohexene to 4-Ethylcyclohexene Catalyzed by Homogeneous and Resin-Anchored $Ir(CO)(Cl)(PPh_3)_2$.*[a]

Run No.	Mode (% P in resin)	Temp. °C	P/Ir[b]	H_2 Pressure psig	Time Hrs.	Conversion[c] %7	%8
1	homogeneous	70	2	100	6.5	59	1
2	homogeneous	82	2	100	2.5	96	4
3	homogeneous	90	2	100	1.0	96	4
4	homogeneous	120	6	100	21.0	83	1
5	homogeneous	130	12	100	21.0	50	1
6	homogeneous	150	12	100	21.0	98	1
7	homogeneous	150	12	320	48.0	93	2
8	anchored (1.91)	120	4	100	0.3	95	5
9	anchored (1.25)	80	7	100	1.3	95	5
10	anchored (4.02)	120	7	100	18.0	46	14
11.	anchored (4.02)	80	7	100	48.0	33	8
12.	anchored (4.02)	120	11	100	52.0	35	8
13.	anchored (2.54)	150	12	100	24.0	44	4

TABLE III. (Cont'd.)

[a]All reactions were carried out in benzene (15ml) employing 0.064 mmol Ir, 27.8 mmol 4-vinylcyclohexene

[b]In the homogeneous reactions P/Ir = 2 means no PPh_3 was added to reaction medium

[c]The conversions are based on 4-vinylcyclohexene charged to the reaction. Where the %7 + %8 are <100% the remainder was unreacted 1.

TABLE IV. *Typical Examples of the Rate Enhancement Found After the First Cycling of Anchored $Ir(CO)(Cl)(PPh_3)_2$ in 4-Vinylcyclohexene Hydrogenations*

Run	Cycle	%P	P/Ir	Temperature °C	Time Hrs.	Conversion %7	%8
1a	1st	4.02	7	120	18	46	14
1b	2nd	4.02	7	80	48	33	8
2a	1st	2.54	12	150	24	44	4
2b	2nd	2.54	12	150	4	52	7
3a	1st	1.25	7	80	2.25	61	3
3b	2nd	1.25	7	80	1.30	95	5
4a	1st	1.91	4	120	5	82	0
4b	2nd	1.91	4	120	0.3	95	5

Sequential Cyclooligomerization-Reduction Sequences Using Anchored Catalysts

The cyclooligomerization of butadiene, by resin-anchored $Ni(CO)_2(PPh_3)_2$, to 4-vinylcyclohexene, 1,5-cyclooctadiene, and 1,5,9-cyclododecatriene was previously described (Pittman et al, 1975). This group also showed these oligomers were quantitatively reduced to their saturated analogs in the presence of anchored $Rh(PPh_3)(Cl)$ and selectively reduced to monoenes in the presence of anchored $Ru(CO)(Cl)(PPh_3)_2$. If the cyclooligomerization is conducted in the presence of 10

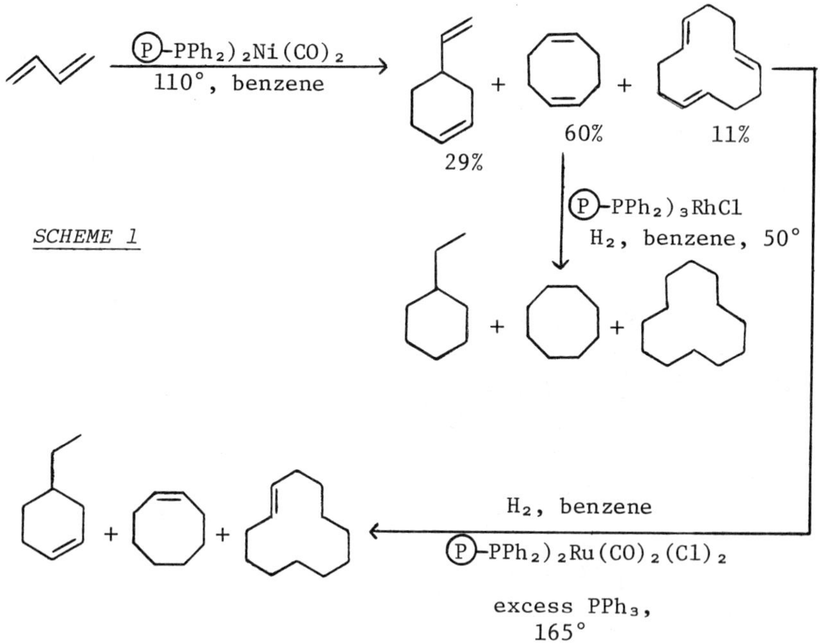

SCHEME 1

psig of CO, a 98% selectivity to 4-vinylcyclohexene is achieved.

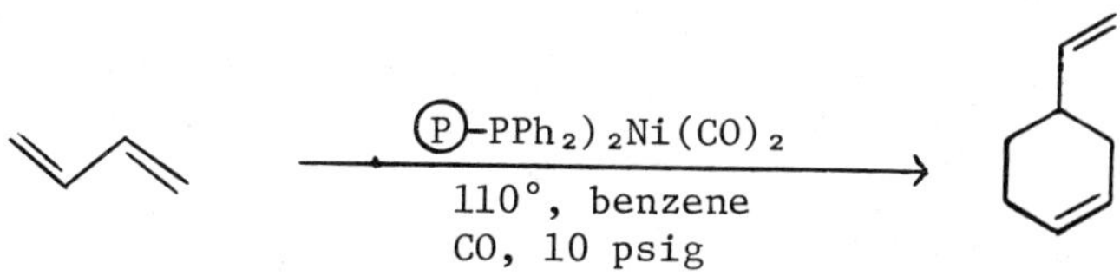

(5)

The same reaction sequences can be carried out sequentially. Thus, $Ni(CO)_2(PPh_3)_2$ and $Rh(PPh_3)_3Cl$ or $Ru(CO)_2(Cl)_2$-$(PPh_3)_2$ may be anchored to the same resin or to separate resins, which are then charged to the same reactor, and the reactions are carried out in one pot. This route is shown in scheme 2.

Butadiene was oligomerized in benzene for 24 hrs. at 110-115° using the anchored catalysts. The reaction bomb was then cooled to 50° and pressurized to 350 psig with hydrogen for 24 hrs. High yields of alkanes (see Table V) were obtained from the cyclooligomers in all cases (85-89%), and the overall yields from butadiene were high until the molar turnover limit of $(PPh_3)_2Ni(CO)_2$ was reached (about 1100). The mixed anchored catalysts could be recycled without substantial loss of activity until this turnover limit was reached. This limit was defined as the point where the rate fell sharply to a point where further conversion was impractical. Separate experiments demonstrated that $(PPh_3)_2Ni(CO)_2$ does not catalyze hydrogenation, nor does $(PPh_3)_3RhCl$ catalyze oligomerization under the conditions employed. These catalysts do not interfere with each other when employed bound to the same polymer (see Scheme 2).

The homogeneous reactions exhibited faster hydrogen pressure drops. Using equal molar amounts of catalyst and cyclooligomers in the same volume of benzene, the hydrogenation using (P)$-PPh_2)_3$ - RhCl, A, was about 0.8 as fast as the homogeneous reaction while the rate using $(CO)_2Ni_2(Ph_2P$ (P)$-PPh_2)_3RhCl$, B, was about 0.15 as fast. These results compare to Grubbs' and Kroll's (1971) observation that $(PPh_3)RhCl$ bound to 2% cross-linked styrenedivinylbenzene resins were only 0.06 as active as using the homogeneous catalyst. The reason that B exhibited slower rates than A is due to the fact the mole fraction of metal in the polymer was higher. Since Ni was almost completely dicoordinated and Rh di- and tricoordinated with polymer-bound phosphine groups, the effective cross-link density of resin B was greater than A since the chelated metals now become cross-linking sites. This was substantiated by examining the P/Ni + Rh ratio in the polymer and by recovering displaced PPh_3 (as PPh_3O) from resin preparations. Both A and B were prepared from 1% divinylbenzene resins while Grubbs employed 2% divinylbenzene resins. Furthermore, we employed a smaller particle size distribution (37-74μ) than did Grubbs (74-149μ). Thus, diffusion retardation of rate would be expected to be greater in Grubbs' work.

Fahey (1973 a,b) demonstrated 1,5-COD and 1,5,9-CDT were selectively hydrogenated to (Z)-cyclooctene and (E)-cyclododecene in 92 and 97% yields using $(PPh_3)_2RuCl_2(CO)_2$ homogeneously with excess PPh_3 present. We extended these studies to polymer anchored systems and achieved similar results. However, excess resin anchored PPh_2 groups (up to a 15 mol excess) were not able to enhance selectivity to monoene in the manner that PPh_3 did. Studies were undertaken to demonstrate that selective hydrogenation could be effected sequentially with cyclooligomerization.

TABLE V. *Sequential Reactions of 1,3-Butadiene with Homogeneous and Polymer-Bound $(Ph_3P)_2Ni(CO)_2$ and $(Ph_3P)_3RhCl$*

Run	Catalyst	mmol	Catalyst	mmol	mmol of butadiene	% conversion	% 4-VCH	%1,5-COD	%1,5,9-CDT	% alkane[d]
1	$(Ph_3P)_2Ni(CO)_2$	0.157					23.6[a]			
2[e]			$(Ph_3P)_3RhCl$	0.054	128	0	0	0	0	0
3	$(Ph_3P)_2Ni(CO)_2$[b]	0.157	$(Ph_3P)_3RhCl$[c]	0.108	187	98.3	20.8	66.4	12.8	99.3
4	Resin 1[f]	0.179	Resin 2[f]	0.083	181	84.7	22.3	54.9	20.5	86.4
5	Resin 1[f]	0.179	Resin 2[f]	0.083	55	92.2	22.7	61.3	16.0	86.3
6	Recycle	0.179	Recycle	0.083	55	91.8	23.1	60.8	16.1	84.5
7	Recycle	0.179	Recycle	0.083	55	81.3	22.8	59.7	17.5	83.2
8	Recycle	0.179	Recycle	0.083	55	21.6	23.9	58.6	17.5	81.9
9	Resin 3[f]	0.413	Resin 3[f]	0.300	55	92.3	24.9	60.7	14.4	84.1
10	Recycle	0.413	Recycle	0.300	55	92.1	25.1	60.5	14.5	82.8

[a]mmol of 4-VCH added to reaction at 350 psi H_2. [b]Homogeneous oligomerizations 90°, 24 hr., polymer-bound oligomerization 110-115°, 24 hr. [c]Hydrogenations at 50°, 350 psi H_2. [d]Total percent of the cyclooligomerization products which were hydrogenated to alkanes. [e]No H_2 pressure. [f]Polymer 3 analyzed for 2.49% Ni and 3.09% Rh while 2 contained 1.83% P and 1.70% Rh and 1 analyzed for 2.39% P and 2.10% Ni.

SCHEME 2

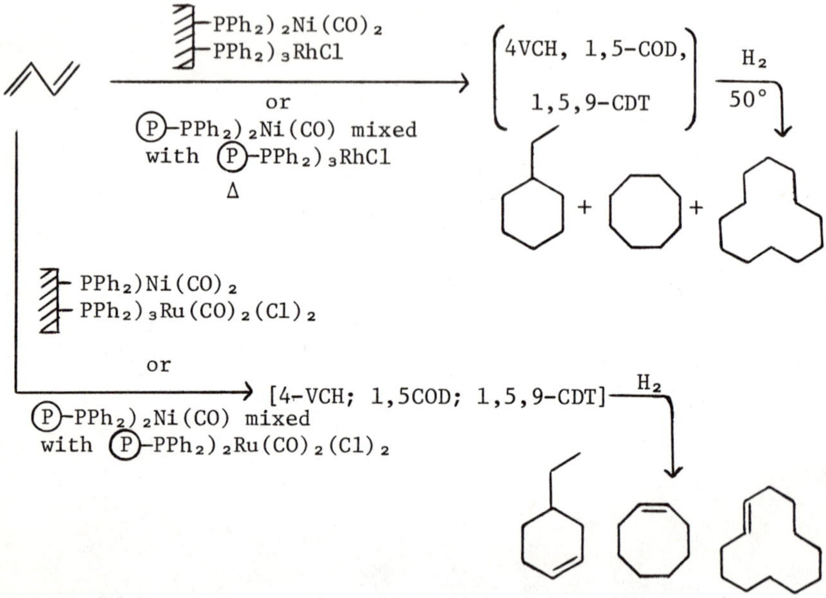

Butadiene was efficiently converted to cyclooligomers 4-VCH, 1,5-COD, and 1,5,9-CDT followed by sequential selective hydrogenation to ethylcyclohexane, cyclooctene, and cyclododecene. The sequential reactions could be carried out (1) using $(PPh_3)_2Ni(CO)_2$ and $(PPh_3)_2RuCl_2(CO)_2$ together homogeneously (2) using a mixture of resins and (3) using a resin to which both the nickel and ruthenium species were anchored. The latter system was not pursued in great detail because the nickel catalyst was destroyed at the temperature which the resin-anchored ruthenium complex was used. Thus studies were concentrated on the first two modes. As summarized in Table VI, high yields of the monoenes were obtained from butadiene in all cases in which a large molar excess (15-20) of PPh_3, relative to Ru, was added to the reaction solutions. Furthermore, the product distribution in the sequential homogeneous and sequential resin catalyzed reactions were very similar (Table VI).

The homogeneous and resin catalyzed oligomerizations were carried out at 90° for 24 hr and 115° for 24 hr, respectively. The homogeneous hydrogenations were conducted at 140°, 4 hr, and 150 psig of hydrogen while those employing anchored catalyst were performed at 165°, 18 hr, and 150 psig. Thus butadiene was first oligomerized and then the reactor temperature

TABLE VI. *Sequential Reaction of 1,3-Butadiene with Homogeneous and Polymer-Bound $(Ph_3P)_2Ni(CO)_2$ and $(Ph_3P)_2RuCl_2(CO)_2$*

No.	Catalyst[a]	mmol	Catalyst	mmol	mmol of butadiene	Total Yield	% 4-VCH	%1,5-COD	%1,5,9-CDT	%ECH[d]	%CO[d]	%CD[d]
1[b]			$L_2RuCl_2(CO)_2$	0.090	100	0	0	0	0	0	0	0
2[c]	$L_2Ni(CO)_2$	0.157	$L_2RuCl_2(CO)_2$	0.090	162	97.2	21.8	68.8	9.4	85.3	89.1	98.2
3	$L_2Ni(CO)_2$	0.157	$L_2RuCl_2(CO)_2$	0.090	55	98.1	23.6	64.1	13.3	86.1	88.7	94.6
4	Recycle	0.157	Recycle	0.090	55	91.2	22.9	65.3	11.8	85.8	84.9	91.7
5	Recycle	0.157	Recycle	0.090	55	56.2	23.9	62.8	13.3	86.4	87.2	92.2
6[e]	Resin 1[f]	0.079	Resin 6[f]	0.053	55	92.3	24.1	56.8	19.1	84.2	88.9	91.8
7	Recycle	0.179	Recycle	0.053	55	0	0	0	0	0	0	0

[a]L represents Ph_3P. [b]No hydrogen pressure. [c]Homogeneous oligomerization at 90°, 24 hr., hydrogenations at 140°, 4 hr., 150 psi of H_2, in 20 ml of benzene with 0.262 gm of Ph_3P added. [d]Percent of respective polyene hydrogenated to monoene: ECH is 4-ethylcyclohexene, CO is Z-cyclooctene, and CD is E-cyclododecene. [e]Polymer-Bound oligomerization at 115°, 24 hr., hydrogenation at 165°, 18 hr. [f]Polymer 1 analyzed for 2.39% P and 2.10% Ni while 6 analyzed for 1.88% P and 1.27% Ru.

was raised and a hydrogen pressure applied. The slower rates observed for the heterogeneous selective hydrogenations is probably a diffusion effect. Comparison to homogeneous hydrogenations is complicated because, in addition to hydrogen and polyene diffusion into the catalyst sites, one must consider the concentration of PPh_3 within the resin's matrix. Using 165° in the hydrogenation step caused $(PPh_3)_2Ni(CO)_2$ or its bound analog to decompose. Thus, a mixture of resins containing the Ni and Ru catalysts could not be recycled because of the decomposition of Ⓟ$-PPh_2)Ni(CO)_2$ during hydrogenation. However, Ⓟ$-PPh_2)Ni(CO)_2$ + $(PPh_3)_2RuCl_2(CO)_2$ or the all homogeneous system could be recycled, after product removal by vacuum distillation, because 140° was employed in the hydrogenation step in those cases. As with all systems employing $(PPh_3)_2Ni(CO)_2$, or dual anchored catalysts, recycling was possible until the maximum molar turnover ratio (1100-1200) of the nickel complex was reached. It should be noted that independent hydrogenation experiments showed Ⓟ$-PPh_2)_2Ru(Cl)_2$-$(CO)_2$ or $(PPh_3)_2RuCl_2(CO)_2$ had very high ($\cong$ 5,000) turnover numbers.

Sequential Linear Oligomerization-Hydrogenation of Butadiene

Butadiene was sequentially oligomerized and hydrogenated to a mixture of saturated normal and branched acetates in the presence of polymer-anchored $PdCl_2$, Et_3N, HOAc, and THF at 100° . The unsaturated oligomers readily hydrogenated at 200 psig (hydrogen) at temperatures >60° . The ratio of normal to branched products was a function of the P/Pd ratio and was largest when P/Pd was low (<1.5). Some Pd leaching occurred on the first cycle when P/Pd ≈ 1 and the resin turned black. However, the resin remained active and could be recycled repeatedly. When P/Pd = 3, the yield of the normal product dropped to 64% while the branched product increased to 36%.

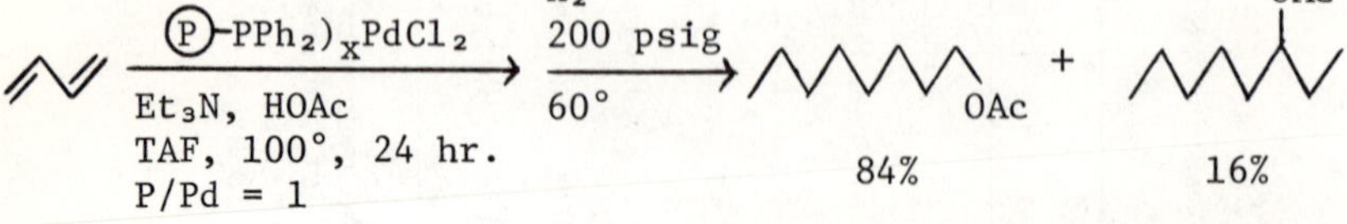

(6)

In conclusion, the application of more than one resin-bound transition metal catalyst in multistep sequential synthesis has been successfully demonstrated. Extension of this concept throughout the field of homogeneous catalysis should be possible. For example, single pot, multistep, homogeneously catalyzed, syntheses of styrene and (E)-1,9-non-4-endial from butadiene are under study in this laboratory using resin anchored catalysis such as (Ⓟ—$PPh_2)_2Pd(C_4H_2O_3)$, (Ⓟ—$PPh_3)_2RuCl_2(PPh_3)$, and (Ⓟ—$PPh_2)_2IrCl(CO)$.

Multistep catalysts are well known in heterogeneous catalyses. One only need consider zeolites or hydrocracking catalysts (such as Co-Mo-Al_2O_3-silica). The dual "heterogenized" homogeneous enzyme studies of Mosbach (1971) represent the thrust of these ideas into biochemistry. In dual anchored catalysts, some critical considerations include: proximity effects of the two catalysts, catalyst compatibility when heterogenized, ability to operate the sequential reactions simultaneously (i.e., low intermediate concentration) to give rate enhancements, and rate enhancement via the generation of a high concentration of coordinatively unsaturated active sites.

REFERENCES

1. Bailar, Jr., J.C., Catal. Rev., 10 (1), 17 (1974).
2. Burnett, M.G., Morrison, R.J., and Stregnell, C.J., J. Chem. Soc. Dalton, 701 (1973).
3. Evans, G.O., Pittman, Jr., C.U., McMillan, R., Beach, R.T., and Jones, R., J. Organometal. Chem., 67, 295 (1974).
4. Fahey, D.R., J. Org. Chem., 38, 80 (1973a); Fahey, D.R., J. Org. Chem., 38, 3348 (1973b).
5. Frankel, E.N. and Butterfield, R.O., J. Org. Chem., 34, 3930 (1969).
6. Grubbs, R.H., and Kroll, L.C., J. Amer. Chem. Soc., 93, 3062 (1971).
7. Grubbs, R.H., Kroll, L.C., and Sweet, E.M., J. Macromol. Sci. Chem., 7, 1047 (1973).
8. Grubbs, R.H., Gibbons, C., Kroll, L.C., Bonds, Jr., W.D., and Brubaker, Jr., C.H., J. Amer. Chem. Soc., 96, 2373 (1974).
9. James, R., "Homogeneous Hydrogenation", Wiley, New York, N.Y., 1973.
10. Haag, W.O. and Whitehurst, D.D., Abstracts of the Vth International Congress on Catalysis, Palm Beach, Fla., Aug. 21, 1972, pp 30-31.
11. Lyons, J.E., J. Catalysis, 30, 490 (1973).
12. Mosbach, K., Sci. Am., 26, 225 (1971).

13. Pittman, Jr., C.U., Evans, G.O., Chem. Technol., 560 (1973).
14. Pittman, Jr., C.U., Kim, B.T., and Douglas, W.M., J. Org. Chem., 40, (1975a).
15. Pittman, Jr., C.U. and Hanes, R., Ann. N.Y. Acad. Sci., 239, 76 (1974).
16. Pittman, Jr., C.U. and Smith, L.R., J. Amer. Chem. Soc., 97, 341 (1975b).
17. Pittman, Jr., C.U. and Smith, L.R., J. Amer. Chem. Soc., 97, 1743 (1975c).
18. Strohmeier, W. and Onada, T., Z. Naturforsch, 24b, 1493 (1969).
19. Szonyi, G., Adv. Chem. Ser., No. 70, 53 (1968).

SYNTHESIS BY
CARBON MONOXIDE INSERTION AT CARBON-CHLORINE BONDS

JOHN A. SCHEBEN AND IRVING L. MADOR†

U. S. Industrial Chemicals Co., Division of
National Distillers and Chemical Corp.
Cincinnati, Ohio 45237

Results of the catalytic carbonylation of a variety of vinylic, allylic and other organic chlorides by either a gas or liquid phase reaction are described. In the presence of a supported palladium metal catalyst, vinyl chloride selectively yields acryloyl chloride. Catalytic activity gradually decreases with reaction time in the gas phase, while activity is maintained at higher levels in the liquid phase. Acrylate esters can be produced by operating in an alcoholic medium. Tertiary amines function as promoters for these reactions. Palladium chloride catalyzes the carbonylation of various allylic chlorides to unsaturated acyl chlorides, while carbon tetrachloride yields trichloroacetyl chloride and dichloromethane forms chloroacetyl chloride. The dual insertion of carbon monoxide and acetylene or butadiene at the C-Cl bond of allyl chloride was also effected.

At the U. S. Industrial Chemicals Company Research Department in Cincinnati, Ohio, various groups have been working in different aspects of noble metal catalysis for a number of years. One of the fruits of this work is an ethylene-based process for vinyl acetate. Another area of industrial importance is that involving the use of noble metals as catalysts in carbonylation reactions. This article will review some of the developments resulting from our work on the insertion of carbon monoxide at carbon-chlorine bonds.

EXPERIMENTAL SECTION

Materials

The gaseous reactants were better than 99% pure, while the liquid reactants were the best grade available. Both labora-

†Present address, Allied Chemical Co., Morristown, N. J. 07960

tory compounded and commercially available catalysts were used. The laboratory catalysts were prepared by impregnating either an alumina, carbon or silica support with aqueous solutions of the appropriate metal salts. Reduction of the metal salts was accomplished with either hydrogen or carbon monoxide gas at 100-400°C.

Apparatus and Procedures

Most of the catalyst screening and testing for the liquid phase work was conducted in small (30 ml. volume) 316 SS or Alloy 20 reactors, which could be readily attached to the arm of a shaker unit for agitation, and then heated at the desired temperature by insertion in an oven. The liquid phase development studies with allyl chloride were conducted in either a 300-ml, 1-gallon or 4-gallon, 316 SS, stirred autoclave of conventional design.

The gas phase investigation was made in a fixed bed flow reactor. The reactor was fabricated from a piece of 316 SS tubing 0.625 OD x 0.395 ID x 24 inches, and could accommodate up to 70 ml. of catalyst. Usually 10 ml. of catalyst was tested and the unused portion of the reactor was filled with a 316 SS rod which decreased the reactor dead volume and acted as a preheater section. During operation the reactor was mounted vertically in an annular metal block which was heated by a dual set of Nichrome heating wires. The reactants passed downwards through the catalyst bed. Appropriate fittings were attached to each end of the reactor, and the system pressure was maintained by a control valve and pressure controller. A high pressure pump was used to feed the liquid reactants. The incoming gas flow was controlled by a very fine needle valve. The reactor effluent passed from the control valve to a sample valve, then through several traps maintained at dry-ice temperature and a wet test meter. The sample valve was attached to a gas chromatograph for on-stream analysis.

The normal operating procedure for carbonylating vinyl or allyl chlorides in the gas phase was as follows. Add 10 g of catalyst and heat to the selected operating temperature. Then purge the system and regulate the carbon monoxide pressure. After adjusting the organic chloride feed pump, allow the system to equilibrate at the desired conditions for at least an hour and sample the reactor effluent every 5 minutes for an hour. The hourly molar acyl chloride production was determined from the hourly average product peak height. Analysis was performed on a 0.125 x 72 inch column containing 25% SE-30 on 60/80 Chromosorb W, HMDS at 45°C, using helium as the carrier gas.

RESULTS AND DISCUSSION

Carbonylation of Ethylene

In the early 1960's, Blackham (1964) observed that the following reaction takes place in aprotic solvents.

$$CH_2{=}CH_2 + CO + PdCl_2 \longrightarrow ClCH_2CH_2COCl + Pd \qquad (1)$$

The product, β-chloropropionyl chloride readily reacts with alcohols to yield the corresponding β-chloroester, eq. 2; dehydrohalogenation of β-chloroesters to acrylate esters is accomplished by treatment with basic compounds, eq. 3. The initial carbonylation reaction can be made catalytic with respect to the palladium compound by the use of the copper chloride redox system eq. 4.

$$ClCH_2CH_2COCl + ROH \longrightarrow ClCH_2CH_2COOR + HCl \qquad (2)$$

$$ClCH_2CH_2COOR + Base \longrightarrow CH_2{=}CHCOOR + Base{\cdot}HCl \qquad (3)$$

$$CH_2{=}CH_2 + CO + 2CuCl_2 \xrightarrow{PdCl_2} ClCH_2CH_2COCl + 2CuCl \qquad (4)$$

These reactions have industrial potential since they represent a novel route to acrylate esters based on ethylene and carbon monoxide. However, the number of steps required is a drawback to commercialization.

A more direct route to acrylates, in which the carbonylation and dehydrohalogenation steps are combined, was then found, eq. 5, Scheben and Mador (1969).

$$CH_2{=}CH_2 + CO + 2CuCl_2 + Base \xrightarrow{PdCl_2} CH_2{=}CHCOCl + 2CuCl + Base{\cdot}HCl \qquad (5)$$

Thus, acryloyl chloride can be prepared in a single step reaction from ethylene and carbon monoxide by employing a weak base as an acid acceptor, and cupric chloride as the agent responsible for reoxidizing the Pd metal.

Various inorganic metal oxides, calcium oxide, or basic salts, such as disodium hydrogen phosphate, were found to perform as acceptors. Organic compounds such as nitriles and acid anhydrides also function as acid acceptors, and it was found that the carbonylation reaction could be conveniently carried out with acetonitrile functioning as both the solvent and acid acceptor.

Styrene and carbon monoxide similarly react to form β-phenyl acryloyl chloride. Although these syntheses represent useful techniques for the preparation of acryloyl halides, there are difficulties in contacting the reacting gases with the solid inorganic acid acceptors as well as the copper and palladium salts. The use of a liquid acid acceptor, such as acetonitrile alleviates part of this problem, but at the same time, obviously, organic by-products are introduced.

Carbonylation of Vinylic Halides

Another approach to the synthesis of acryloyl chloride is the insertion of carbon monoxide at the carbon-chlorine bond of vinyl chloride, Scheben, et al (1971a).

$$CH_2{=}CHCl + CO \xrightarrow{Pd/Al_2O_3} CH_2{=}CHCOCl \qquad (6)$$

Vinyl chloride is a commercially available and reasonably priced starting material. The catalytic carbonylation of vinyl chloride to acryloyl chloride takes place in the presence of a supported palladium metal catalyst. Product selectivity in this reaction was found to be in the high nineties. The best vinyl chloride conversion in the vapor phase continuous flow reactor was around twelve percent, with catalyst turnover numbers approaching 7 moles C_2H_3COCl/mole Pd. Turnover numbers reaching 10 moles/mole Pd were found under liquid phase conditions.

VAPOR PHASE SYSTEM

A variety of metals, metal salts or metal combinations were deposited on a number of different types of supports and tested as catalysts for this reaction. Activity was experienced only with the supported noble metal catalysts. Rhodium, ruthenium and iridium metals form volatile carbonyls under the carbonylation conditions and display limited activity. Supported palladium catalysts were more active than those containing platinum. Alumina was a better support than either carbon or silica.

Mixtures of palladium metal with other metals were also tested. The palladium-gold combination offered increased catalyst stability and improved vinyl chloride reaction rates. A comparison of the activities of the supported palladium and palladium-gold catalysts is shown in Table I. Increasing the support diameter from 1.98 to 3.18 mm had little effect on the conversion level, thereby indicating that diffusion problems within the pellets were not rate limiting.

Figure 1 shows the effect of temperature on the carbonylation of vinyl chloride. Based on observed rates two hours after start-up, the acryloyl chloride production rate increased with increases in temperature up to 180°C and then leveled out. The rate observed after eight hours reaction indicates that

TABLE I. *Catalysts and Supports*

Catalyst	C_2H_3COCl STY*
2% Pd-Al_2O_3	10
4% Pd-Al_2O_3	26
2% Pd-0.9% Au-Al_2O_3	37
1% Pd-0.5% Au-Al_2O_3	38
1% Pd-0.7% Au-Carbon	0
4% Pd-1.8% Au-SiO_2	8

Conditions = 40.8 atm., 180°C., mole ratio. C_2H_3Cl:CO = 0.26:1, 170 seconds contact time. *STY = g. C_2H_3COCl/liter of catalyst/hour.

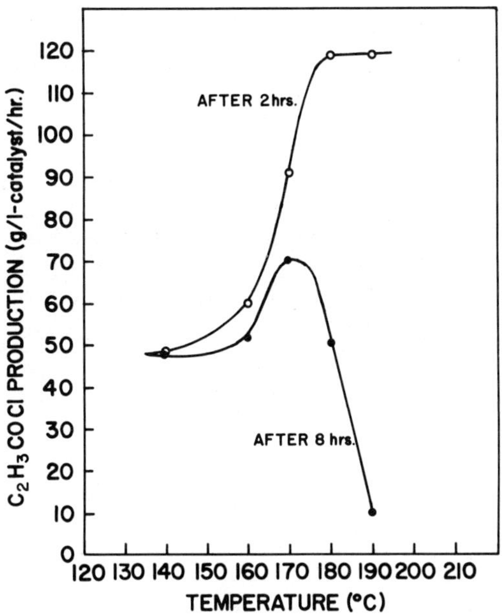

Figure 1. Effect of reaction temp. on acryloyl chloride productivity; catalyst 10 g. 1% Pd-Al_2O_3, C_2H_3Cl:CO=1, 40.8 atm, 20 sec. contact time.

above 170°C there was a rapid reduction of product formation with time. Compared to higher temperatures, in the range 140°-160°C the catalyst displays a somewhat reduced activity, but greater stability. Most of the carbonylation reaction rates were obtained at 160°-180°C, after relatively short operating periods.

On increasing the carbon monoxide pressure from 27.2 to 122 atmospheres as shown in Figure 2, the yield of acryloyl chloride reaches a plateau around 54 to 68 atmospheres. The gradual loss in conversion with the increase in pressure beyond 68 atmospheres may be due to the formation of stable palladium carbonylation intermediates having no free coordination site available for the olefin moiety. Displacement of these carbonyls may be difficult at high carbon monoxide pressures.

As can be seen in Table II, the molar feed ratio of vinyl chloride to carbon monoxide was varied between 0.19-2.32:1. Optimum activity was observed with the reactant ratio near unity. This ratio agrees with the reaction stoichiometry.

The average contact time for the reactants in the reaction zone was varied between 29 and 341 seconds by changing the volume of the catalyst charge. The effects of this change are shown in Table III. Maximum reaction rate was observed at a contact time of around 46 seconds.

The effect of linear gas velocity on the reaction rate was checked by varying the linear velocity of the reactant gas mixture in the range 1 to 3 cm/sec while keeping the contact time constant. Essentially the same reaction rates were obtained. Thus, the diffusion of products or reactants has very little effect on the reaction rate.

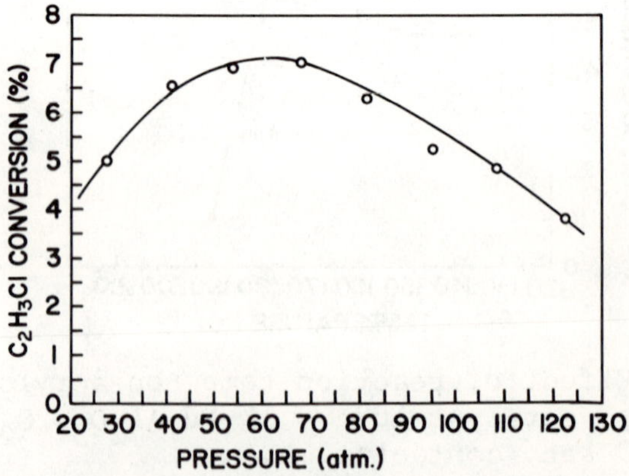

Figure 2. Effect of pressure on vinyl chloride conversion catalyst 10 g. 2% Pd-0.9% Au-Al_2O_3, C_2H_3Cl:CO=0.3, contact time = 85-286 sec, 180°C.

TABLE II. *Molar Ratio of Vinyl Chloride to Carbon Monoxide*

Hourly Molar Feed Ratio C_2H_3Cl/CO	C_2H_3COCl STY
0.19, t_1	23
0.31, t_1	47
0.67, t_2	50
0.77, t_2	62
1.19, t_2	87
2.32, t_1	61

Conditions: 180°C., 40.8 atm., 1% Pd-0.5% Au-Al_2O_3. Contact time t_1 = 108 sec., t_2 = 79 sec.

TABLE III. *Effects of Contact Time*

Contact Time, sec.*	C_2H_3COCl STY
29	43
46	91
67	52
121	38
208	33
341	24

Conditions: 180°C., 40.8 atm., 2% Pd-0.9% Au-Al_2O_3, mole ratio C_2H_3Cl:CO about 0.3:1.

*Contact time = $\frac{3600}{LHSV}$, LHSV = volumes of charge/volume of catalyst/hr.

Limited data indicates a gradual deactivation of the catalyst with time on stream at temperatures around 180°C. In a 48 hour test at this temperature and a system pressure of 40.8 atm., with a 2% Pd-0.9% Au-Al_2O_3 catalyst and a 130 sec. contact time, the vinyl chloride conversion dropped from 7 to 3.5% (STY 34 to 22), and the catalyst showed a 3.7% weight gain. At a reaction temperature of 150°C and 20 sec contact time at 40.8 atm pressure, a 1% Pd-Al_2O_3 catalyst displayed essentially no loss in activity (STY 47, 5% conversion) during a 30-hour test, and only a 1% weight gain. This catalyst deactivation is probably due to active sites being covered with polymeric materials, since the original catalyst activity can be restored by washing with dilute alkali.

Several kinds of organic and inorganic compounds were tested as promoters of this reaction. The vinyl chloride feed was diluted with varying amounts (1-66% by volume) of pyridine, dimethylformamide, triethylamine, xylene, methanol or acryloyl chloride. Pyridine and dimethylformamide were deleterious at a one percent level, whereas, triethylamine at the same concentration had no immediate effect on the reaction. With 33 volume percent methanol in the vinyl chloride feed, the normal hourly STY value increased from 38 to 45 and methyl acrylate was produced. Dilution of the vinyl chloride feed with large amounts of xylene (66% by volume) resulted in a loss of activity and failed to prevent catalyst deactivation.

The addition of 2 to 5% by volume acryloyl chloride to the vinyl chloride feed was harmful; the net hourly acryloyl chloride production dropping about 60%. This observation suggested that the carbonylation reaction under these conditions is in a state of equilibrium. Although we have observed that the decarbonylation of acryloyl chloride takes place at atmospheric pressure in the presence of a supported palladium catalyst at around 190°C, under a slight carbon monoxide partial pressure the decarbonylation reaction is inhibited. It is possible that the reduced catalyst activity is due to the polymerization of the added acryloyl chloride and fouling of the catalyst. Activity of the supported palladium catalysts was not improved by the addition (1.5% by wt of catalyst) of a variety of inorganic salts, e.g., compounds of nickel, cobalt, rhenium, barium, silver, mercury or magnesium.

BATCH LIQUID PHASE SYSTEM

Preliminary experiments showed that acryloyl chloride could also be prepared in a batch type rocker reactor using a heterogeneous catalyst. While one might expect catalyst fouling to be minimized under liquid phase conditions because of the effect of continual catalyst surface washing, the long hold-up times with a very reactive nomomer might be detrimental. In practice, the hourly acryloyl productivity (moles/mole Pd) was better in the batch system than in the continuous flow reactor, (10 vs 7 moles C_2H_3COCl/mole Pd).

Under liquid phase conditions, the carbonylation of vinyl chloride goes well in the presence of methanol, ethanol, chlorinated or ethereal type solvents, but not in aromatic or aliphatic hydrocarbons. A supported palladium metal catalyst was more active than either $PdCl_2$, $Pd(acac)_2$ or $(n\text{-}Bu_3P)_2PdCl_2$. The catalytic activities of palladium supported on carbon and on alumina were equivalent.

The ability of various solvents to coordinate with palladium catalysts and to accelerate or retard their reactions is known. We found that in general, tertiary amines serve to increase the rate of this carbonylation reaction. The promotional effects of a variety of amines are shown in Table IV. Triethylamine is particularly effective with the $Pd\text{-}Al_2O_3$ catalyst. The effects of changing the amine to palladium ratio are shown in Table V. These results show that the function of the tertiary amine is not that of an acceptor for HCl, but rather that of an accelerator for the carbonylation reaction. Specifically, triethylamine may aid the CO insertion reaction by forming an amine coordinated intermediate.

TABLE IV. *Effect of Amine Promoters*

Promoter	Molar Ratio C_2H_3COCl/Pd
2,2'-bipyridine	0
urea	4
N,N'-dimethylpiperazine	10
N,N-dimethylformamide	11
N,N-dimethylacetamide	17
N,N-dimethylaniline	18
N-methylmorpholine	19
triethylamine hydrochloride	18
trihexylamine	24
triethylamine	30

Conditions: 170°C., 47.6 atm., 4 hrs., 0.2×10^{-3} moles Pd as $Pd\text{-}Al_2O_3$, 0.47 moles C_2H_3Cl, 1-2 moles promoter/mole Pd.

TABLE V. *Effect of $(C_2H_5)_3N/Pd$ Ratio*

Molar Ratio $(C_2H_5)_3N/Pd$	Molar Ratio C_2H_3COCl/Pd
0	3
0.15	9
0.25	28
0.5	30
1.0	28
2.0	30
4.0	22
12.5	16
50.0	17

Conditions: 170°C., 47.6 atm., 4 hrs., 0.2×10^{-3} moles Pd as Pd-Al_2O_3, 0.47 moles C_2H_3Cl.

The addition of various inorganic weak bases, e.g., carbonates, acetates, phosphates or borates, was less effective than the addition of triethylamine. The presence of catalytic quantities of sodium methoxide or sodium phenoxide was detrimental.

This carbonylation reaction, promoted by triethylamine and catalyzed by a supported palladium catalyst was strongly influenced by the reaction temperature and reaction time. These effects are shown in Figures 3 and 4. It is apparent from these figures that the optimum reaction temperature is around 170°C. Higher reaction temperatures favor the formation of tars. A 60% increase in catalyst productivity is noted when the reaction period is increased from 2 to 4 hours (18 vs 30 moles C_2H_3COCl/mole Pd). Beyond this time period the production increase is somewhat slower.

Contrary to what was observed in the vapor phase system, a 1:1 C_2H_3Cl to CO ratio is not ideal in the batch reactor; here an excess of vinyl chloride is beneficial. Table VI illustrates this point. During these batch operations the initial C_2H_3Cl to CO ratio was not maintained; thus a gradual drop in reaction rate during a given experiment was experienced.

Methyl acrylate was directly synthesized by carrying out the carbonylation of vinyl chloride in methanol media. The influence of this solvent was dramatic, especially with the Pd-Al_2O_3 catalyst and catalytic amounts of triethylamine as promoter. During a four hour reaction period at 170°C, the methyl

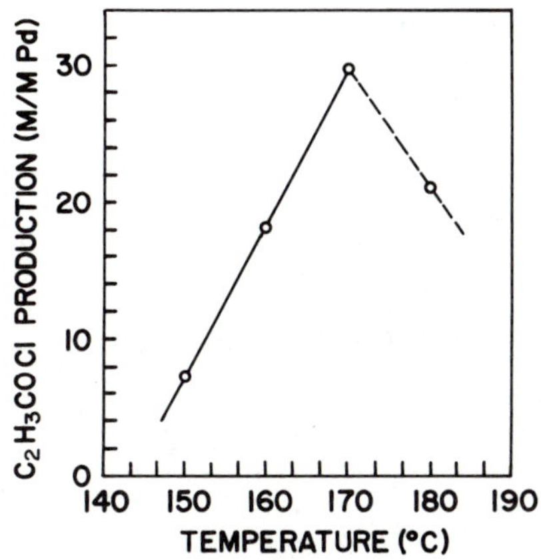

Figure 3. Effect of reaction temp. on acryloyl chloride productivity; catalyst 5% Pd-Al_2O_3, 0.2 x 10^{-3} M Pd, 4 hrs, 0.15 M C_2H_3Cl, 47.6 atm, 0.9 x 10^{-4} M $(C_2H_5)_3N$.

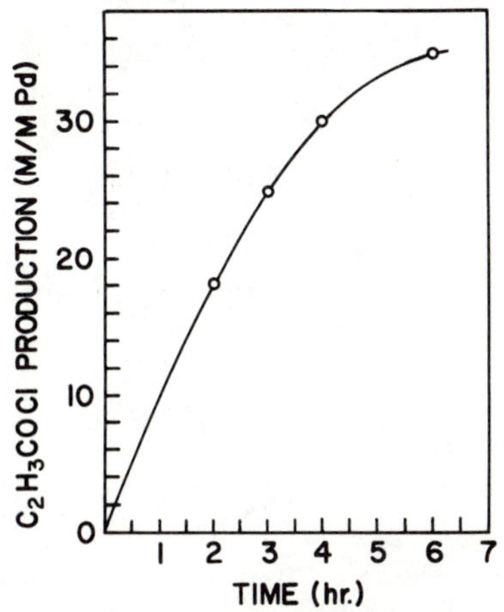

Figure 4. Effect of reaction time on acryloyl chloride productivity; catalyst 5% Pd-Al_2O_3, 0.2 x 10^{-3} M Pd, 0.15 M C_2H_3Cl, 47.6 atm, 170°C, 0.9 x 10^{-4} M $(C_2H_5)_3N$.

TABLE VI. *Effect of Mole Ratio $C_2H_3Cl:CO$*

Feed Molar Ratio C_2H_3Cl/CO	Product Molar Ratio C_2H_3COCl/Pd
0.7	13
0.9	14
1.9	30
2.9	32
3.9	27
4.5	28

Conditions: 170°C., 0.2 x 10^{-3} moles Pd as Pd-Al_2O_3, 0.36 x 10^{-3} moles $(C_2H_5)_3N$, 0.147 moles C_2H_3Cl, 4 hrs.

TABLE VII. *Carbonylation of Vinylic Halides*

Vinylic Halide	Product
$CH_2=C(Cl)-CH_3$	$CH_2=CHCH_2COCl$
$ClCH_2-C(Cl)=CH_2$	$ClCH_2-C(COCl)=CH_2$
$ClCH=CH-CH_2Cl$	$ClCH=CHCH_2COCl$
$CH_2=CCl_2$	$CH_2=C(Cl)-COCl$
C_6H_5Cl	C_6H_5COCl

acrylate production amounted to 122 moles per mole of Pd charged compared with 30 moles per mole without methanol. The catalyst retained its original activity even after several re-uses.

Table VII shows the acyl halide products obtained from a number of other vinylic halides. The formation of 3-butenoyl chloride from the reaction of carbon monoxide with 2-chloropropene indicates that a chlorine atom was transposed. In confirmation, the isomerization of 2-chloropropene to allyl chloride was shown to take place in the presence of palladium catalysts. The product obtained from 2,3-dichloropropene is also of special interest, since the vinylic halide part of the molecule is carbonylated in preference to the allylic halide. 1,3-Dichloropropene is carbonylated in the allylic position only. Vinylidene chloride reacts with CO to form α-chloroacryloyl chloride. Chlorobenzene yielded benzoyl chloride, Mador and Scheben (1969). No carbon monoxide inserted products were found with 1,2-dichloroethylene, 1-chloropropene, or trichloroethylene.

MECHANISM

A detailed study of the mechanism for the carbonylation of vinylic halides with palladium metal catalysts was not made during the course of this work. Recently, Schoenberg, et al (1974) suggested a plausible mechanism for their observed ester formation in the reaction of vinylic bromides and iodides with carbon monoxide, an alcohol, a catalytic amount of a palladium complex and a tertiary amine. It is likely that the carbonylation we describe follows a similar pathway. We suggest that the initial step involves Pd-coordination with the unsaturated species, eq. 7. The π-complex undergoes an oxidative addition of the halide to the metal, eq. 8, and the complex so formed undergoes CO insertion to form an acryloyl palladium species, eq. 9. In the presence of excess CO, alcohol or tertiary amine the acyl complex is converted to the acryloyl chloride or derivative and the Pd catalyst is regenerated, eq. 10.

For additional references to the carbonylation of vinylic halides see Brewis et al (1967); Closson and Ihrman (1969); Tsuji et al (1963) and Tsutsumi et al (1963).

Carbonylation of Allylic Halides

A good deal of research effort has been expended on the preparation, characterization, mechanism of formation and reactions of π-allyl complexes. Reactions of various π-allyl palladium complexes with nucleophiles are well documented, Tsuji (1969). Our work in this area began with a study of the insertion of carbon monoxide into allyl chloride, Mador and Scheben (1967). The reaction is catalytic with respect to the

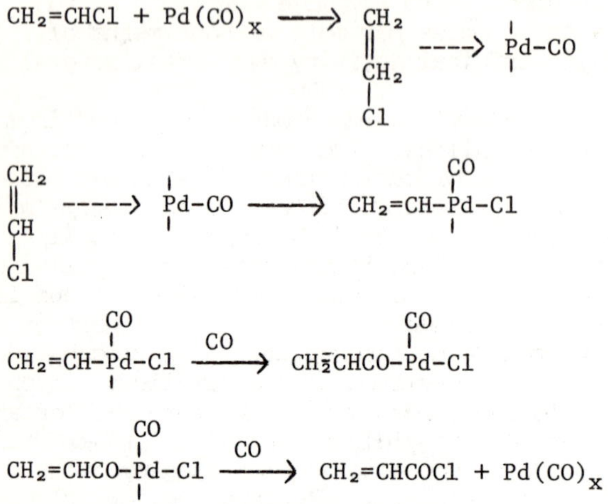

(7) (8) (9) (10)

palladium chloride, and the product 3-butenoyl chloride is produced in high conversion and selectivity. In studying this reaction the effects of pressure, temperature, catalyst concentration and reaction time on the product selectivity and allyl chloride conversion were determined.

$$CH_2{=}CHCH_2Cl + CO \xrightarrow{PdCl_2} CH_2{=}CHCH_2COCl \qquad (11)$$

3-Butenoyl chloride is readily isomerized by basic compounds to its thermodynamically more stable isomer, 2-butenoyl chloride. Temperature also promotes this isomerization. Generally, if the reaction temperature is kept below 100°C and the reaction period held to 7 hours or less, the amount of isomerization will be less than 5%. Figure 5 illustrates the effects of temperature and pressure on the conversion of allyl chloride to 3-butenoyl chloride. The maximum conversion to 3-butenoyl chloride is obtained around 95°C. Increasing the temperature to 115°C at 68 atm pressure causes the conversion to drop from 70 to 55%. This loss is due to the formation of high boiling tars at the higher reaction temperature. Also, as indicated in Figure 5, increasing the carbon monoxide pressure leads to increases in the conversion to acid chloride product, the increase being approximately linear, in the temperature range 60°-95°C.

As can be seen in Figure 6, the reaction of allyl chloride with carbon monoxide and palladium chloride is about 70% com-

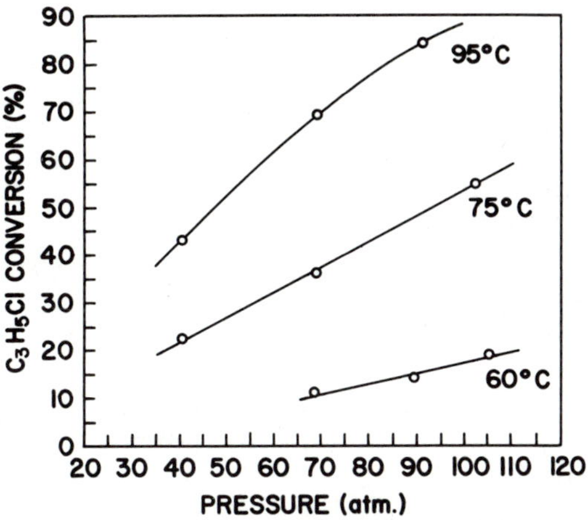

Figure 5. Effect of pressure on allyl chloride conversion to 3-butenoyl chloride; catalyst 8 x 10^{-3} M $PdCl_2$, 1 M C_3H_5Cl, 6.5 hrs.

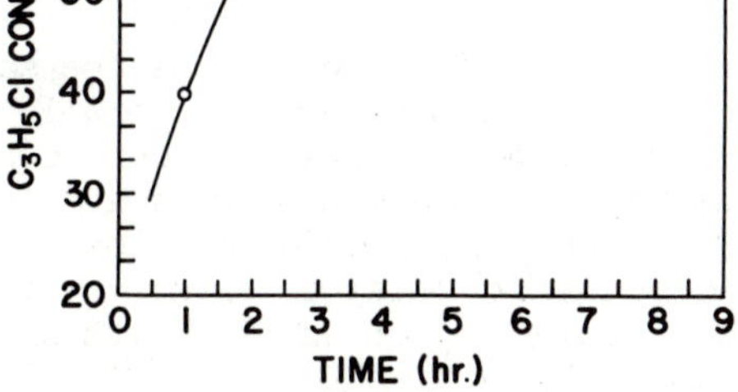

Figure 6. Effect of reaction time on allyl chloride conversion to 3-butenoyl chloride; catalyst 8 x 10^{-3} M $PdCl_2$, 1 M C_3H_5Cl, 95°C, 68 atm.

plete in 3 hours at 68 atmospheres pressure with a reaction temperature of 95°C. The allyl chloride conversion rate does not change appreciably with longer reaction times, suggesting that the reaction is approaching equilibrium. With higher carbon monoxide pressure the reaction rate is increased and the limiting conversion value is reached more rapidly.

When the palladium chloride catalyst concentration was varied between 3-12 millimoles/mole allyl chloride the conversion was only slightly affected, see Figure 7, suggesting that catalytic activity at these concentrations is mainly dependent on the palladium chloride solubility and on the rate of formation of the active intermediate species.

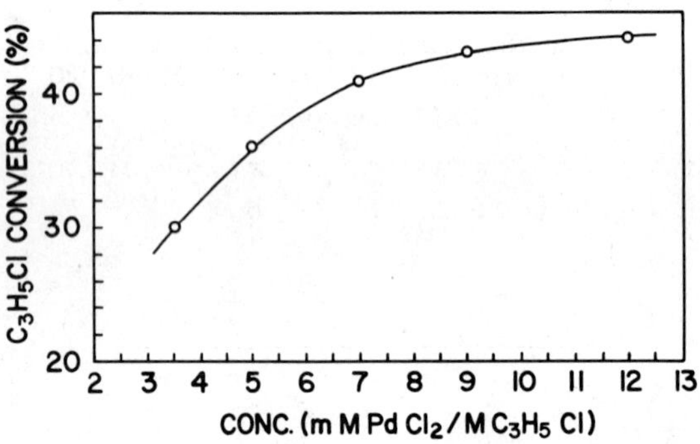

Figure 7. Effect of catalyst concentration on allyl chloride conversion to 3-butenoyl chloride; 1 M C_3H_5Cl, 95°C, 68 atm, 1 hr.

Allyl chloride was usually employed as the solvent as well as the reactant. The reaction rate was generally slower in the presence of added inert solvents. In our development work palladium chloride was the catalyst of choice. It was as active as the π-allyl-palladium chloride complex, other palladium salts or supported palladium metal catalysts. Other noble metal complexes or salts were not as effective as the palladium compounds.

During the course of this work large quantities of 3-butenoyl chloride were required for experimental purposes. The scale-up was performed in a 4-gallon batch type stirred reactor. About 12 lbs of crude 3-butenoyl chloride was prepared every 24 hours in this system. The average conversion and selectivity numbers were as good as those obtained in a 300 ml stirred reactor. Table VIII illustrates some typical experimental conditions and results found in this bench scale work.

TABLE VIII. *Large Scale Synthesis 3-Butenoyl Chloride*

Run	Moles $PdCl_2$	Moles C_3H_5Cl	Temp. °C	Time hrs.	% Conv.
1	0.197	125	86	62	78
2	0.197	130	96	38	77
3	0.225	128	96	38	85

Conditions: CO pressure 98 to 200 atmospheres, shaft speed 600 rpm, 4-gallon reactor.

Runs 1 and 2 show that a 10° rise in reaction temperature increases the reaction rate by about 50 percent. The effect of changing the allyl chloride to palladium chloride ratio from 660:1 to 567:1 is shown in runs 2 and 3, respectively. Under these conditions the isomerization reaction amounts to less than 9%. It was demonstrated that the rate of carbonylation of allyl chloride in the bench scale equipment is related to the reactor geometry, the stirrer shaft speed and especially the reactor surface to volume ratio. Better overall reaction rates were found in a one-gallon stirred reactor (600 RPM), than in a four-gallon reactor having the same inside diameter and operated under similar conditions.

Substituted allylic halides also react with carbon monoxide in the presence of catalytic amounts of palladium(II) compounds. The three monomethyl or monophenyl-allyl chlorides react similarly with carbon monoxide, eq. 12, 13 and 14. It should be noted that the same reaction product is obtained by carbonylating either the I and III allylic halide isomers. This leads to the assumption that a common π-allyl-palladium complex is formed as an intermediate in these reactions. Carbonylation of II ($R=CH_3$) yields 3-methyl-3-butenoyl chloride. This isomer reacts much more slowly with carbon monoxide than does either I or III ($R=CH_3$).

$$\underset{\text{I}}{CH_2{=}CH\overset{R}{\overset{|}{C}}HCl} + CO \xrightarrow{\text{Pd salts}} R{-}CH{=}CHCH_2COCl \qquad (12)$$

$$\underset{\text{II}}{CH_2{=}\underset{R}{\underset{|}{C}}CH_2Cl} + CO \xrightarrow{\text{Pd salts}} CH_2{=}\underset{R}{\underset{|}{C}}{-}CH_2COCl \qquad (13)$$

$$\underset{\text{III}}{R{-}CH{=}CHCH_2Cl} + CO \xrightarrow{\text{Pd salts}} R{-}CH{=}CHCH_2COCl \qquad (14)$$

$$R = CH_3 \text{ or } C_6H_5$$

A 70:30 mixture of I and III ($R{=}CH_3$) was heated at 73°C under either nitrogen or carbon monoxide pressure in the presence of palladium chloride. It was observed that the isomerization of I to III under nitrogen takes place more readily than the reverse reaction (ratio I:III now about 1:2). Since III is the more thermodynamically stable isomer, the observed product distribution is not unexpected. The nearly complete disappearance (ratio now about 1:10) of I on treatment with CO suggests that I is carbonylated faster than III.

As extensions of this catalytic carbonylation reaction other allylic halides were treated with carbon monoxide. Allyl bromide yielded 3-butenoyl bromide. Although 1,4-dichlorobutene-2 was reacted under a variety of conditions, only the mono-carbonylated product, 5-chloro-3-butenoyl chloride was recovered. The highly substituted allylic halide, 1-chloro-5,5,7,7-tetramethyl-octene-2 reacted slowly with carbon monoxide, insertion taking place exclusively at the terminal position to yield the linear product. Benzyl chloride yielded phenylacetyl chloride, Scheben et al (1971b).

Carbonylation of Polyhalogenated Compounds

In extending these insertion reactions, a number of polyhalogenated compounds were found to react with carbon monoxide in the presence of a catalytic quantity of a palladium compound, Mador and Scheben (1969c). Carbon tetrachloride reacts according to eq. 15 yielding trichloroacetyl chloride.

$$CCl_4 \text{ to } CO \xrightarrow{PdCl_2} CCl_3COCl \qquad (15)$$

Methylene chloride gives chloroacetyl chloride under similar conditions. These reactions afford convenient routes to chloroacetyl chloride type products. The selectivity to the mono-acyl product is generally excellent, but the conversion of the polyhalomethane measures less than 5%.

Carbonylation of Allyl Chloride-Olefin Mixtures

Carbonylations of allyl chloride in the presence of acetylene or butadiene lead to the formation of polyunsaturated acyl halide products, eq. 16 and 17.

$$CH_2{=}CHCH_2Cl + CO + C_2H_2 \xrightarrow{PdCl_2} CH_2{=}CHCH_2CH{=}CHCOCl \quad (16)$$

$$CH_2{=}CHCH_2Cl + CO + C_4H_6 \xrightarrow{PdCl_2} CH_2{=}CHCH_2CH_2CH{=}CHCH_2COCl \quad (17)$$

Thus 2,5-hexadienoyl chloride was prepared in 35% conversion from allyl chloride, acetylene and carbon monoxide, Scheben (1971). For a review of analogous insertion reactions, though catalyzed by nickel, see Chiusoli (1973). In the reaction of allyl chloride, carbon monoxide and butadiene, 3,7-octadienoyl chloride is the major product, 25% conversion. The other products consist of 3-butenoyl and 3-pentenoyl chlorides. The carbonylation of butadiene (via crotyl chloride) gives rise to the 3-pentenoyl chloride product. The product distribution varies with the diolefin to allyl chloride ratio; thus with a butadiene to allyl chloride ratio greater than two, the butadiene carbonylation reaction is favored, Scheben (1970). Medema and van Helden (1969) reported similar types of products. With a mixture of ethylene and allyl chloride, the palladium chloride catalyst is reduced and the products are 3-butenoyl chloride and β-chloropropionyl chloride.

Vapor Phase Synthesis of 3-Butenoyl Chloride

A vapor phase synthesis of 3-butenoyl chloride from allyl chloride, carbon monoxide and a supported palladium catalyst also was demonstrated, Mador and Scheben (1969a). The advantages of a fixed bed flow system over a batch liquid phase pressure reactor are the simplicity of operation, shorter contact times and ease of catalyst recovery. Table IX shows representative results of the carbonylation of allyl chloride in a fixed bed reactor with a 3% $Pd\text{-}Al_2O_3$ catalyst. These results indicate that 3-butenoyl chloride production is favored at high pressures and short contact times at reaction temperatures around 110-112°C. Supported catalysts containing various palladium salts were not as active as those containing palladium metal. Generally, better catalyst utility was observed in the liquid phase system.

TABLE IX. *Vapor Phase Carbonylation of Allyl Chloride*

Temp. °C	CO Press. atm.	Flow Rate l./min. S.T.P.	C_3H_5COCl STY
75	96	0.32	3
75	96	0.60	2
85	64	0.13	2
110	68	0.33	7
111	69	0.14	12
114	102	0.11	17
115	102	0.22	34

Conditions: Molar ratio $CO:C_3H_5Cl$, 32:1, residence time 0.5-7 seconds, 9 g. 3% $Pd-Al_2O_3$.

ACKNOWLEDGEMENT

The authors acknowledge the cooperation of National Distillers and Chemical Corp. for permission to publish this work. We also wish to extend our appreciation to J. Kwiatek for his comments and criticisms about the manuscript, and to J. Hinnenkamp, R. Weiner, J. M. Fisher and P. Buzzi for their assistance in this work.

REFERENCES

1. Blackham, A. U. (to National Distiller and Chemical Corp.), U.S. Patent 3,119,861 (Jan. 28, 1964).
2. Brewis, S., Dent, W., Owen, R. (to Imperial Chemical Industries, Ltd.), British Patent 1,091,042 (Nov. 15, 1967).
3. Chiusoli, G., Accounts of Chemical Research 6, 422 (1973).
4. Closson, R., Ihrman, K. (to Ethyl Corp.), U.S. Patent 3,457,299 (July 22, 1969).
5. Mador, I. L., Scheben, J. A. (to National Distillers and Chemical Corp.), U.S. Patent 3,309,403 (March 14, 1967).
6. Mador, I. L., Scheben, J. A. (to National Distillers and Chemical Corp.), U.S. Patent 3,423,456 (Jan. 21, 1969a).
7. Mador, I. L., Scheben, J. A. (to National Distillers and Chemical Corp.), U. S. Patent 3,452,090 (June 24, 1969b).

8. Mador, I. L., Scheben, J. A. (to National Distillers and Chemical Corp.), U.S. Patent 3,454,632 (July 8, 1969c).
9. Medema, D., van Helden, R., Abstracts, Vol. 14, No. 2, B92, Division of Petroleum Chem. 157th National Meeting of the American Chem. Soc., Minneapolis, Minn., April 1969.
10. Scheben, J. A. (to National Distillers and Chemical Corp.), U.S. Patent 3,536,739 (Oct. 27, 1970).
11. Scheben, J. A. (to National Distillers and Chemical Corp.), U.S. Patent 3,627,827 (Dec. 14, 1971).
12. Scheben, J. A., Mador, I. L. (to National Distillers and Chemical Corp.), U.S. Patent 3,468,947 (Sept. 23, 1969).
13. Scheben, J. A., Fisher, J. M., Mador, I.L. (to National Distillers and Chemical Corp.), U.S. Patent 3,626,005 (Dec. 7, 1971a).
14. Scheben, J. A., Mador, I. L., Orchin, M. (to National Distillers and Chemical Corp.), U.S. Patent 3,560,561 (Feb. 2, 1971b).
15. Schoenberg, A., Bartoletti, L., Heck, R. F., J. Org. Chem., 39, 3318 (1974).
16. Tsuji, J., Accounts of Chemical Research, 2, 144 (1969).
17. Tsuji, J., Morikawa, M., Kiji, J., Tetrahedron Letters, No. 16, 1061 (1963).
18. Tsutsumi, Nagao, Hiromatsu, Ishii, Japan Chem. Soc. 16th Annual Report, 458 (1963).

ASYMMETRIC HOMOGENEOUS HYDROGENATION WITH CHIRAL RHODIUM-PHOSPHINE CATALYSTS

JAMES D. MORRISON, WILLIAM F. MASLER AND SUSAN HATHAWAY
Department of Chemistry, University of New Hampshire,
Durham, New Hampshire 03824

During the past several years, there have been revolutionary advances in the methodology of asymmetric synthesis. A particularly significant development has been the utilization of chiral metal complexes as catalysts for a number of asymmetric transformations. Using such catalysts asymmetric hydrogenations have been caried out with greater than 95% asymmetric bias. That is, starting with achiral olefins, saturated products that are almost optically pure have been synthesized directly, without recourse to classical resolution (except, of course, resolutions that are required to prepare the chiral catalyst). This is a significant development in the history of synthetic chemistry, for it has long been the dream of chemists to be able to mimic the 100% stereoselectivity of enzymes. Now we are close to accomplishing this goal. Some may feel we are, in fact, about as close as we will get with any system of significant scope. It is requiring ever larger disparities in free energies of activation between competing diastereomeric transition states. However, it is also true that synthetic chemists are often not unduly intimidated by foreboding theoretical considerations. Indeed they frequently respond as fighting bulls to the taunts of the toreador. So perhaps there is even now on the horizon some cleverly designed system that will lift us to the Elysian fields of 100% stereoselectivity.

In this paper, we will explore the development of the new technology embodied in the phrase asymmetric homogeneous hydrogenation, particularly as practiced with Rh(I)-chiral phosphine catalysts. We will trace the method from its conception at the interface of other developing methodologies, through early stages of unspectacular yet germinal testing, to its recent major accomplishments. However, it is desirable to first put the general phenomenon of asymmetric synthesis, and the specific application of asymmetric catalytic hydrogenation, in perspective.

The basic principles of asymmetric synthesis have been discussed in a number of recent reviews (Morrison and Mosher, 1971; Scott and Valentine, 1974; Eliel, 1974). The critical part of an asymmetric synthesis scheme is some reaction in which an achiral unit in an ensemble of substrate molecules undergoes a reaction which converts it to a chiral unit in such a manner that the stereoisomeric products are produced in unequal amounts. A more succinct and generally equivalent statement is that in an asymmetric reaction a prochiral unit is converted into a chiral unit.

Usually an asymmetric synthesis is carried out in one of two ways. Either a second chiral center is created in a molecule under the influence of an existing chiral center in that same molecule (intramolecular chiral mediation) or a chiral reagent interacts with a prochiral substrate to mediate the creation of a new chiral center (intermolecular chiral mediation).

For example, when (-)-benzoin is allowed to react with an ethyl Grignard reagent, the addition of the ethyl group occurs preferentially on one of the diastereotopic faces of the carbonyl group to give predominantly the optically active R,S diastereomer (Fig. 1).

FIGURE 1. *Addition of a Grignard reagent to a chiral ketone (Morrison and Mosher, 1971, pp. 86-87).*

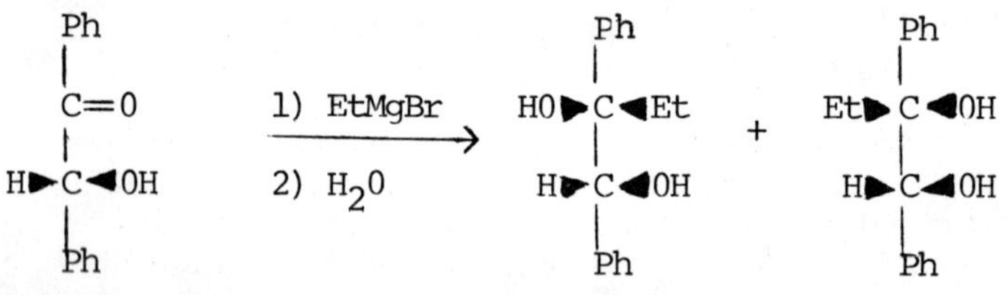

R-(-)-Benzoin R,R R,S

MAJOR DIASTEREOMER

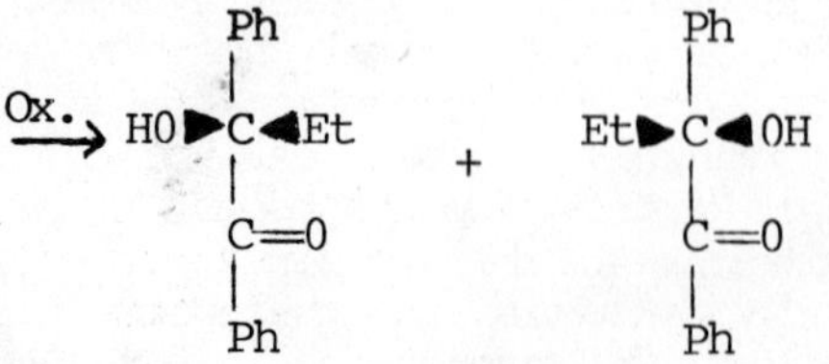

MAJOR ENANTIOMER

(+)-R (-)-S

The original chiral center in the substrate can be converted to an achiral center by oxidation to prove that the prochiral carbonyl group in the benzoin was indeed converted to a chiral center. When this is done, a levorotatory product is isolated. This reaction is an example of intramolecular chiral mediation; the chiral center in the benzoin substrate mediates the production of the new chiral center.

Intermolecular chiral mediation is illustrated by the reduction of acetophenone with lithium aluminum hydride that has been modified with an optically active additive like (-)-quinine. In this instance, hydride is added to one of the enantiotopic faces of the ketone preferentially to produce more of one enantiomer than the other (Fig. 2). The chiral centers in the quinine mediate the process.

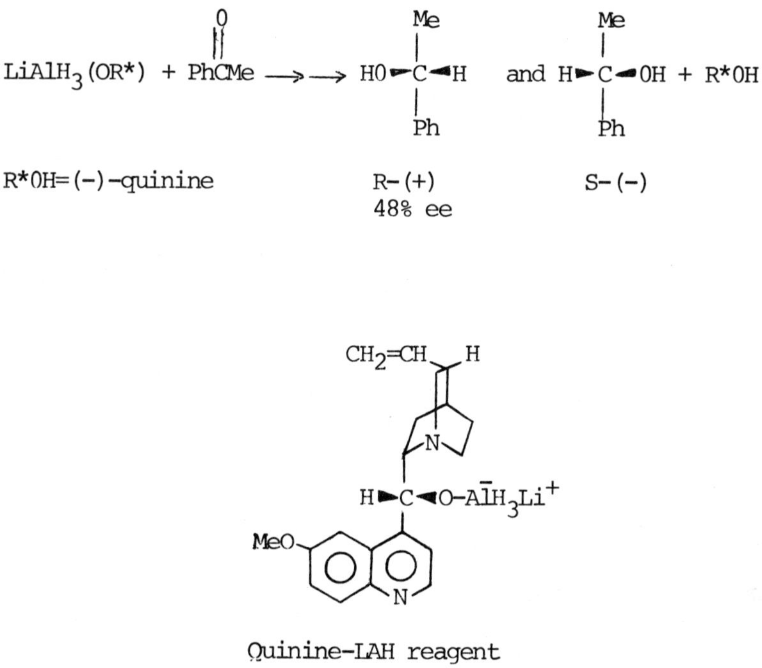

FIGURE 2. *Asymmetric reduction of acetophenone using a reagent prepared from lithium aluminum hydride and a chiral carbinol (Cervinka and Belovsky, 1967).*

Asymmetric reactions involving chiral reagents (such as lithium aluminum hydride modified with (-)-quinine) have been used with great success. Their principle shortcoming is that stoichiometric quantities of chiral compounds must be invested and only rarely can these compounds be recycled. An asymmetric reaction involving a catalyst that is chiral would be a better way to accomplish an asymmetric synthesis since with only a small amount of chiral material large quantities of optically active product could be produced. The idea of chiral catalysis has fascinated chemists for a long time, in part because enzymes are chiral catalysts. The greatest amount of work has been done in the area of chiral hydrogenations.

Heterogeneous catalysts modified by the addition of chiral substances have been used to asymmetrically hydrogenate olefins, however, only a few effective systems have been found. Palladium deposited on silk fibroin was used to asymmetrically hydrogenate 4-benzylidene-2-methyl-5-oxazolone to give, after hydrolysis, optically active phenylalanine (Fig. 3). The optical purity* of the product was found to be dependent upon the origin of the fibroin and its chemical pretreatment.

FIGURE 3. *Asymmetric heterogeneous hydrogenation using palladium supported on a chiral protein material as catalyst (Akabori et al, 1956, 1959, 1960).*

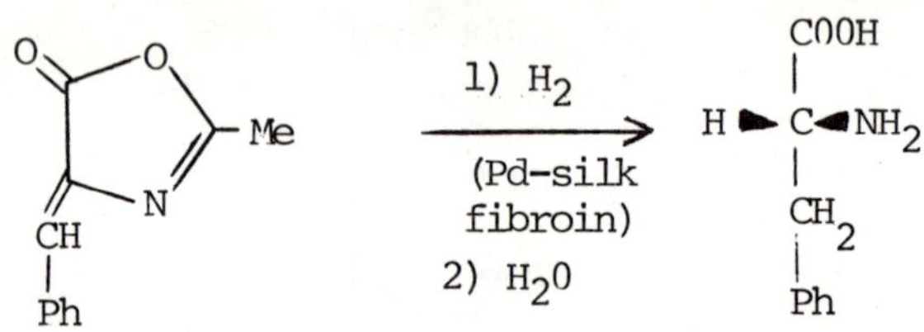

R(+), 30-70% ee

*We will use the terms optical purity and per cent enantiomeric excess (%ee) to express the extent to which one enantiomer is produced in excess over the other in an asymmetric reaction; i.e., optical purity or %ee = %R - %S (or vice versa). Normally %ee is determined by measuring the rotation of the product and expressing it as a percentage of the maximum rotation for that product ([δ] obs./[δ] max. x 100), although increasingly %ee is being determined by absolute nmr methods involving diastereomeric derivatives or chiral shift reagents.

Raney nickel has been modified with amino acids and other chiral reagents to give catalysts that have been used to effect asymmetric reductions (Izumi, 1971). However, these catalysts suffer from some of the same kinds of vagaries that have been observed for the palladium on silk fibroin catalysts. For example, the optical purities of the products were found to be very dependent upon pH and the method of catalyst preparation.

The generally low percent asymmetric synthesis in asymmetric heterogeneous hydrogenations may be due, in part, to a non-uniform distribution of chiral modifying agents over the catalytic surfaces. In the case of silk fibroin, metal clumping on the chiral support or dissociation of the metal from the fibroin may allow some reduction to occur in an achiral local environment.

A homogeneous system in which some chiral metal complex would activate hydrogen and an alkene for hydrogenation would perhaps offer a better chance for high asymmetric bias. Several groups saw an opportunity for developments in this area a few years ago, shortly after the first reports on the catalytic activity of solutions of Rh(I) triphenylphosphine complexes that have come to be known as "Wilkinson's Catalyst". It was not the discovery of such homogeneous hydrogenation systems alone, however, which sparked ideas about asymmetric hydrogenation.

For at about the time of Wilkinson's discovery, new schemes were developed by others for the preparation and configurational correlation of chiral phosphines. The combination of advances in homogeneous catalysis and chiral phosphine technology prompted research on chiral phosphine complexes of the Wilkinson type. Horner, Buthe, and Siegel (1968a) were the first to hypothesize in print that rhodium complexes containing optically active tertiary phosphine ligands should effect the asymmetric hydrogenation of unsymmetrically substituted olefins, but it is now clear that several other groups had the same idea at about the same time.

The idea is attractive for several reasons. In a metal complex catalyst that bears chiral ligands the metal "active site" should always be chiral; it cannot dissociate from the chiral influence as a metal layered onto a chiral support might be able to. Also, in a homogeneous reaction there is a better chance that a mechanism can be rather clearly defined and that the rational design of a highly effective system might be contemplated.

Examples of asymmetric hydrogenation based on the "Horner idea" were reported by Knowles and coworkers (the Monsanto group) in 1968. Rhodium complexes of the type RhL_3Cl_3 (where L was a chiral phosphine) were used in the hydrogenation of α-phenylacrylic acid (atropic acid) and itaconic acid under the conditions indicated in Fig. 4. When L was (R)-(−)-methylphenyl-n-propylphosphine (phosphorus is the chiral center), 15% optically pure (S)-(+)-α-phenylpropionic acid and 3% optically pure methylsuccinic acid (configuration unreported) were asymmetrically synthesized (Fig. 4)*.

FIGURE 4. *Early examples of homogeneous asymmetric hydrogenation (Knowles et al, 1968).*

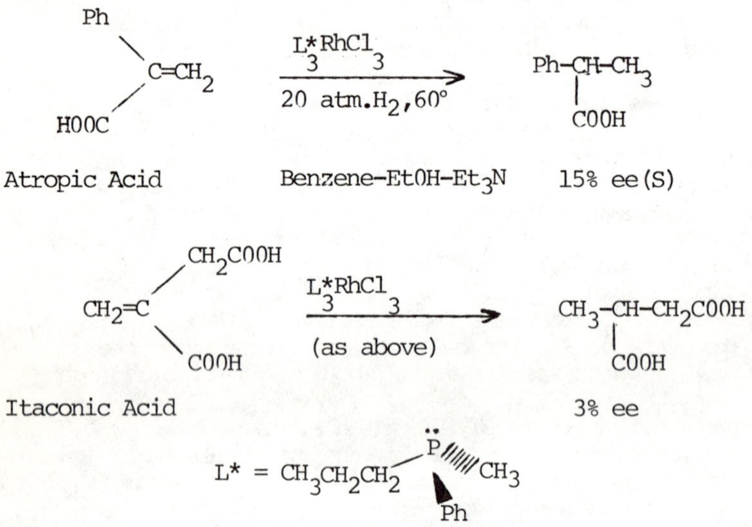

Although the structure of the active catalyst obtained in solution was uncertain in these early experiments, the Monsanto group suggested at the time that Wilkinson-type Rh(I) complexes could be involved. They did not speculate on how the reduction of Rh(III) to Rh(I) might be accomplished, but one possibility is shown in Eqns. 1 and 2. Presumably, added base (such as triethylamine) promotes metal reduction by removing acid.

$$Rh^{III}L_3Cl_3 + H_2 \rightleftharpoons HRh^{III}L_3Cl_2 + HCl \quad (1)$$

$$HRh^{III}L_3Cl_2 \rightleftharpoons Rh^{I}L_3Cl + HCl \quad (2)$$

---

*The phosphine ligand ($PPhMePr^n$) had an optical purity of about 70%. In the original publication (Knowles et al., 1968), there is a misprint which shows the chiral phosphine as $PPhMePr^i$.

Almost simultaneously with Knowles and Sabacky, Horner and coworkers (1968b) tested their own hypothesis using a catalyst prepared in situ from μ-dichlorobis(1,5-hexadiene)dirhodium-(I) and (S)-(+)-methylphenyl-n-propylphosphine (Fig. 5).

FIGURE 5. *Early examples of asymmetric homogeneous hydrogenation with a Wilkinson-type catalyst (1,5-HD = 1,5-hexadiene) (Horner et al, 1968b).*

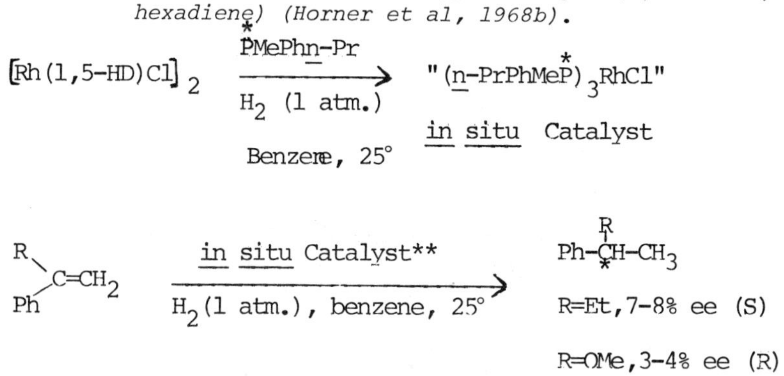

** (S)-Methylpropylphenylphosphine

In this case, the structure of the catalyst was more certain since related reactions of $Rh(olefin)_2Cl_2$ compounds and tertiary phosphines were known to give neutral, square planar Rh(I) complexes of the type RhL_3Cl. Therefore, for their hydrogenations of α-substituted styrenes, Horner and coworkers assumed a mechanism which paralleled that proposed for Wilkinson's catalyst and involved the intermediate shown in Fig. 6. Horner and Siegel (1972a,b) also investigated the reduction of a number of other α-substituted styrenes, $C_6H_5C(R){=}CH_2$ (R = Et, n-Pr, i-Pr, OEt, α-naphthyl, $CH_2C_6H_5$, Br), using a number of chiral phosphines, MeP̊PhR' (R' = n-Pr, i-Pr, n-Bu, t-Bu). The optical purities of the hydrogenation products varied between 2-19%.

Likewise, the Monsanto group (Knowles et al., 1970) extended their studies to other substrates (particularly α,β-unsaturated acids) and other phosphines. In these later experiments, they generated the catalyst in situ, as had the Horner group. A methanol solution containing an α-substituted acrylic acid and a trace amount of triethylamine was then added and hydrogenations were carried out at 30 atmospheres and 60° C. For various combinations of olefin and chiral ligand, optical purities ranged from 3-21% (Fig. 7.).

FIGURE 6. *Presumed intermediate in asymmetric homogeneous hydrogenation (Horner et al, 1968b).*

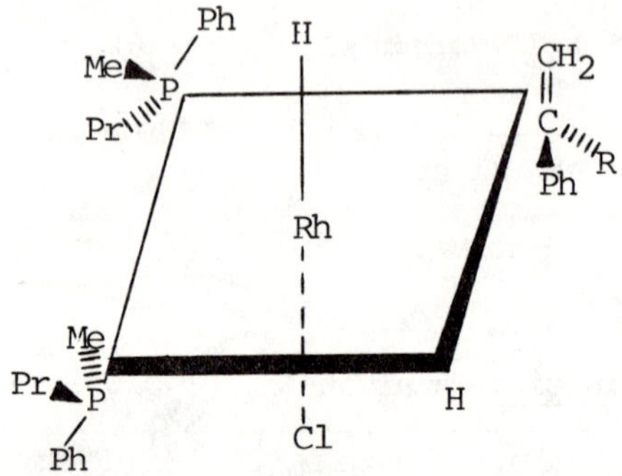

FIGURE 7. *Asymmetric homogeneous hydrogenation of α-substituted acrylic acids with a Wilkinson-type catalyst under "extended" conditions (higher temperature and pressure than usual) (Knowles et al, 1970).*

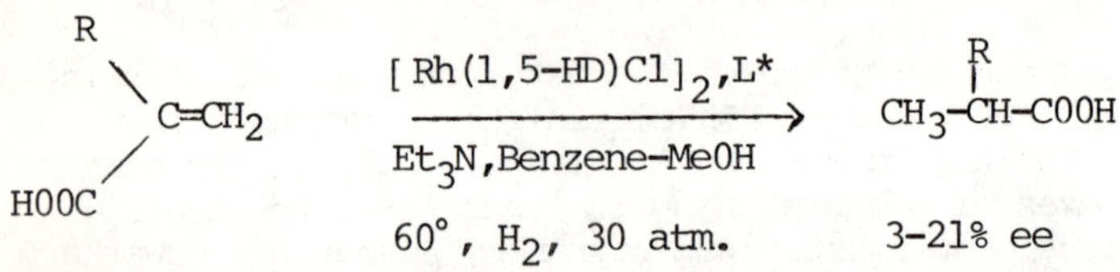

L*=PMePhR' (chiral)
R'=n-Pr,i-Pr, Cyclohexyl
1,5-HD=1,5-Hexadiene

As described above, the rhodium-phosphine catalyst used extensively by the Horner and the Monsanto groups contained phosphines which were asymmetric at phosphorus. A negligible optical yield was obtained in the one reported reaction involving an optically active phosphine ligand with a chiral carbon atom somewhat remote from the metal ($PhP(CH_2\overset{*}{C}HMeEt)_2$). However, in 1971 Morrison and coworkers (the New Hampshire group) showed that a successful asymmetric hydrogenation could be carried out with at least one phosphine not asymmetric at phosphorus, neomenthyl-diphenylphosphine (NMDPP). In fact, using a catalyst believed to be Rh(NMDPP)$_3$Cl, these workers

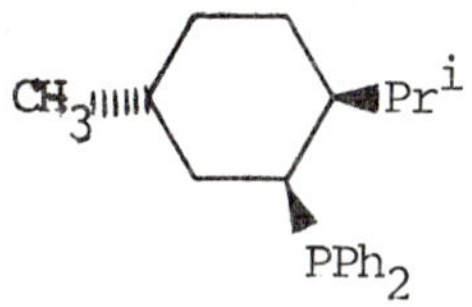

NMDPP

achieved what was, at the time, the highest degree of asymmetric induction ever accomplished with a chiral hydrogenation catalyst, homogeneous or heterogeneous (Morrison et al., 1971). A 61% enantiomeric excess of (S)-(+)-3-phenylbutanoic acid was obtained by hydrogenation of (E)- β-methylcinnamic acid in the presence of Rh(NMDPP)$_3$Cl (Fig. 8).

FIGURE 8. *Asymmetric homogeneous hydrogenation with a neomenthyldiphenylphosphine catalyst (Morrison et al, 1971).*

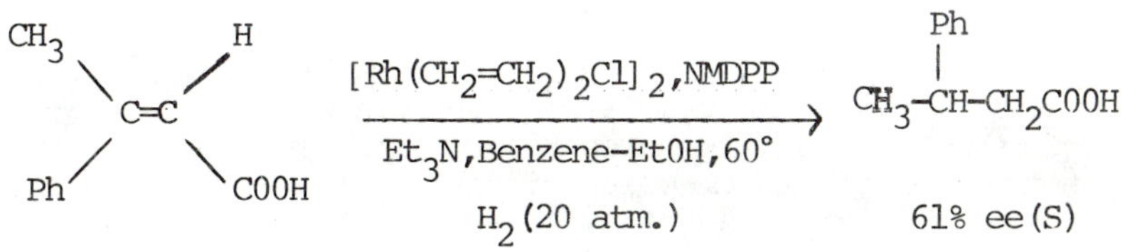

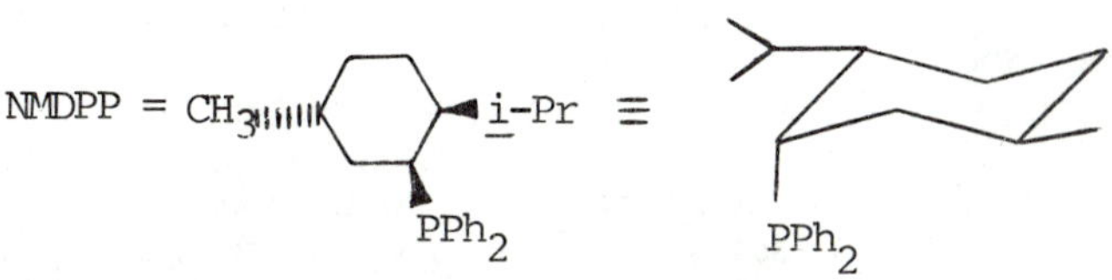

The New Hampshire group noted that in their work α,β -unsaturated carboxylates uniformly yielded products of higher optical purity than simple olefins. For example, the product enantiomeric excesses for (E)- β-methylcinnamic acid (61% e.e.), (E)-α -methylcinnamic acid (52% e.e.), and α -phenylacrylic acid (28% e.e.) were in the range of 4-9 times greater than that for α-ethylstyrene (7% e.e.). In agreement with the Monsanto group, it was suggested that these observations might indicate a mechanism involving bifunctional substrate coordination through both the carboxylate anion and olefinic bond.

The relatively high optical purities obtained with the Rh-NMDPP system are particularly interesting from a practical viewpoint since the NMDPP ligand* is prepared from an inexpensive, commercially available, chiral precursor, (-)-menthol (Morrison and Masler, 1974). Tertiary phosphines chiral at phosphorus, on the other hand, are much less accessible and require a classical resolution step.

Almost coincident with the New Hampshire group's report of dramatic success with NMDPP, Kagan and coworkers (The Paris group) reported that optical purities of 60-70% could be obtained with a new chiral diphosphinerhodium(I) catalyst system (Dang and Kagan, 1971). Like NMDPP, the new phosphine ligand (-)-2,3-O-isopropylidene-2,3-dihydroxy-1,4-bis(diphenylphosphino)butane (called (-)-DIOP)* could be prepared from a readily available chiral compound, (+)-tartaric acid. The DIOP

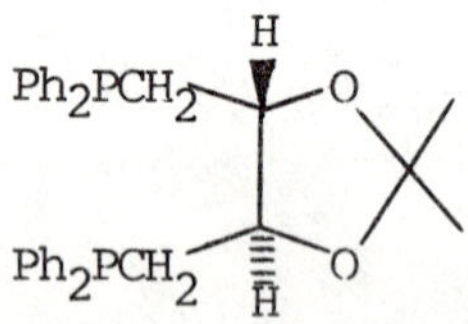

(-)-DIOP

catalyst, often represented as [Rh(DIOP)Cl(solvent)] was generated in situ as shown in Fig. 9. The substrates used in initial experiments and hydrogenated to products having up to 72% e.e. were two α-acylaminoacrylic acids and α-phenylacrylic acid.

The high stereoselectivities observed with the DIOP catalyst have been attributed to the appreciable conformational rigidity of the DIOP ligand, arising from the presence of a bicyclic system (the chelating diphosphine ring and the fused dioxolane ring) and also to the presence of a trans ring junction. Stereochemical control through participation of the carboxylic acid function of the substrate also seemed to be indicated since in contrast to the result shown in Fig. 9 for free α-phenylacrylic acid, hydrogenation of methyl α-phenylacrylate gave methyl-2-phenyl-propanoate of the R configuration, and only 7% e.e. Later experiments, however, showed

*Both NMDPP and DIOP (+ and -) are now commercially available from Strem Chemical Co., 150 Andover St., Danvers, Mass. 01923.

FIGURE 9. *Asymmetric homogeneous hydrogenations with the DIOP catalyst. Hydrogenation with a (+)-DIOP catalyst would, of course, give enantiomeric products.*

$$[Rh(Cyclooctene)_2Cl]_2 + 2\ DIOP \xrightarrow[\text{Argon, r.t.}]{\text{Benzene-EtOH}} 2\ [Rh(DIOP)\text{-}Cl(S)]$$

in situ DIOP catalyst

(S) = solvent

(–)-DIOP =

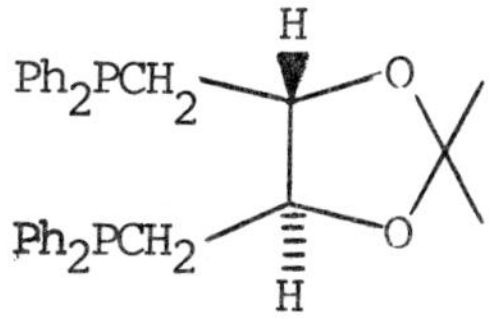

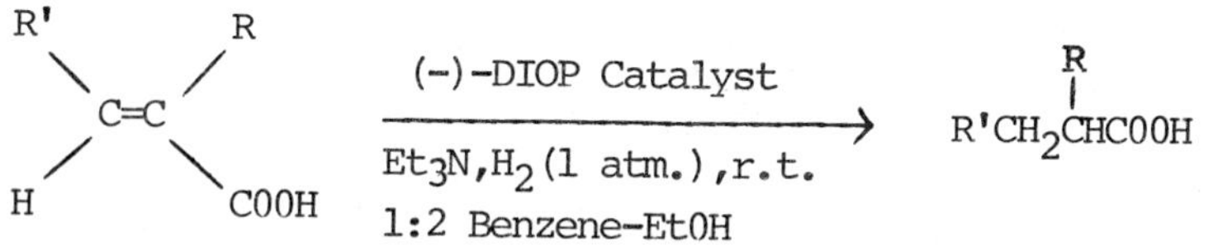

A: R'=Ph, R=$NHCOCH_3$
B: R'=H, R=$NHCOCH_2Ph$
C: R'=H, R=Ph

A: 72% ee (R)
B: 68% ee (R)
C: 63% ee (S)

that although the carboxyl group was important in the case of α-phenylacrylic acid, it was not crucial for a successful asymmetric hydrogenation when the substrate also contained the enamide function (Kagan and Dang, 1972). For example, compound I was hydrogenated with high asymmetric bias (78% e.e.). This result and others were taken as evidence that coordination through the enamide group may influence the stereochemical course of the reaction.

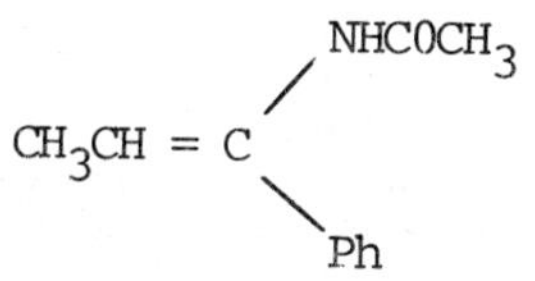

I

The favorable affect of the enamide function on asymmetric induction is indicated not only by the result with compound I, but also by later results summarized in Table I, where optical purities in the range of 70-80% were generally obtained for various derivatives of alanine, phenylalanine, tyrosine, and 3,4-dihydroxyphenylalanine (DOPA). The Paris group found that the Rh-(-)-DIOP catalyst yielded the "unnatural" R or D-amino acid derivatives, whereas L-amino acid derivatives could be obtained with a (+)-DIOP catalyst. Since the optical purity of the N-acylamino acids can often be considerably increased by a single recrystallization (fractionation of pure enantiomer from racemate) and the N-acetyl group can be removed by acid hydrolysis, this scheme provides an excellent asymmetric synthesis route to several amino acids.

TABLE I. *Asymmetric Hydrogenations of α-Acylaminoacrylic Acids with the Soluble DIOP Catalyst*[a]

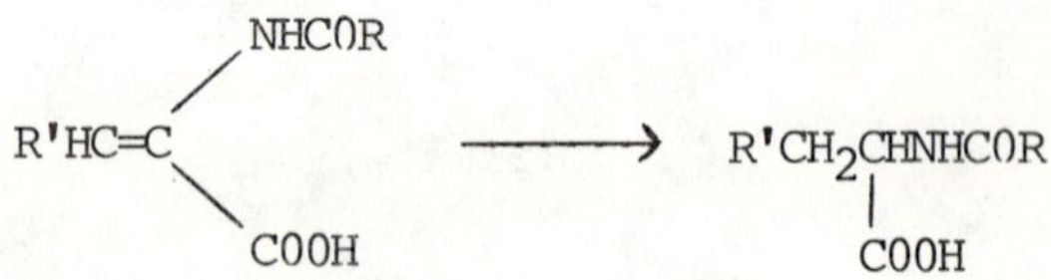

R'	R	Yield(%)	%ee[b]
H	CH_3	96	73
Ph	CH_3	95	72
p-OH-Phenyl	CH_3	92	80
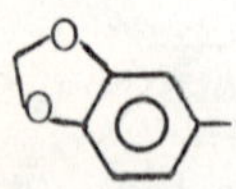	CH_3	97	79
p-OH-Phenyl	Ph	95	62

[a]Reaction conditions as in Fig. 9, but without NEt_3.
[b]All products have the R (or D) configuration.

An even more efficient asymmetric synthesis of α-amino acid derviatives has been described by the Monsanto group. They have found that chiral o-anisylcyclohexymethylphosphine (ACMP), like DIOP, exerts an extraordinary effective chiral influence in the reduction of α-acylaminoacrylic acid substrates.

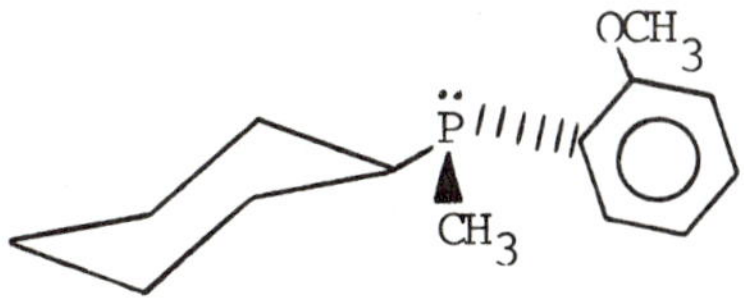

(+)-(R)-ACMP

Catalysts prepared from (+)-ACMP give L-amino acid derivatives and those containing the (-)-phosphine give derivatives of the D-series. Many instances of 85-90% e.e. have been observed (Table II) (Knowles et al., 1971, 1972, 1973, 1974).

TABLE II. *Asymmetric Homogeneous Hydrogenations of α-Acylaminoacrylic Acids with the Monsanto (+)-(R)-ACMP Catalyst*

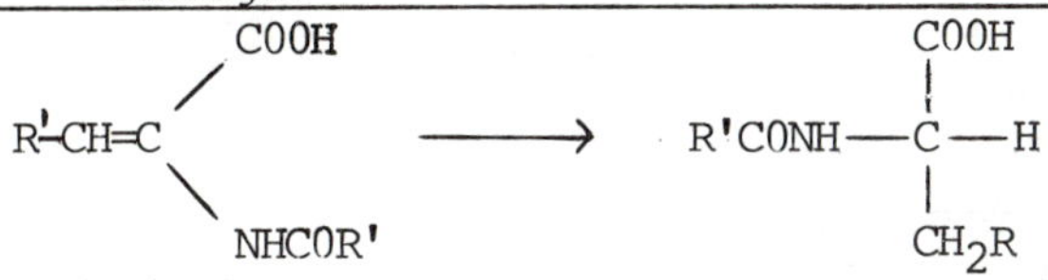

R	R'	Product %ee*
3-OMe, 4-OH-Phenyl	Ph	90
3-OMe, 4-OAc-Phenyl	Me	88
Phenyl	Me	85
Phenyl	Ph	85
p-Cl-Phenyl	Me	77
3-(1-acetylindolyl)	Me	80
H	Me	60

*With (+)-(R)-ACMP catalyst, the products all have the S (or L) configuration.

The ACMP ligand was deliberately designed to create the opportunity for secondary bonding between the substrate and the ligand. It was felt that α-acylaminoacrylic acid substrates might possibly act as "tridentate ligands" toward the catalyst; the olefinic and carboxylate groups interacting with the rhodium and the acylamino groups being hydrogen bonded to the methoxy groups of the ACMP ligands.

It should be pointed out that asymmetric reactions other than hydrogenation have been carried out with chiral phosphine complexes of rhodium (and a few other metals). For example, asymmetric hydrosilylations (addition of Si-H across C=C, C=O and C=N bonds) have been catalyzed by such complexes (Yamamoto et al, 1971, 1972, 1973; Langlois et al., 1973). When a ketone or imine is hydrosilylated, the intermediate silyloxy or silylamino compound can be hydrolyzed to an alcohol or amine. Thus the overall result is equivalent to direct hydrogenation (Eqn. 3). Asymmetric hydroformylations (Tanaka et al., 1972; Solomon et al., 1973; Stern et al., 1973; Botteghi et al., 1973; Pino et al., 1973) and a variety of other chiral reactions have also been observed with metal complexes made from chiral phosphines (Corriu and Moreau, 1973; Kiso et al., 1974; Bogdanovic et al., 1972; Trost and Dietsche, 1973).

$$\gt C{=}O \xrightarrow[\text{cat.}]{-\overset{|}{Si}-H} -\underset{H}{\overset{|}{C}}-O\underset{|}{\overset{|}{Si}}- \xrightarrow{H_2O} -\underset{H}{\overset{|}{C}}-OH \qquad (3)$$

The Synthesis of Chiral Phosphine Ligands for Use in Chiral Catalysts of the Wilkinson Type

As we have seen, two kinds of chiral tertiary phosphine ligands have been used in asymmetric hydrogenation experiments involving rhodium complexes. The Horner and Monsanto groups have concentrated upon ligands whose chirality is centered at an asymmetric phosphorus atom. The New Hampshire and Paris groups have focused their attention mainly on phosphines which carry chiral carbon moieties.

A synthetic approach to phosphines chiral at phosphorus that overcomes some limitations inherent in earlier methods has been developed by Mislow and coworkers (Korpium et al., 1967, 1968; Farnham et al., 1970). When unsymmetrically substituted phosphinyl halides are esterfied with (−)-menthol, the resulting diastereomeric phosphinates can be separated by fractional crystallization.* (Fig. 10). Displacement of the menthyloxy group by an appropriate Grignard reagent gives chiral tertiary phosphine oxides. The chiral tertiary phosphine oxides can be reduced to chiral tertiary phosphines by one of several methods; trichlorosilane (retention of config-

*Alternative methods of preparation of chiral phosphinates have also been reported (Emmick and Letsinger, 1968; Nudelman and Cram, 1968).

uration), trichlorosilane and a weakly basic amine (retention), trichlorosilane and a strongly basic amine (inversion), hexachlorodisilane (inversion) or phenylsilane (retention). While it does not circumvent a classical resolution step, the Mislow approach does introduce greater flexibility since a number of chiral phosphines can be obtained from a single resolved precursor.

FIGURE 10. *The synthesis of chiral phosphine oxides by reaction of Grignard reagents with diastereomerically pure menthyl phosphinates. Deoxygenation of chiral phosphine oxides gives chiral phosphines.*

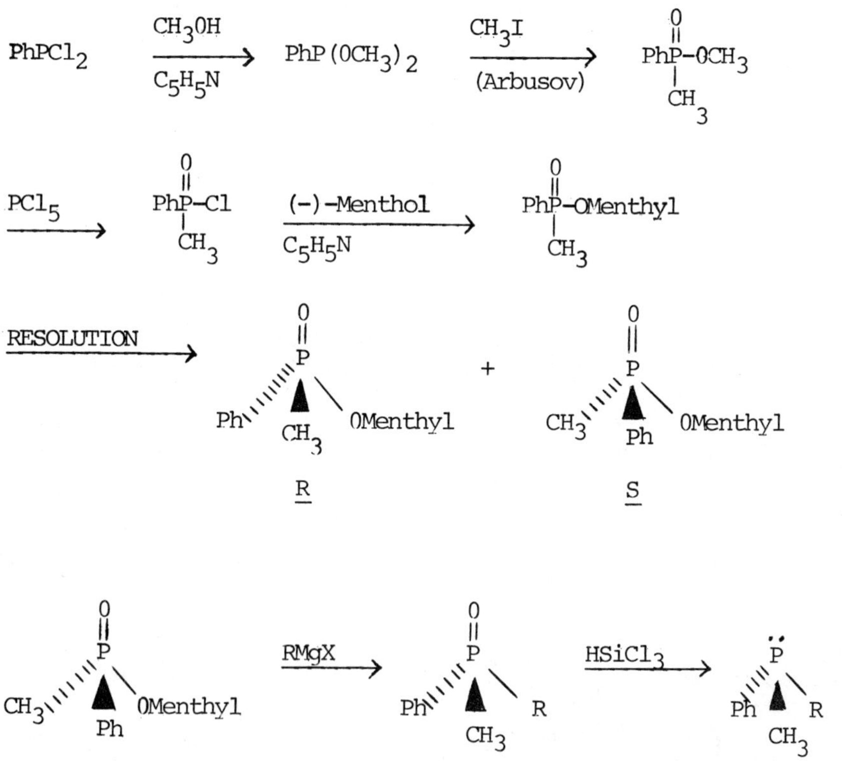

The Monsanto group has applied the Mislow synthetic scheme to the synthesis of the ACMP ligand. Figure 11 shows the reaction sequence starting with the menthyl phosphinate that has the R configuration at phosphorus. The Grignard displacement reaction gave a 70-90% yield of phosphine oxide. The selective reduction of the phenyl group in the chiral phosphine oxide was accomplished in about 60% yield.

FIGURE 11. *Synthesis of (S)-ACMP. The (R)-ACMP ligand is prepared from the phosphinate that is epimeric at phosphorus (see Fig. 10).*

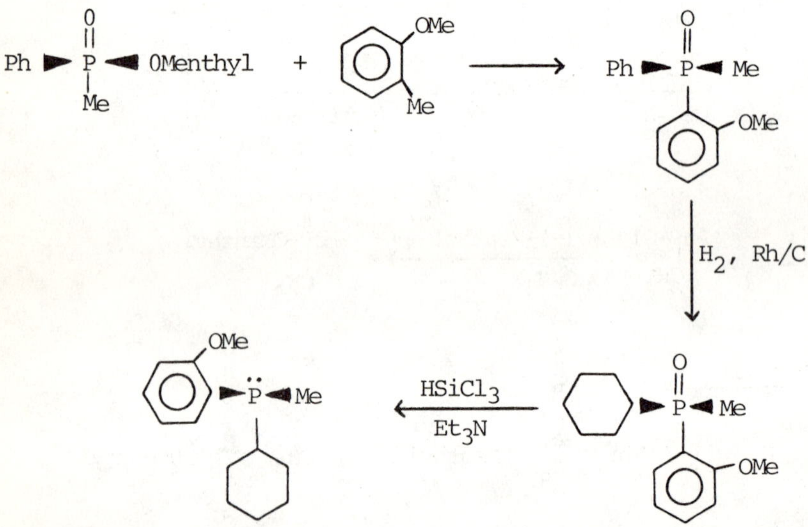

Deoxygenation of the o-anisylcyclohexymethylphosphine oxide was carried out with Si_2Cl_6 or $HSiCl_3$-Et_3N (inversion of configuration at phosphorus) in about 50% yield.

The ACMP ligand can be used in situ with a soluble Rh(I)-alkene complex to produce a catalytically active system but normally it is converted to a stable crystalline complex of the type $(ACMP)_2Rh$ (diene) $^+X^-$ where the diene is, for example, 1,5-cyclooctadiene and X^- is BF_4^-, PF_6^- or $B(Ph)_4^-$.

The Monsanto application of the Mislow scheme has also

produced other ligands of the ACMP type, for example with i-propyl, i-butyl or benzyl groups in place of cyclohexyl and i-propyl, ethyl or benzyl in place of methyl in the ether function. Ligands with these structural variations gave catalysts that were less effective than ACMP in terms of the optical purities of hydrogenation products (W.S. Knowles et al., 1974).

Hydrogenation of an α-acylaminoacrylic acid with the Monsanto ACMP catalyst may be carried out with either the free acid or its triethylamine salt. Free acid substrates give the best %ee results at low pressures and %ee is not especially sensitive to temperature. The anion substrate form is best reduced at lower temperatures (typically 0°) and %ee values do not show much pressure sensitivity. Almost any alcohol solvent may be used and small amounts of water are tolerable. Hydrogenations are usually carried out in methanol solution (or with some substrates a 20-25% slurry is used) being careful to purge oxygen which is a serious poison. At 40 psig, 50°, and about 0.01% metal levels, the reduction of most acylaminoacrylic substrates is complete in about 4 hrs (often less). The amino acid product may be crystallized from methanol, often with an increase in optical purity due to fractional crystallization of the pure enantiomer from the racemate. Rhodium may be recovered, but the ligand is oxidized during work-up.

The New Hampshire group has prepared a number of chiral phosphine ligands from chiral alkyl halides (Fig. 12) and tosylates via displacements with diphenylphosphide anion. For example, in some early experiments, lithium diphenylphosphide was used to prepare (+)-S-2-methylbutyldiphenylphosphine, (-)-R-3-phenylbutyldiphenylphosphine, (+)-R-2-octyldiphenylphosphine (configuration presumed but not rigorously proved), (+)-neomenthyldiphenylphosphine (NMDPP) and (-)-menthyldiphenylphosphine (MDPP) from the appropriate chloride or bromide (Morrison and Burnett, 1970; Morrison and Masler, 1974). The (+)-NMDPP ligand proved to be especially effective in hydrogenation experiments (Morrison et al., 1971; Masler, 1974).

In more recent studies of the (+)-NMDPP synthesis, the relative effectiveness of lithium, sodium, and potassium diphenylphosphides was determined. Under a standard set of conditions, the reaction of (-)-menthyl chloride with sodium diphenylphosphide (Fig. 13) gave the highest yield of (+)-NMDPP. The ratios of the yields of (+)-NMDPP were 1.0:1.55:1.16 for lithium, sodium, and potassium diphenylphosphide, respectively (Morrison and Masler, 1974).

The chiral diphosphine ligand, (+)-(1R,3S)-1,2,2-trimethyl-1,3-bis(diphenylphosphinomethyl) cyclopentane, commonly called (+)-CAMPHOS, has also been prepared by the New Hampshire group (Masler, 1974). The synthesis of this ligand posed special challenges and ultimately resulted in some new synthetic approaches that may be useful in other ligand synthesis.

FIGURE 12. *Synthesis of (S)-2-phenylbutyldiphenylphosphine. Catalysts prepared from this ligand, and structurally related ligands, typically give products of low optical purity.*

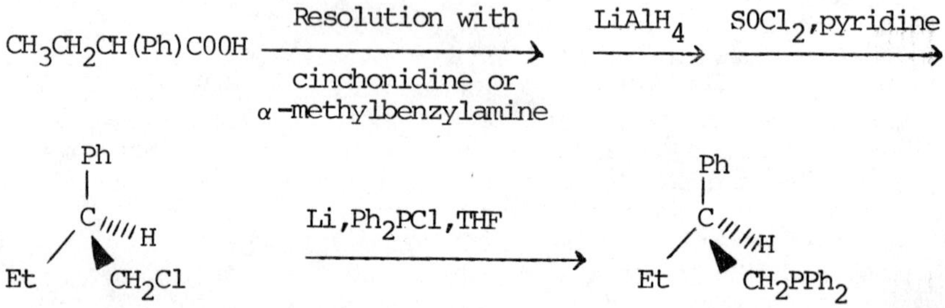

FIGURE 13. *Synthesis of Neomenthyldiphenylphosphine (NMDPP).*

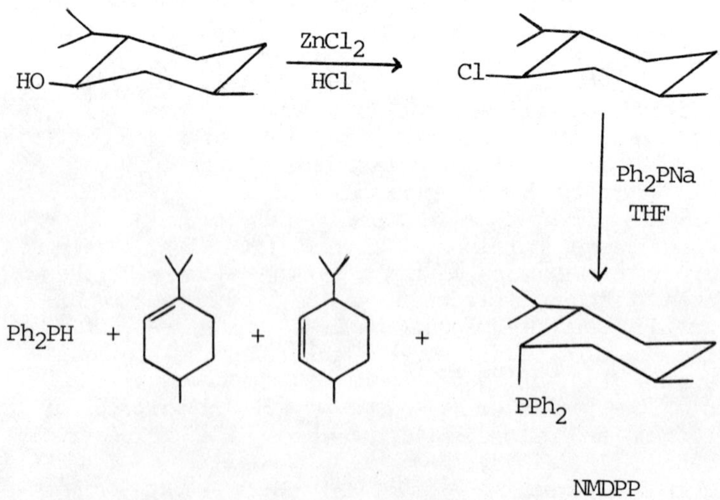

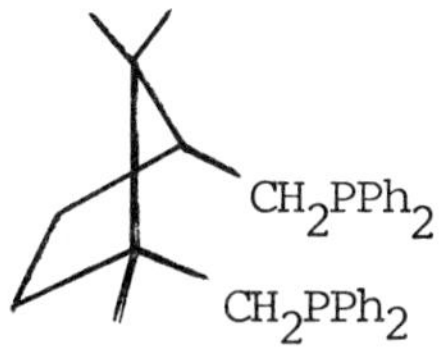

(+)-CAMPHOS

The synthesis of CAMPHOS by displacement on its ditosylate precursor (prepared from the alcohol obtained by reduction of commercially available (+)-camphoric acid) with the diphenylphosphine anion appeared promising on paper but, initially at least, was dismal failure in practice. The reaction of lithium diphenylphosphide (from Ph_2PCl and Li) gave no CAMPHOS. However, when potassium diphenylphosphide (from Ph_2-PH and K) in tetrahydrofuran was used, in place of the lithium reagent, (+)-CAMPHOS was formed*. An alternative and more circuitous route to (+)-CAMPHOS starting with (+)-camphor was also developed (Fig. 14).

FIGURE 14. *Syntheses for CAMPHOS (Masler, 1974).*

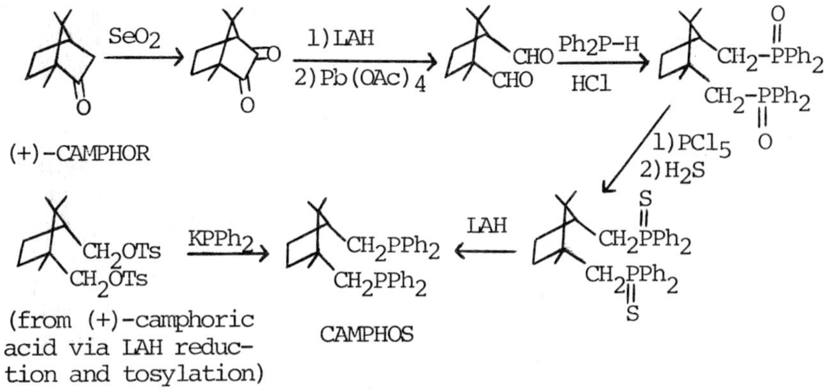

*The observations made by the New Hampshire group concerning the variable reactivity of metal phosphides with alkyl halides and tosylates should be kept in mind when planning ligand syntheses by these routes. It appears that for any particular halide or tosylate substrate, the best metal phosphine for displacement can be determined only by experiment.

The key step in this synthesis was the HCl catalyzed addition of diphenylphosphine to the dialdehyde. This rather obscure addition reaction which vis subsequent rearrangement gives camphos dioxide may prove to be a very useful approach to other chiral phosphine ligands also.

Addition occurs readily with aldehydes and also with unhindered ketones. Since many chiral aldehydes and ketones are available from natural sources, the possibilities for chiral ligand synthesis by this route are virtually unlimited.

As has been noted, the Paris group has achieved much success with diphosphine ligands derived from chiral tartaric acid, both enantiomers of which are commercially available. The "parent ligand" in the Paris arsenal is DIOP which is prepared as shown in Fig. 15.

FIGURE 15. *Synthesis of (-)-DIOP.*

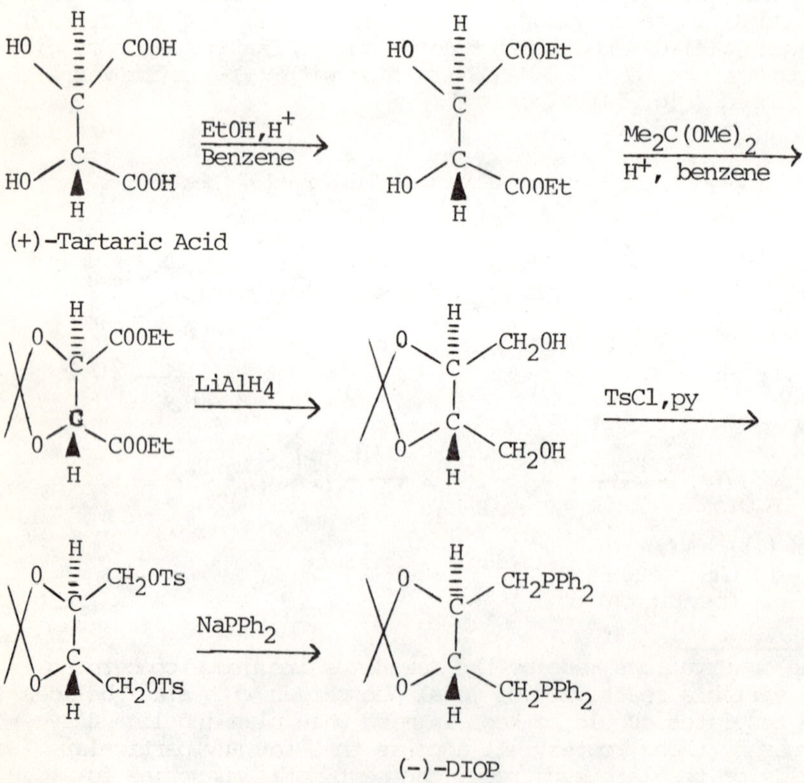

Modified DIOPs such as II and III have likewise been synthesized.

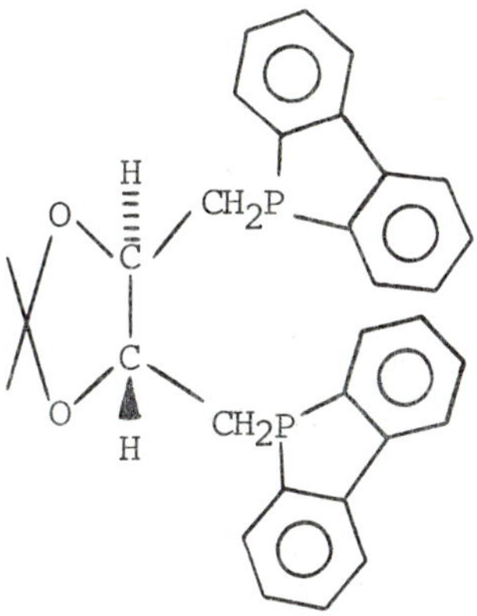

II

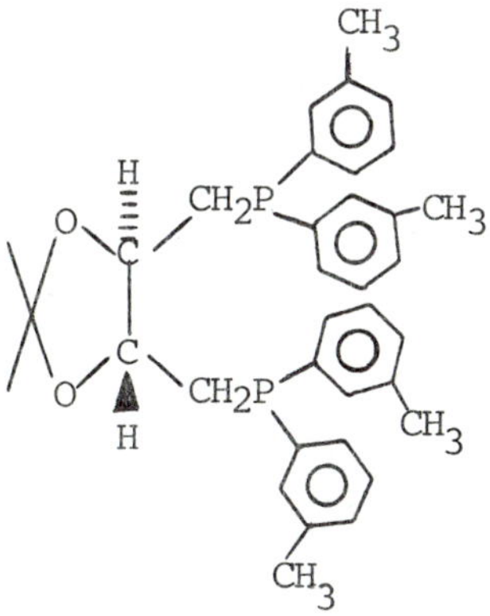

III

Chiral, insolubilized, catalytically-active, transition metal complexes that incorporate DIOP moieties (Fig. 16) have also been developed by the Paris group (Poulin et al., 1973; Dumont et al., 1973). Insolubilized complexes exhibit some features of both homogeneous and heterogeneous catalysts and other speakers have extolled their virtues.

The insolubilized DIOP catalyst was found to be rather ineffective for the asymmetric hydrogenation of olefinic substrates; the hydrogenation of α-ethylstyrene proceeded readily, but gave (−)-R-2-phenylbutane with an optical purity of only 1.5%. Methyl atropate was hydrogenated to (+)-S-methylhydratropate (2.5% ee). The soluble DIOP catalyst gave 15 and 17% ee, respectively, for the same reduction. The optical purity of the products was lower when recovered insolubilized catalyst was used. There was no reduction of α-acetamidocinnamic acid in ethanol-benzene with the insolubilized catalyst, presumably due to the hydrophobic nature of the polymer support which causes it to shrink in hydroxylic solvents.

More impressive results were obtained for the asymmetric hydrosilyation of ketones over the insolubilized catalyst. Acetophenone, for example, was hydrosilylated with phenylnaphthylsilane and the intermediate was hydrolyzed with HCl to give (−)-S-phenylmethylcarbinol (58% ee). Similar results were obtained with the homogeneous DIOP catalyst. Using diaryl

FIGURE 16. *Synthesis of insolubilized DIOP*

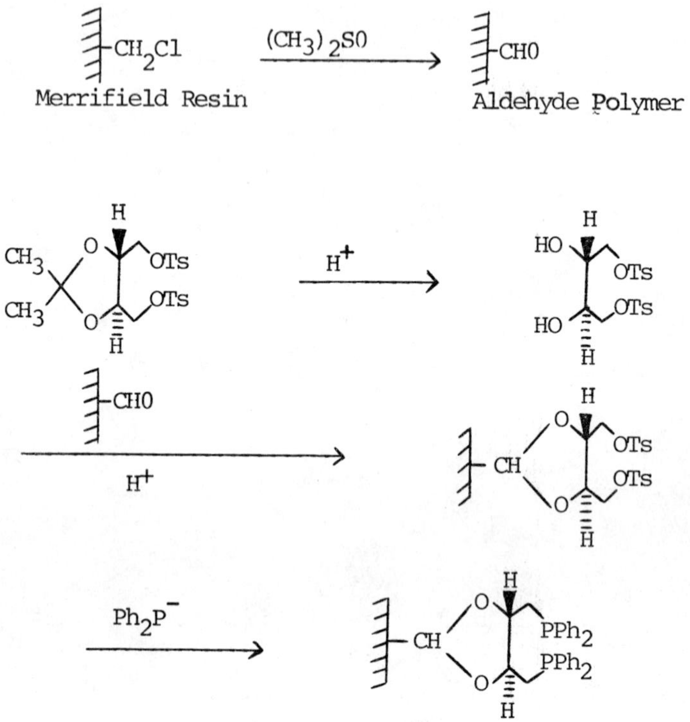

Insolubilized DIOP

or arylalkyl silanes, yields from 52-100% and optical purities of from 7-59% were realized (Table III).

Typically, the soluble DIOP catalyst is prepared in situ from the reaction of 2 moles of DIOP with one mole of $[Rh\ Cl(COD)_2]_2$ or a similar alkene complex. Excess phosphine greatly reduces or destroys entirely the catalyst activity. Hydrogenations have most often been carried out in a 1:2 benzene:ethanol solution, 2-5 mM in catalyst, at room temperature and one atmosphere hydrogen pressure. At the conclusion of the hydrogenation, the reaction solution is evaporated to dryness and isolation procedures appropriate for the particular product obtained are then used.

The New Hampshire group has compared the effectiveness of

various catalysts toward α,β-unsaturated carboxylic acid substrates. Reductions were carried out in a 1:1 ethanol-benzene solvent, at 300 psi H_2, for 24 hrs. at 60°. The solvent was deoxygenated immediately prior to use. The catalyst was formed in 100 ml of solvent from 34μ moles of $[Rh(COD)Cl]_2$ and 0.5 mmole of phosphine, was prereduced for 0.5 hr. at 3.5 atm H_2, and then the substrate (most often 25 mmole, a substrate: rhodium ratio of 375*) was added in 100 ml of solvent containing 4 mmole of triethylamine.

TABLE III. *Asymmetric Synthesis of Alcohols by Hydrosilylation of Ketones.*[a] *A comparison of an Insolubilized DIOP Catalyst vs. a Solution DIOP Catalyst.*

$$R_1R_2C{=}O \longrightarrow R_1R_2CH{-}O{-}\overset{|}{\underset{|}{Si}}{-} \xrightarrow{H+} R_1R_2CHOH$$

Ketone	Silane	Ketone/ Rh	Time (Hr.)	Product %ee Insolubilized Catalyst	Product %ee Soluble Catalyst
Acetophenone	$H_2SiPhCH_3$	25	42	12	13
Acetophenone	H_2SiPh_2 H_2SiPh_2	35	24	29	28
Acetophenone	$H_2SiPh(C_{10}H_7)$	33	45	59	58
Acetophenone	$H_2SiPh(C_{10}H_7)$	20	24	55	58
Phenyl-isopropyl ketone	$H_2SiPhCH_3$	25	48	7	20
Phenyl-isopropyl ketone	H_2SiPh_2	25	48	28	35

[a]In all cases, the hydrosilylated product was hydrolyzed to the alcohol. The (-)-S-carbinol was obtained in every case.

*Substrate: Rh ratios between 185 and 435 have been used in comparative studies.

Table IV shows the structures of the substrates that have been used in comparative studies by the New Hampshire group. Some of these substrates are quite hindered, but with the NMDPP, CAMPHOS, DIOP, and ACMP catalysts quantitative yields are usually obtained under the "extended" reaction conditions described above.** The MDPP catalyst does not give a good yield in most instances even under extended conditions.

At the conclusion of a hydrogenation reaction, solutions were evaporated to dryness and the residue was partitioned between 10% NaOH solution (50 ml) and methylene chloride (50 ml). The aqueous layer was separated, washed with ether, and then was acidified with hydrochloric acid.* The organic acid product was extracted with ether. The ether solution was dried ($MgSO_4$) and concentrated to give a "crude" product. Itaconic, citraconic, mesaconic and (E) and (Z) - α -phenylcinnamic acids give solid products; the other cinnamic acid derivatives and atropic acid give liquids. "Crude" solid products were analyzed by nmr and optical rotations were taken on the unrecrystallized "crude" material. "Crude" liquid products were analyzed by nmr, then distilled, and optical rotations were taken on distilled material.

It is possible to perceive a number of interesting stereochemical relationships in the comparative data of Table V. It is clear that the phosphorus-chiral ACMP catalyst, which is so outstandingly effective for the asymmetric reduction of α-acylaminoacrylic acids, does not compete favorably with catalysts made from carbon-chiral ligands in terms of the %ee values obtained with α, β -unsaturated carboxylic acids that do not contain an acylamino group. This is not to say that other ligands that are chiral at phosphorus will also be less effective than those that are chiral at carbon. An important point, however, is that <u>the match-up of ligand and substrate is a critical, specific and as yet unpredictable feature of such reactions. A good ligand for one kind of substrate is not necessarily best for another kind.</u>

Some dramatic differences are also apparent between NMPP and MDPP ligands. These ligands are diastereomers, more precisely, they are epimers since they differ only in configuration at C-3. It is quite reasonable that these ligands should

---

*Substrate: Rh ratios between 185 and 435 have been used in comparative studies.

**The extended" reaction conditions involve higher pressure and temperature than are normally used for Wilkinson-type catalysts and unhindered substrates.

behave differently, since diastereomers have different chemical and physical properties, albeit sometimes only slightly different. However, NMDPP and MDPP generate considerably disparate behavior both in terms of the activity and the chiral influence of the catalyst derived from them. Toward every substrate examined thus far, the MDPP catalyst has had a very low activity, much lower activity than the NMDPP catalyst. Also, the MDPP catalyst generally gave much lower asymmetric bias than the NMDPP catalyst, and was the only chiral catalyst to give an achiral product* (two examples).

If one inspects molecular models, it is possible to envisage a possible rationalization for the lower activity of the MDPP catalyst compared to the NMDPP catalyst. It appears that the MDPP ligand is less hindered around phosphorus than is the NMDPP ligand. It may be that MDPP more effectively completes for unsaturated coordination sites on the metal (especially under the high ligand loading conditions we have used). This is equivalent to the proposition that a MDPP ligand is less easily dissociated and consequently catalysis is retarded. Of course, other explanations are also possible, one being that the MDPP more effectively hinders the unsaturated site of the catalyst and in this way reduces its effectiveness.

There appears to be no general stereocorrelation model that can be perceived for the NMDPP and MDPP ligands. As has been pointed out, these ligands are epimeric, being "locally enantiomeric" at C-3. One might be tempted to presume that catalysts prepared from them would produce enantiomeric products since the C-3 chiral carbons are closest to the metal. However, such an intuitively comfortable presumption is as dangerous as the equally satisfying premise that the better ligands will always be those that are chiral at phosphorus, rather than at some more remote carbon atom. It is clear from the comparative data that NMDPP and MDPP do sometimes induce the production of enantiomeric products from the same substrate; but just as often they give products with the same chiralities. There appears to be no general relationship on the basis of comparative data collected thus far.

The comparative data can also be used to provide insight on another point. The first six substrates listed in Table V

*In principle, all chiral catalysts should give chiral products. However, the energy difference between diastereomeric transition states can be so slight that the product does not have an observable rotation.

TABLE IV. *The α,β-Unsaturated Carboxylic Acid Substrates Used in Comparative Studies of Chiral Rh-Phosphine Catalysts.*

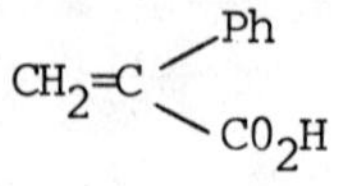

Atropic Acid

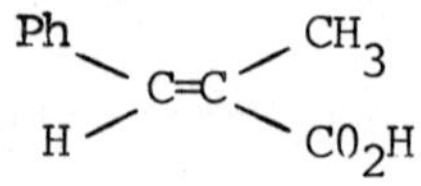

(E)- α-Methylcinnamic Acid

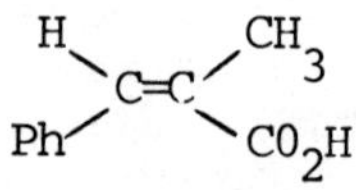

(Z)- α-Methylcinnamic Acid

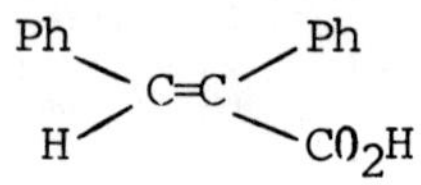

(E)- α-Phenylcinnamic Acid

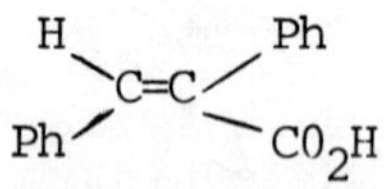

(Z)- α-Phenylcinnamic Acid

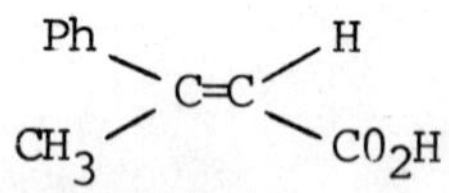

(E)- β-Methylcinnamic Acid

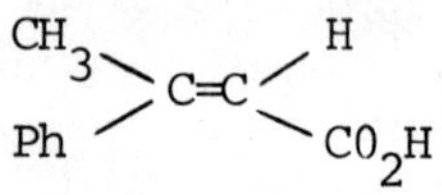

(Z)- β-Methylcinnamic Acid

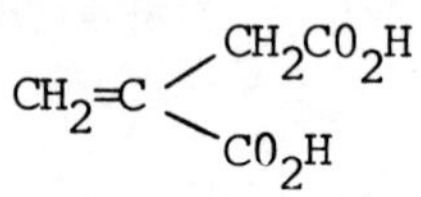

Itaconic Acid

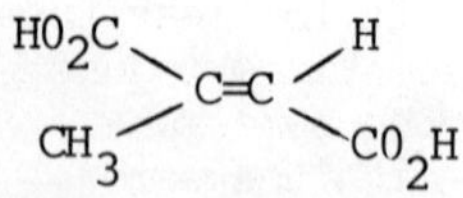

Mesaconic Acid

TABLE V. *The Asymmetric Hydrogenation of α,β-Unsaturated Carboxylic Acids. A Comparison of Product %ee Values for Several Ligands.*

Substrates	ACMP	DIOP	NMDPP	MDPP	CAMPHOS
(E)-α-Methylcinnamic Acid	12(R)	25(S)	60(R)	17(S)	15(R)
(Z)-α-Methylcinnamic Acid	24(R)	33(R)	25(R)	racemic	11(S)
(E)-α-Phenylcinnamic Acid	24(S)	15(R)	34(S)	27(R)	12(S)
(Z)-α-Phenylcinnamic Acid	1.5(R)	1.0(S)	9.1(S)	3.2(S)	14(S)
(E)-β-Methylcinnamic Acid	37(S)	14(R)	62(S)	1.2(S)	9.7(S)
(Z)-β-Methylcinnamic Acid	13(R)	28(S)	31(R)	31(S)	11(R)
Atropic Acid	1[a]	44(S)	30(S)	racemic	6.1(S)
Itaconic Acid	-	-	8.1(R)	18(R)	11(R)
Mesaconic Acid	-	-	5.9(R)	7.2(S)	1.8(R)

[a]Result taken from Knowles et al., 1973, where it appears without configurational notation; all other data from unpublished research of the New Hampshire group.

comprise a set of three diastereomeric* (geometrically isomeric) pairs. The question is, with the same catalyst do E and Z isomers give enantiomeric products? The answer is that from the data no generality that covers this situation is apparent when all catalysts are considered. With DIOP, enantiomers are obtained from diastereomeric substrates in each instance, but with the other catalysts there is no regularity. There is almost an equal number of examples of each of the two possible patterns. This is not too surprising if one remembers that diastereomeric substrates, like diastereomeric ligands, can be thought of as simply different compounds. There is no reason to presume that diastereomers must display enantiomeric patterns of behavior, but neither is there any stereochemical principal that prevents them from doing so. In another chiral hydrogenation system (Abley and McQuillin, 1971) the fact that diastereomeric olefinic substrates gave products of the same configuration and almost the same optical purity with the same chiral catalyst has been taken as a possible indication that hydrogen transfer has occurred after a loss of diastereomeric identity. It is important to recognize that whereas this is a sufficient explanation, it is not a necessary one.

Concluding Remarks

There are other asymmetric homogeneous hydrogenation systems that we have not had time to mention in this brief presentation. However, the chiral rhodium-phosphine systems and related systems involving rhodium, cobalt and ruthenium have been reviewed in detail in a soon to be published volume of Advances in Catalysis (Morrison et al., 1975).

Chiral hydrogenation with soluble metal complexes is a very new area of research. Only a few chiral ligands have been tested. The possibilities for valuable contributions in this area are vast. In the short-term future, we will hope for many new ligands that will induce high asymmetric bias. For the long-term, we envision some totally new concepts for chiral catalysis and perhaps some major theoretical insights that will make it possible to perceive structure-efficiency relationships leading to the rational design of chiral catalytic systems.

---

*The olefinic substrates that are cis-trans isomers are by modern stereochemical nomenclature more generally termed diastereromers. That is, they are stereoisomers that are not enantiomers. The fact that they contain no asymmetric carbons is irrelevant to this classification.

REFERENCES

1. Akabori, S., Sakurai, S., Izumi, Y., and Fujii, Y., Nature, 178, 323 (1956).
2. Akabori, S., Izumi, Y., Fujii, Y., and Sakurai, S., Nippon Kagaku Zasshi, 77, 1374 (1956); Chem. Abst., 53, 51496 (1959).
3. Akabori, S., Izumi, Y., and Fujii, Y., Nippon Kagaku Zasshi, 78, 886 (1957); Chem. Abst., 54, 9889c (1960).
4. Abley, P., and McQuillin, F.J., J. Chem. Soc., (C), 844 (1971).
5. Bogdanovic, B., Henc, B., Meister, B., Pauling, H., and Wilke, G., Angew. Chem. Internat. Edit., 11, 1023 (1972).
6. Botteghi, C., Consiglio, G., and Pino, P., Chimia, 27, 477 (1973).
7. Burnett, R.E., Ph.D. Thesis, Univ. of N.H., 1971.
8. Červinka, O. and Belovsky, O., Coll. Czech. Chem. Commun., 32, 3897 (1967).
9. Corriu, R.J.P., and Moreau, J.J.E., Tetrahedron Lett., 4469 (1973).
10. Dumont, W., Poulin, J.C., Dang, T.P., and Kagan, H.B., J. Amer. Chem. Soc., 95, 8295 (1973).
11. Dang, T.P., and Kagan, H.B., Chem. Comm., 481 (1971).
12. Eliel, E.L., Tetrahedron, 30, 1503 (1974).
13. Emmick, T.L., and Letsinger, R.L., Ibid., 90, 3459 (1968).
14. Farnham, W.B., Murray, R.K., Jr., and Mislow, K., Ibid., 92, 5809 (1970).
15. Horner, L., Buthe, H., and Siegel, H., Tetrahedron Lett., 4023 (1968a).
16. Horner, L., Siegel, H., and Buthe, H., Angew. Chem. Int. Ed. Engl., 7, 942 (1968b).
17. Horner, L., and Siegel, H., Phosphorus, 1, 199 (1972); Chem. Abst., 76, 85238 a (1972) ; Phosphorus, 1, 209 (1972) Chem. Abst., 77, 48585m (1972) .
18. Izumi, Y., Angew. Chem. Int. Ed. Engl., 10, 871 (1971).
19. Kagan, H.B., and Dang, T.P., J. Amer. Chem. Soc., 94, 6429 (1972); German Pat., 2,161,200 (1972) Chem. Abst., 77, 114567k (1972) .
20. Kiso, Y., Yamamoto, K., Tamao, K., and Kumada, M., ibid., 94, 4373 (1972).
21. Kiso, Y., Tamao, K., Miyake, N., Yamamoto, K., and Kumada, M., Tetrahedron Lett., 3 (1974).
22. Knowles, W.S., and Sabacky, M.J., Chem. Comm., 1445 (1968).
23. Knowles, W.S., Sabacky, M.J., and Vineyard, B.D., Ann. N.Y. Acad. Sci., 172, 232 (1970); Chem. Eng. News, 48, (29), 41 (1970).
24. Knowles, W.S., Sabacky, M.J., and Vineyard, B.D., Chem. Comm. 10 (1972a).

25. Idem, Chem. Eng. News, 50, (6), 4 (1972b).
26. Idem., Chem. Tech., 591 (1972c).
27. Knowles, W.S., and Sabacky, M.J., German Pat. 2,123,063 (1971) Chem Abst., 76, p60074 f (1972d) .
28. Knowles, W.S., Sabacky, M.J., and Vineyard, B.D., German Pat. 2,210,938 (1972) Chem. Abst., 77, 165073d (1972e) .
29. Idem., Ann. N.Y. Acad. Sci., 214, 119 (1973).
30. Knowles, W.S., Sabacky, M.J., and Vineyard, B.D., Adv. in Chemistry Series, No. 132, Homogeneous Catalysis-II, Amer. Chem. Soc., (1974).
31. Korpium, O., and Mislow, K., J. Amer. Chem. Soc., 89, 4784 (1967).
32. Korpium, O., Lewis, R.A., Chickos, J., and Mislow, K., Ibid., 90, 4843 (1968).
33. Langlois, N., Dang, T.P., and Kagan, H.B., Tetrahedron Lett., 4865 (1973).
34. Masler, W.F., Ph.D. Thesis, Univ. of N.H., 1974.
35. Morrison, J.D., and Burnett, R.E., Abstracts of Papers, 159th Natl. Mtg. of the American Chemical Society, Houston, Texas, Feb. 1970, NO. ORGN 85.
36. Morrison, J.D., and Mosher, H.S., Asymmetric Organic Reactions, Prentice-Hall, Englewood Cliffs, N.J., 1971.
37. Morrison, J.D., Burnett, R.E., Aguiar, A.M., Morrow, C.J., and Phillips, C., J. Amer. Chem. Soc., 93, 1301 (1971).
38. Morrison, J.D., Masler, W.F., J. Org. Chem., 39, 270 (1974).
39. Morrison, J.D., Masler, W.F., and Neuberg, M.K., Asymmetric Homogeneous Hydrogenation, in Advances in Catalysis, Vol. 25, Academic Press (In Press) (1975).
40. Nudelman, A., and Cram, D.J., J. Amer. Chem. Soc., 90, 3829 (1968).
41. Ogata, I., and Ikeda, Y., Chemistry Letters, 487 (1972).
42. Pino et al., Angew. Chem. Internat. Edit., 12, 669 (1973).
43. Poulin, J.C., Dumont, W., Dang, T.P., and Kagan, H.B., C.R. Hebd. Seances, Acad. Sci. Ser. C Sci. Chim., 277, 41 (1973).
44. Salomon, C., Cosiglio, G., Botteghi, C., and Pino, P., Chimia, 27, 215 (1973).
45. Scott, J.W., and Valentine, D., Jr., Science, 184, 943 (1974).
46. Stern, R., Hirschauer, A., and Sajus, L., Tetrahedron Lett., 3247 (1973).
47. Tanaka, M., Watenabe, Y., Mitsudo, T., Yamamoto, K., and Takegami, Y., Chemistry Letters, 483 (1972).
48. Yamamoto, K., Hayashi, T., and Kumada, M., J. Amer. Chem. Soc., 93, 5301 (1971).
49. Yamamoto, K., Uramoto, Y., and Kumada, M., J. Organometal. Chem., 31, C9 (1971).

50. Yamamoto, K., Hayashi, T., and Kumada, M., ibid., 46, C65 (1972).
51. Idem. ibid., 54, C45 (1973).

STEREOCHEMICAL ASPECTS OF METAL-CATALYZED FORMATION AND HYDROGENOLYSIS OF EPOXY ALCOHOLS

JAMES E. LYONS

Sun Oil Company

P.O. Box 1135, Marcus Hook, Pennsylvania 19061

ABSTRACT

The oxidation of cyclohexene in the presence of $[C_5H_5V(CO)_4]$ gives cis-1,2-epoxycyclohexane-3-ol in 65% yield after 10% conversion of the olefin. Evidence is presented for the existence of a reaction pathway which involves the stereoselective epoxidation of 2-cyclohexene-1-ol by cyclohexenyl hydroperoxide. Both cis- and trans-1,2-epoxycyclohexane-3-ol are hydrogenolyzed over several transition metal catalysts in a regioselective and stereoselective manner to cis- and trans-1,2-cyclohexanediol respectively. The stereochemistry of the catalytic hydrogenolysis differs markedly from published accounts of $LiAlH_4$ reduction of these epoxy alcohols. Homogeneous catalytic hydrogen transfer in the presence of Rh(I) and Ru(II) complexes resulted in geometrical isomerization of cis-1,2-epoxycyclohexane-3-ol to the trans-isomer. Palladium catalysts disproportionate cis-1,2-epoxycyclohexane-3-ol in the absence of hydrogen to catechol and predominately 1,2-cyclohexanediol.

INTRODUCTION

The oxidation of olefins in the presence of catalytic quantities of transition metal complexes usually gives mixtures of allylic hydroperoxides, allylic alcohols, α,β-unsaturated ketones, epoxides, peroxide dimers, polymers and various products of oxidative cleavage of the double bond (for reviews of this subject see: Lyons, 1974a; Saito, 1968; and Yasui, 1969). Allison et al (1966) reported that good yields of a class of unusual oxidation products, 1,2-epoxy - 3-alcohols, are formed when complexes of vanadium(III) are present during autoxidation of several substituted olefins. These authors recognized the potential of this reaction for making precursors of useful diols and triols, and investigated the formation and reactions of epoxy alcohols in some detail. An allylic hydroperoxide was identified as a reaction intermediate and Allison et al (1966) concluded that the epoxy alcohol arose via a vanadium catalyzed rearrangement of this species. The stereochemistry of the oxidation of 4-methyl-2-pentene to four stereoisomeric epoxy alcohols was shown to proceed through two isomeric allylic hydroperoxides (Allison et al, 1966).

In an effort to further probe this interesting reaction in simpler systems, we decided to study oxidations in which a single epoxy alcohol might be formed from one allylic hydroperoxide precursor. We examined the oxidation of two symmetrical olefins, tetramethylethylene, TME, and cyclohexene in the presence of a vanadium (I) complex $[C_5H_5V(CO)_4]$, which was found to be both an efficient initiator of oxidation and a catalyst for epoxy alcohol formation (Lyons, 1973, 1974a,b). The oxidation of TME was found to proceed almost completely through a single allylic hydroperoxide, I, to give only one epoxy alcohol, II, eq. 1 (Lyons and Turner, 1972; Lyons, 1973, 1974a). The vanadium(I) - catalyzed oxidation of cyclohexene also gives a single allylic hydroperoxide, (III), from which cis-1,2-epoxycyclohexane - 3-ol, IV, is formed stereoselectively (Lyons, 1973, 1974b), eq. 2. Since: (a) allylic alcohols are epoxidized up

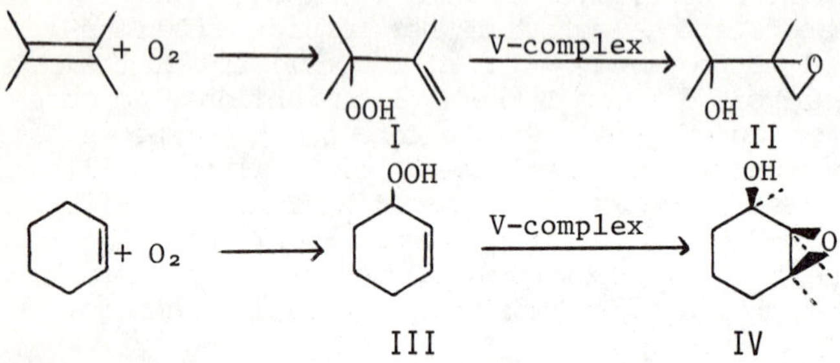

(1)

(2)

to 100 x faster than their olefin precursors in these systems (Lyons, 1974a; Sheng and Zajacek, 1970); (b) the conversion of allylic hydroperoxides I and III to the corresponding epoxy alcohols in the presence of allylic alcohols V and VI, proceeds via intermolecular epoxidation, eq. 3,4, rather than intramolecular rearrangement, and (c) the allylic alcohols V and VI are present in significant quantities during oxidation; this suggested (Lyons 1973, 1974a), that the major pathway for epoxy alcohol formation is an intermolecular epoxidation involving oxidation intermediates rather than an intramolecular rearrangement of allylic hydroperoxide.

This paper considers the stereochemistry of this reaction and shows that the stereoselectivity which is observed for ole-

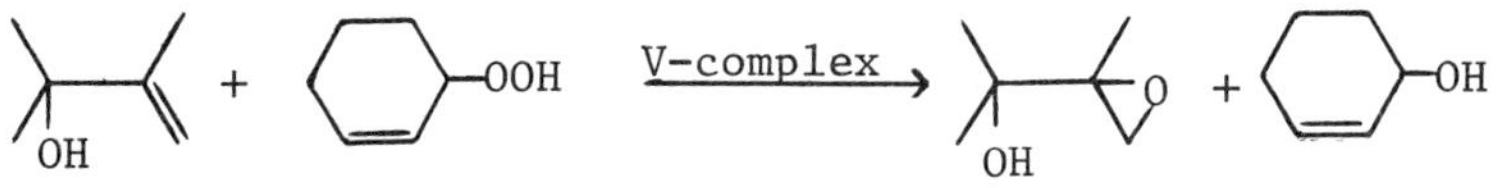

(3)

V III II VI

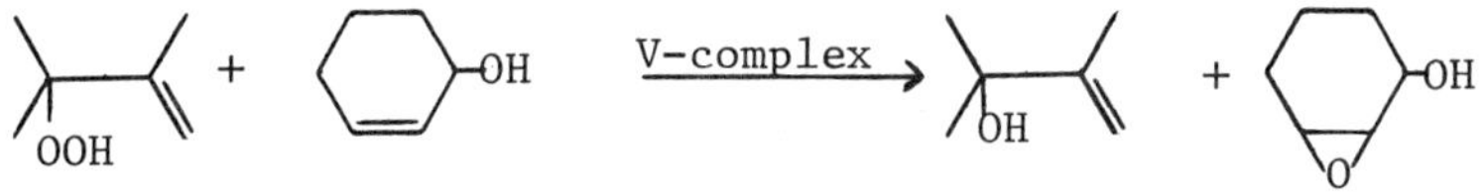

(4)

I VI V IV

fin oxidation is consistent with that of epoxidation of an allylic alcohol (Lyons, 1973; Sharpless and Michaelson, 1973). Mechanistic implications of the stereochemistry of epoxyalcohol formation during the oxidation of cyclohexene are discussed.

Catalytic hydrogen transfer reactions of the cis- and trans- 1,2-epoxycyclohexane - 3-ol are also reported. The heterogeneous catalytic hydrogenolysis of the isomeric epoxyalcohols was found to be both regioselective and stereoselective and exhibits marked stereochemical differences from reductions by lithium aluminum hydride (Hartman and Rickborn, 1972) in solution. Although homogeneous transition metal hydrogenation catalysts do not rapidly hydrogenolyze the cyclic epoxyalcohols, phosphine complexes of Ru(II) and Rh(I) smoothly equilibrate the cis- and trans-isomers. Regioselective dehydrogenation of the isomeric epoxyalcohols over several solid catalysts gives catechol as the predominant product.

RESULTS AND DISCUSSION

Stereoselective Cyclohexene Oxidation

The usual products of the oxidation of cyclohexene in the presence of transition metal complexes are mixtures of cyclohexenyl hydroperoxide, 2-cyclohexene-1-ol, 2-cyclohexene-1-one, cyclohexene oxide, dimers and polymers (Arzoumanian et al, 1974 a,b; Kaneda et al, 1973; Fusi et al, 1971, 1974; Gould and Rado, 1969; Kurkov et al, 1968; Collman et al, 1967). Allison et al, 1966, have reported that 1,2-epoxycyclohexane-3-ol was formed in 24% yield at 15% conversion of cyclohexene during oxidation in the presence of vanadium naphthenate; however, the geometrical configuration of the product was not reported in this case.

We have found that the complex, $[C_5H_5V(CO)_4]$, (C_5H_5 = cyclopentadienyl) is an efficient catalyst for the stereoselective oxidation of cyclohexene to cis-1,2-epoxycyclohexane-3-ol,

IV, in good yield. The cis-epoxy alcohol, IV, accounts for 65% of the products of reaction after 10% of the cyclohexene has reacted and 55% after 30% of the cyclohexene has been oxidized (Table I). The stereoselectivity of this reaction is ~99%.

TABLE I. *cis-1,2-Epoxycyclohexane-3-ol Via Oxidation of Cyclohexene*[a] *in the Presence of Cyclopentadienylmetalcarbonyl Complexes,* $[C_5H_5M(CO)_x]_y$

Metal Complex	Solvent	Reaction Time, hrs.	Conversion of Cyclohexane, %	Yield[b] of IV, %
$[C_5H_5V(CO)_4]$	CH_2ClCH_2Cl[c]	3	10	65
		5	30	55
$[C_5H_5V(CO)_4]$	--	3	11	51
		5	20	47
		9	48	42
$[C_5H_5Fe(CO)_2]_2$	--	5	20	--
$[C_5H_5Mo(CO)_3]_2$	--	5	24	1

[a]Oxygen was bubbled, 2.0 l/hr., through 80 ml of cyclohexene containing 0.25 gram of the catalyst with stirring at 65°C. Products were vacuum transferred at 65°C and 0.01 mm from the catalyst and residue and analyzed by glpc. [b]Yield based on cyclohexene which has reacted. For complete product analyses see Lyons, 1974a. [c]Oxygen was bubbled, 0.5 l/hr., through 6 ml of 1,2-dichloroethane and 6 ml cyclohexene containing 0.05 gram of the catalyst at 65°C. Analytical procedure identical to (a).

The formation of IV as the predominant product of cyclohexene oxidation contrasts sharply with the more conventional results obtained using molybdenum and iron complexes having a similar ligand system. These complexes formed the usual products (Lyons, 1974a) when used as catalysts for cyclohexene oxidation.

The low-valent vanadium complex $[C_5H_5V(CO)_4]$, also gives superior yields of III compared with $VO(acac)_2$, $V(acac)_3$ and vanadium naphthenate, Allison et al, 1966. Furthermore, cyclohexene oxidations run in the presence of $[C_5H_5V(CO)_4]$ do not exhibit the long induction periods which occur when vanadium (III) and vanadium(IV) complexes are used in the absence of radical initiators. Evidence has been presented (Lyons, 1974a) which suggests that the function of the vanadium(I) complex is to facilitate the decomposition of traces of hydroperoxides into free radical species which initiate olefin autoxidation. In so doing the low valent vanadium complex may be oxidized to a V(III) or a V(IV) complex which can efficiently catalyze the conversion of the primary oxidation product--an allylic hydroperoxide--into an epoxy alcohol. Indeed, when $[C_5H_5V(CO)_4]$ solutions are oxidized with t-butyl hydroperoxide prior to use, the resulting vanadium complex smoothly catalyzes reaction of the allylic hydroperoxide, I, to the epoxy alcohol, II (Lyons, 1974a). If the vanadium(I) complex is not oxidized prior to addition of I, rapid decomposition of I occurs during the initial stages of reaction and low yields of epoxy alcohols are formed.

Prior work (Lyons, 1973, 1974a) had established that formation of cis-1,2-epoxycyclohexane-3-ol from the oxidation of cyclohexene in the presence of vanadium complexes was most probably the result of the epoxidation of 2-cyclohexene-1-ol by cyclohexenyl hydroperoxide. We therefore compared the products of reaction of a cyclohexene solution of cyclohexenyl hydroperoxide and 2-cyclohexene-1-ol catalyzed by the oxidized vanadium complex, with those of cyclohexene oxidation. We found that IV was formed smoothly and stereoselectivity, and little or no epoxidation of cyclohexene occurred, eq. 5. It should be noted that this reaction, eq. 5, does not result in the depletion of the allylic alcohol, VI, which is present at any point during oxidation.

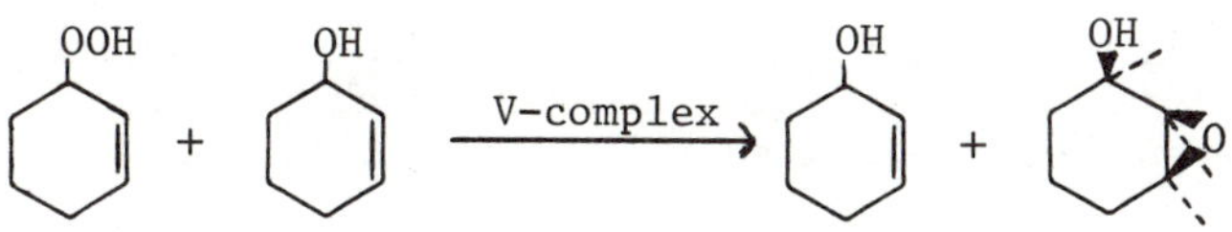

III VI VI IV (5)

The remarkable ability of vanadium to catalyze the stereospecific oxidation of cyclohexene to cis-1,2-epoxycyclohexane-3-ol was not matched by a similar molybdenum complex $[C_5H_5Mo(CO)_3]_2$. Instead, the molybdenum complex catalyzed the oxidation of cyclohexene to a mixture of 2-cyclohexene-1-ol and cyclohexene oxide (Lyons, 1974a). Both reaction, however, had the same intermediate--cyclohexenyl hydroperoxide, III. The molybdenum complex catalyzed reaction of III via a completely different pathway, eq. 6. Clearly the molybdenum complex catalyzes preferential epoxidation of the unreacted olefin which is present in excess whereas the vanadium complex selects the small amount of allylic alcohol, VI, present in the mixture for preferential epoxidation to the epoxy alcohol. A study of the stereochemistry of the epoxidation of the allylic alcohol, VI,

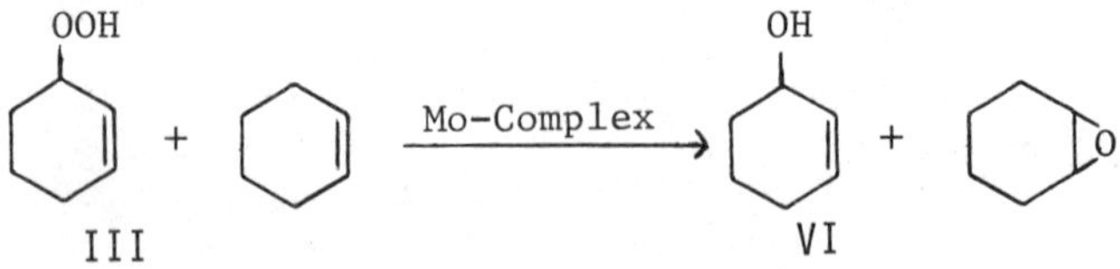

(6)

was undertaken in order to obtain additional insight into the differences in behavior of vanadium, molybdenum and tungsten catalysts.

Stereochemistry of the Epoxidation of 2-Cyclohexene-1-ol. A Comparison Between Vanadium, Molybdenum and Tungsten

The use of transition metal complexes as catalysts for the epoxidation of olefins by alkyl hydroperoxides is a useful synthetic procedure which has been extensively investigated during the past decade (reviewed by Metelitsa, 1972). Most of the olefins which have been studied are simple hydrocarbons. It has been shown (Sheng and Zajacek, 1970), however, that the presence of a hydroxyl group in the allylic position exerts a substantial influence on the rate of the metal-catalyzed epoxidation of an adjacent double bond. We have found that an allylic -OH group can direct the stereochemical course of epoxidation and that the extent to which this directive influence is operative depends on the metal complex which is used as a catalyst.

The oxygen transfer step in metal-catalyzed epoxidation is believed to involve the interaction between a hydroperoxide complex and the olefin, (Metelitsa, 1972), eq. 7. Sheng and

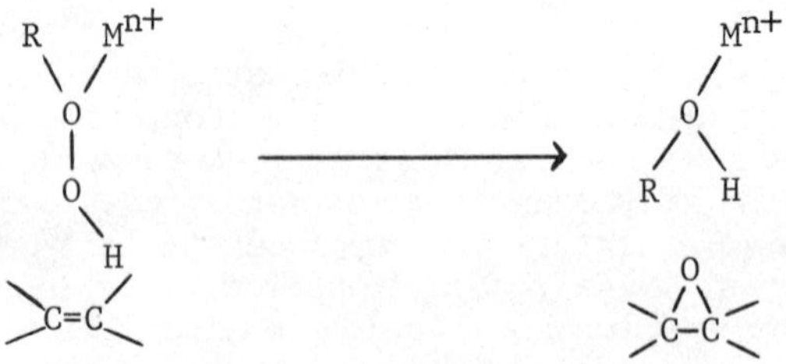

(7)

Zajacek, 1970, have reported that although molybdenum complexes are usually more efficient catalysts than vanadium complexes for olefin epoxidation, allylic alcohols are epoxidized more readily with vanadium complexes than with molybdenum. These authors also noted that while the presence of alcohols caused a marked retardation of the rate of vanadium-catalyzed epoxidation of unsubstituted olefins, the retardation is far less pronounced when molybdenum catalysts are used. It appeared that alcohols complexed much better with the vanadium catalysts than with molybdenum. It was concluded, therefore, that the allylic alcohol was brought into the coordination sphere of the vanadium by complexing with the hydroxyl group. This would place the coordinated hydroperoxide in the vicinity of the double bond, thus facilitating epoxidation, eq. 8.

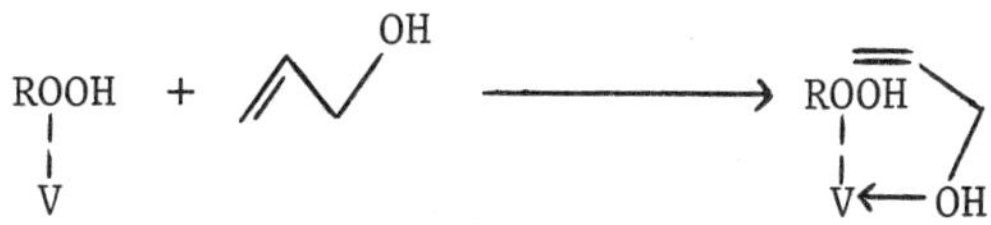

(8)

We have investigated the stereochemistry of the epoxidation of 2-cyclohexene-1-ol by tert-butyl hydroperoxide in the presence of complexes of vanadium, molybdenum and tungsten. The predominant product of reaction is cis-1,2-epoxycyclohexane-3-ol; however, the stereoselectivity of this reaction is dependent on the metal which is used, Table II. When vanadium complexes are used, (Lyons, 1973), the reaction is almost completely stereoselective, eq. 9, whereas complexes of molybdenum and tungsten produce both the cis- and trans-isomers in a ratio of 2/1, Table II. Furthermore, the rate of reaction is faster and the epoxy alcohol yield is higher in the vanadium-catalyzed reaction than with molybdenum complexes. Since both cis- and trans-1,2-epoxycyclohexane-3-ol completely retain their stereochemical integrity under reaction conditions, it would appear that the stereoselectivity of these reactions reflect mechanistic characteristics of the metal-catalyzed epoxidation process.

An independent study by Sharpless and Michaelson (1973) resulted in similar findings for the vanadium catalyzed epoxidation of 2-cyclohexene-1-ol. These authors showed that cis-epoxy alcohols were formed in the epoxidation of several allylic alcohols. Sharpless and Michaelson (1973), however, reported much higher stereoselectivity for the molybdenum catalyzed formation of cis-1,2-epoxycyclohexane-3-ol than we observed.

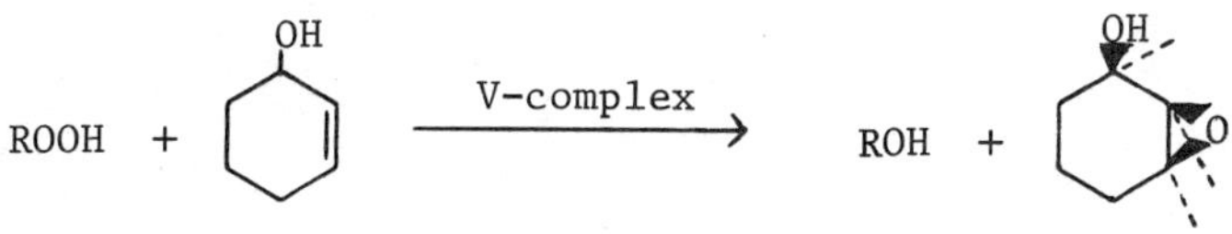

(9)

TABLE II. *Epoxidation of 2-Cyclohexene-1-ol by tert-Butyl Hydroperoxide*[a]

Catalyst	Reaction Time, hrs	Conversion of t-BuOOH, %	Yield of t-BuOH, %[b]	Conversion of ene-ol	Yield of Epoxyol, %[c] cis-	trans-	Ratio cis-/trans-
$VO(acac)_2$	2.5	83	98	69	99	1	99
$V(acac)_3$	2.0	100	89	70	96	1	99
$MoO_2(acac)_2$	2.0	63	76	45	40	19	2.1
$Mo(acac)_3$	3.0	75	61	49	39	21	1.9
$Mo(CO)_6$	1.5	52	87	36	64	29	2.2
$W(CO)_6$	9.5	45	82	40	54	23	2.3

[a]tert-Butyl hydroperoxide, 5.0 grams, was added to a solution of 0.10 gram of the catalyst in 6.0 grams of 2-cyclohexene-1-ol. The reaction mixture was stirred at 75°C for the designated time and analyzed by glpc. [b]Based on moles of t-BuOOH converted. [c]Based on moles of 2-cyclohexene-1-ol converted.

It should be pointed out, however, that their reaction conditions were considerably different and that these authors carried out the reactions to less than 5% conversion while our selectivities were determined at conversions from 45-100% of 2-cyclohexene-1-ol. Sheldon et al (1974) have shown that the nature of the catalyst changes considerably during the early stages of an epoxidation reaction.

The mechanism suggested by Sheng and Zajacek (1970) is consistent with our stereochemical findings using vanadium complexes. An intermediate in which oxygen enters from the side of the molecule to which the -OH group is attached will result in the formation of the cis-epoxy alcohol in a cyclic system, eq. 10. If there is no strong attraction between the metal center and the allylic-OH group, as may be the case in these systems with molybdenum and tungsten complexes, some trans-epoxy alcohol could be formed via eq. 11.

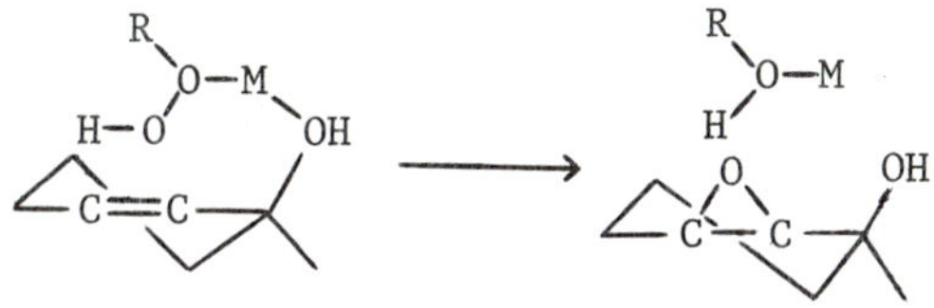

(10)

cis-

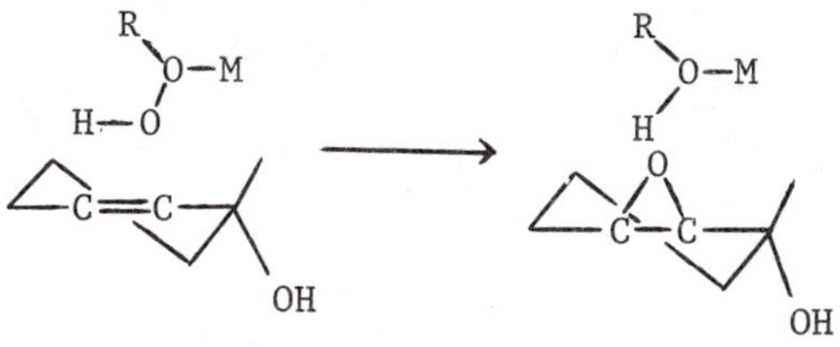

(11)

trans-

Alternative pathways could also account for cis-epoxy alcohol formation. Hydrogen bonding between the allylic-OH group and the epoxidizing agent has been postulated to explain the formation of cis-1,2-epoxycyclohexane-3-ol in the reaction of 2-cyclohexene-1-ol with peracids (Henbest and Wilson, 1957, 1966; Chamberlain et al, 1970). An analogous intermediate, VII, would also account for predominant formation of cis-epoxy alcohol but would not adequately explain the marked dependence of stereoselectivity on the metal center. Another alternative,

VIII, which might involve both hydrogen bonding between the allylic-OH group and coordinated hydroperoxide and metal participation in the epoxidation process is the 1,3-dipolar addition (Kaloustian et al, 1971; Kwart et al, 1967; Mimoun et al, 1970) of a metal hydroperoxide complex to the double bond. Such a pathway has been proposed for peracid epoxidation of olefins (Kwart et al, 1967) and for metal catalyzed epoxidation reactions (Kaloustian et al, 1971; Mimoun et al, 1970) and is consistent with the observed stereochemistry.

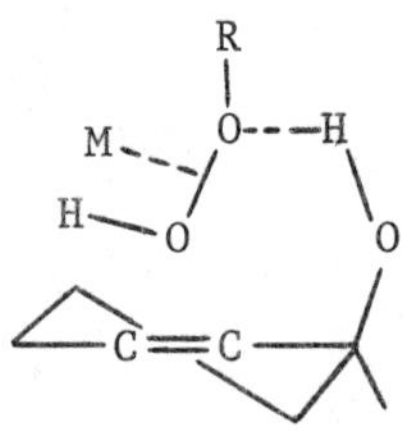

VII

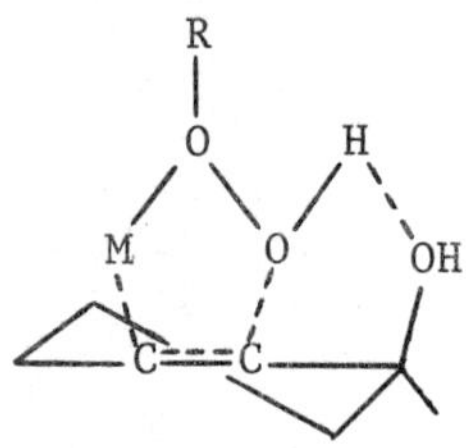

VIII

Stereochemistry of Transition Metal Catalyzed Hydrogenolysis of cis- and trans-1,2-Epoxycyclohexane-3-ol

The stereochemical course of metal hydride reduction of 3-oxygen substituted cyclohexene oxides has been the subject of extensive investigation (reviewed by Zable and Buchanan, 1972). The cleavage of the oxirane ring by catalytic hydrogenolysis has also been studied in considerable detail (reviewed by Cherniphkova and Mushenko, 1972), but few reports concerning the effect of 3-oxygen substituents on this reaction have appeared in the literature (Allison et al, 1966; Mitsui et al, 1965). We have examined the effect of a 3-hydroxyl group on the stereochemical course of the transition metal catalyzed hydrogenolysis of the oxirane ring in both cis- and trans-1,2-epoxycyclohexane-3-ol. Hydrogenolyses over several different hydrogenation catalysts are both stereoselective and regioselective. Although hydrogenolyses over platinum black were the most selective reactions, the overall stereochemical course of hydrogenolysis was independent of the metal catalyst which was used, eq. 12, 13 (Table III). Catalytic hydrogenolysis of the 3-hydroxycyclohexene oxides over platinum black, palladium black, Raney nickel and Raney copper gave predominantly 1,2-cyclohexanediols. In every instance the cis-epoxy alcohol yielded predominantly cis-1,2-diol and the trans-epoxy alcohol gave mainly trans-1,2-diol.

It is most interesting to note that the stereochemical course of the transition metal catalyzed hydrogenolysis of the

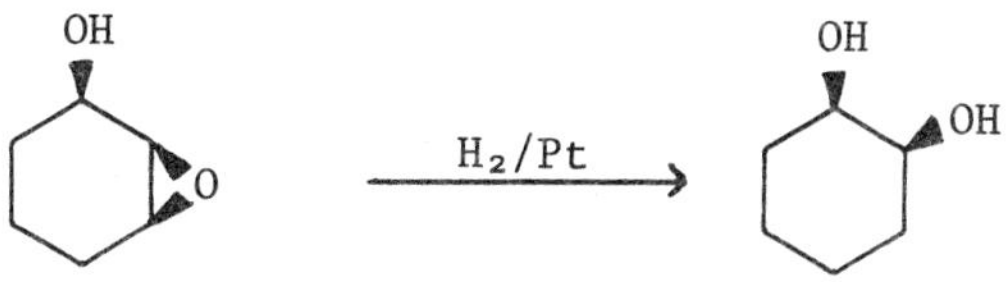

99%

(12)

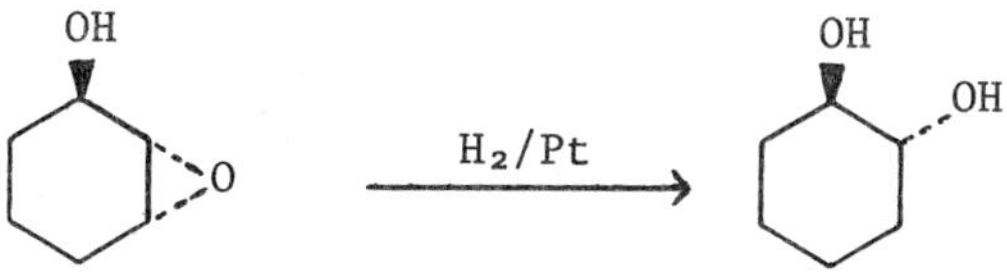

87%

(13)

TABLE III. *The Transition Metal Catalyzed Hydrogenolysis of cis- and trans-1,2-Epoxycyclohexane-3-ol*[a,b]

Substrate, 3-Hydroxy-cyclohexene oxide	Catalyst[c]	Reaction Time, Hrs.	Conversion %	Products, % cis-1,2-cyclohexanediol	trans-1,2-cyclohexanediol	1,3-cyclohexanediol (cis- & trans-)
cis-[a]	Pt-Black	2.0	100	99	trace	1
	Pd-Black	2.0	100	82	1	12
	Raney-Cu	2.0	69	41	2.5	trace
	Raney-Ni	1.0	100	44	12	6
	Raney-Ni	1.7	100	37	18	6
	Raney-Ni	2.0	100	26	21	7
trans-[b]	Pt-Black	2.0	100	9	87	2
	Pd-Black	2.0	100	10	68	16
	Raney-Cu	2.0	100	9	50	1
	Raney-Ni	1.0	100	15	51	11

[a]A 10% solution of the substrate in toluene, 5.0 ml, containing 0.1 g catalyst was stirred under 50 psi of hydrogen in a glass pressure vessel for the designated time. After filtration the clear, colorless product mixture was analyzed by glpc. [b]Same as (a) but P=100 psig H_2. [c]Pt and Pd: Engelhard. Raney catalysts: W. R. Grace and Company.

1,2-epoxycyclohexane-3-ols differs greatly from that of the metal hydride reduction of the same compounds. Hartman and Rickborn (1972) have found that reduction of cis- and trans-1,2-epoxycyclohexane-3-ol exhibited high stereoselectivity and, although regioselectivity was low, 1,2-cyclohexanediols did not always predominate but rather the site of epoxide cleavage was dependent upon the geometry of the reacting substrate, eq. 14, 15. These authors found that the preference for formation of 1,2-diol from the cis-epoxy alcohol and 1,3-diol from the trans-isomer was due to the conformer population of each starting material. A similar dependence upon the conformer population of the epoxy alcohol is clearly not the determining factor in the metal catalyzed hydrogenolysis of these epoxides; however, it is possible that the conformer distribution might be considerably altered at the reactive metal surface.

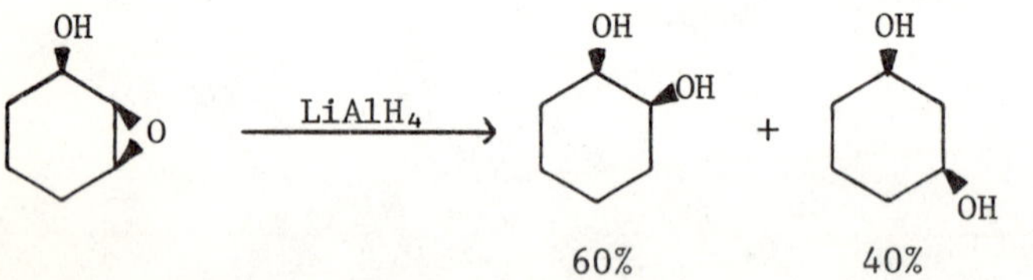

(14)

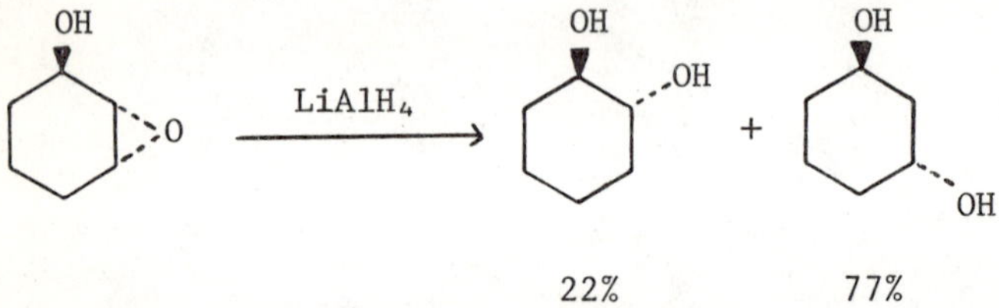

(15)

Perhaps more surprising than the independence of regioselectivity on the geometry of the substrate is the independence of the overall stereochemical pathway on the nature of the transition metal catalyst. For example, Mitsui et al (1965), reported different regioselectivity for the hydrogenolysis of a 3-keto epoxide using Pd/C as the catalyst from that using Raney Ni, eq. 16a,b. Moreover, in the hydrogenolysis of the C-O bond

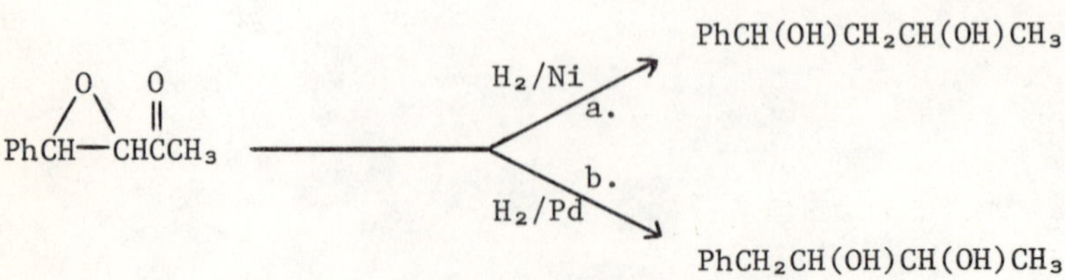

(16)

of optically active benzyl ethers, palladium was shown to catalyze cleavage with inversion of configuration and Raney-nickel gave retention. In contrast to these results we have found no difference in overall stereochemistry between the palladium catalyzed hydrogenolysis of 3-hydroxycyclohexene oxides and the nickel catalyzed reactions.

Hydrogenolyses over Raney nickel were the least stereoselective reactions of this series (Table III). This was largely due to facile metal catalyzed isomerization of cis- and trans-1,2-cyclohexanediol under reaction conditions. By contrast, both cis- and trans-1,2-cyclohexanediol retained their stereochemical integrity in the presence of Pt black under the same conditions. Furthermore, although geometrical isomerization occurred readily over Raney Ni, the 1,2-diol was not converted to the 1,3-isomer under the conditions of hydrogenolysis. In addition to the cyclohexanediols, varying amounts of phenol and cyclohexanol were also produced which in some cases lowered reaction selectivity so that combined yields of diols listed in Table III are often less than 100%.

In summary, the hydrogenolysis of cis- and trans-1,2-epoxycyclohexane-3-ol is both a stereoselective and a regioselective reaction. It does not, however, follow the cleavage pattern for $LiAlH_4$ reduction nor does it respond greatly to differences in the nature of the metal catalyst. Since we are not operating under conditions of equilibrium between cis- and trans-1,2- or 1,3-cyclohexanediols, the observed stereoselectivity and regioselectivity reflect mechanistic characteristics of the metal catalyzed hydrogenolysis reaction. The differences in stereochemistry between $LiAlH_4$ reduction and the metal catalyzed reactions might be due either to differences in the chemical nature of the reactants or merely to geometrical constraints on reactions of surface metal hydrides which are not present in the case of $LiAlH_4$ solutions. For this reason it was of interest to determine the stereochemistry of the homogeneous transition metal catalyzed hydrogenation of these epoxy alcohols in which the reactive metal hydride intermediate was in solution. To date, however, we have not found a soluble transition metal complex which is capable of catalyzing hydrogenolysis of the epoxy alcohol under mild conditions of temperature and pressure. Nonetheless, facile geometrical isomerization of the cis- and trans-epoxy alcohols did occur in the presence of some metal complexes.

Geometrical Isomerization of 1,2-Epoxycyclohexane-3-ol Catalyzed by Rhodium and Ruthenium Complexes

We have found that the selective isomerization of cis-1,2-epoxycyclohexane-3-ol, eq. 17, proceeds smoothly to the trans-isomer in toluene solution at 110°C in the presence of

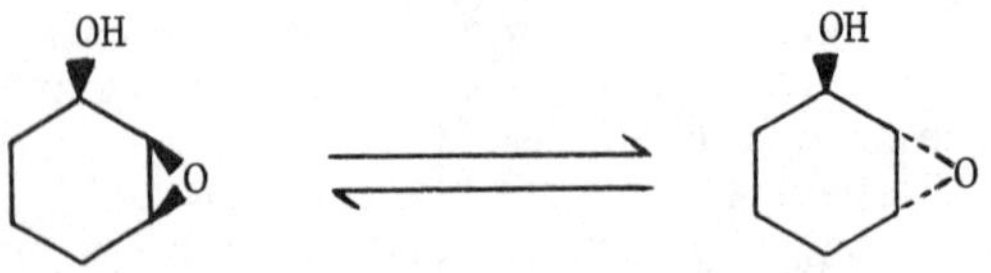

(17)

$RhH(CO)(Ph_3P)_3$, $RhCl(CO)(Ph_3P)_2$ and $RuCl_2(Ph_3P)_3$ (Table IV). It is likely that the oxirane ring remains intact during reaction and that exchange occurs by a metal catalyzed epimerization of the alcohol, eq. 18. Precedent for such a pathway has

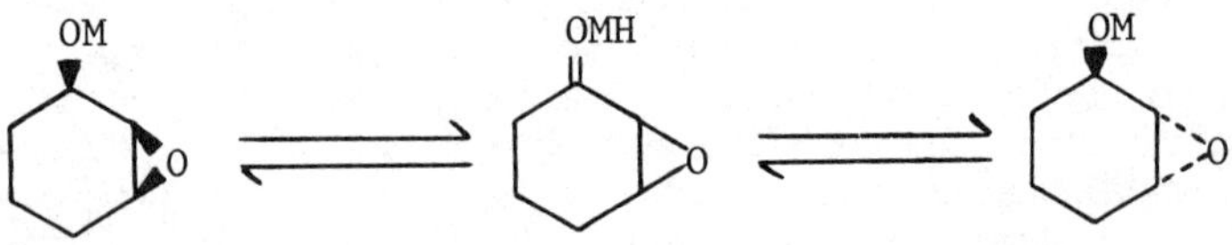

(18)

recently been established by Sasson and Blum (1974) in a report on H-D exchange and racemization of an optically active alcohol by $RuCl_2(Ph_3P)_3$. An alternative pathway involving epoxide ring cleavage does not appear to be as likely as hydroxyl group isomerization since we have found that cyclohexene oxides undergo little ring cleavage under our reaction conditions even under 100 psi H_2 in the presence of complexes listed in Table IV. Furthermore, the relatively few examples of epoxide ring cleavage which are known to be catalyzed by metal complexes result in irreversible formation of carbonyl compounds (Milstein et al, 1974). In addition, we found that cis-1,2-cyclohexanediol is

TABLE IV. *The Isomerization of cis-1,2-Epoxycylohexane-3-ol* [a]

Catalyst, mmoles	Reaction Time Hrs.	2-Cyclo-hexene-1-ol	Products, % cis-1,2-Epoxy cyclohexane-3-ol	trans-1,2-Epoxy cyclohexane-3-ol	1,2-Epoxycyclo-hexane-3-one
$RhH(CO)(Ph_3P)_3(0.25)$	4	---	72.2	24.5	2.2
$RhH(CO)(Ph_3P)_3(1.0)$	2	4.1	52.4	37.4	3.1
$RhH(CO)(Ph_3P)_3(1.5)$	0.5	3.9	46.7	41.2	3.3
$RuCl_2(Ph_3P)_3(0.5)$	1	0.2	62.4	28.4	4.1
	3	5.4	45.9	36.7	3.6
	4	6.3	45.4	37.0	2.9
$RhCl(CO)(Ph_3P)_2(1.0)$	3	1.0	77.5	16.7	0.8

[a] cis-1,2-epoxycyclohexane-3-ol, 44 mmoles, in 10.0 ml toluene was stirred at 110°C for the designated time with the catalyst and then analyzed by glpc. The identity of reaction products was determined by isolation of pure materials by preparative glpc and comparison of ir and nmr spectra with authentic compounds.

readily isomerized to trans-1,2-cyclohexanediol in toluene at 110°C. The formation of small amounts of 1,2-epoxycyclohexane-3-one, Table IV, is consistent with a reaction pathway involving β-hydrogen transfer from ring carbon to the metal, eq. 2.

Similar epimerization processes may be occurring at the metal surface resulting in loss of stereoselectivity in some instances. The selective geometrical isomerization of cyclic alcohols represents another of many recent examples of the utility of homogeneous transition metal catalyzed hydrogen transfer reactions of hydroxylic substrates (Eberhardt et al, 1972; Henbest and Trocha-Grimshaw, 1974; Imai et al, 1974; Regan, 1974; Sasson et al, 1974).

The Regioselective Dehydrogenation of 1,2-Epoxycyclohexane-3-ol to Catechol

Hydrogen transfer reactions carried out over palladium catalysts in the absence of hydrogen result in the formation of catechol and cis-1,2-cyclohexanediol from 1,2-epoxycyclohexane-3-ol, eq. 19 (Table V). Little or no resorcinol is formed in the aromatization reaction, again demonstrating the regioselectivity of surface catalyzed reactions of this epoxy alcohol.

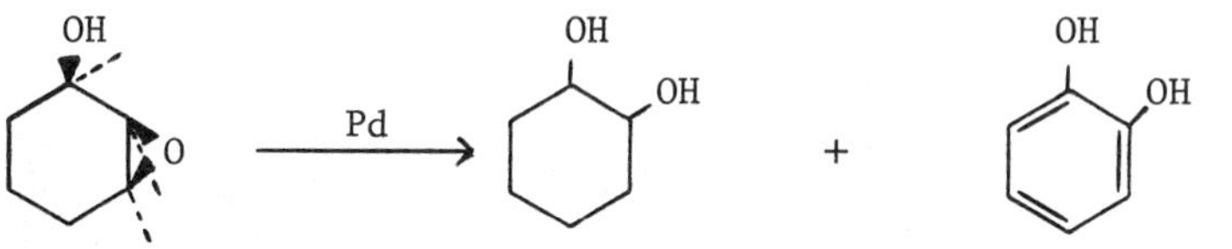

(19)

CONCLUSIONS

The vanadium catalyzed oxidation of cyclohexane to cis-1,2-epoxycyclohexane-3-ol most probably occurs via the stereoselective epoxidation of a reaction intermediate by cyclohexenyl hydroperoxide. The great difference in the product profile of this reaction and that of oxidations carried out in the presence of molybdenum or tungsten epoxidation catalysts appears to be due to the pronounced tendency of the vanadium complex to catalyze the selective epoxidation of an allylic alcohol formed in situ, in preference to the olefinic starting material. The similarity between the stereochemical course of the vanadium catalyzed oxidation of cyclohexene and the epoxidation of 2-cyclohexene-1-ol supports this hypothesis.

TABLE V. *The Liquid Phase Catalytic Disproportionation of cis-1,2-Epoxycyclohexane-3-ol*[a]

Catalyst	Reaction Time Hrs.	Solvent	Conversion, %	Products, %				
				Products of Dehydration[b]	1,2-cyclohexane diol	Catechol	1,3-cyclohexanediol	Others[c]
Pd-Black	0.5	None	100	11	50	13	10	16
Pd-Black	1.0	None	100	9	43	20	11	17
Pd-Black	4.0	None	100	6	37	23	10	24
Pd-Black[d]	1.0	phenyldodecane	70	1	31	6	7	26
	3.0	phenyldodecane	97	5	41	8	3	40
5%Pd-Al_2O_3	2.0	None	83	22	19	12	2	28

[a] cis-1,2-cyclohexane-3-ol, 2.0 g, was stirred at 200±1°C in the presence of the catalyst, 0.20 g, for the designated time under nitrogen. The solution was filtered hot then analyzed by glpc. [b] Cyclohexanol + Cyclohexanone + Cyclohexene-ol + Cyclohexene-one + phenol. [c] No resorcinol found among other by-products. [d] Same as (a) except run in 4.0 ml phenyldodecane as solvent.

Hydrogen transfer reactions over heterogeneous catalysts result in the preferential cleavage of both cis- and trans-1,-2-epoxycyclohexane-3-ol to give 1,2-diols. It is not clear whether this regioselectivity is due to constraints of the metal surface during hydrogen transfer or to other factors such as the formation of stable chelates to surface sites by the 1,2-diols. Attempted catalytic hydrogenolysis of the cis- and trans-epoxy alcohols using transition metal complexes in solution was unsuccessful, but in $LiAlH_4$ reductions (Hartman and Rickborn, 1972) only the cis-isomer exhibits a preference for 1,2-diol formation.

The preparation of epoxy alcohols in high yield by the direct oxidation of readily available olefins is a unique and potentially useful reaction. An understanding of the nature of the catalytic formation and reactivity of these species should ultimately elucidate the scope and limitations of their synthetic applications.

EXPERIMENTAL PROCEDURES

Infrared spectra were determined using Perkin-Elmer infrared spectrophotometers models 137-B and 21. NMR spectra were run using Varian T-60 and A-60 spectrometers. Gas chromatographic analyses were done on a Hewlett-Packard model 5750B instrument. Fractional distillations were performed on a Nester-Faust 18-inch semimicro spinning band column equipped with a stainless steel band.

Materials

Cyclohexene of greater than 99% purity was obtained from Phillips, distilled under nitrogen and then passed through freshly activated silica gel under nitrogen before use. 2-Cyclohexene-1-ol was obtained from Aldrich and distilled under nitrogen prior to use. Tert-butyl hydroperoxide (92% by titration) was purchased from Lucidol. Oxygen cylinders were obtained from Linde Air Products Company. The complexes: $[C_5H_5V(CO)_4]$, $[C_5H_5Mo(CO)_3]_2$, $[C_5H_5Fe(CO)_2]_2$, $[VO(acac)_2]$, $[V(acac)_3]$, $[MoO_2(acac)_2]$, $[Mo(acac)_3]$, $[Mo(CO)_6]$, and $[W(CO)_6]$ were obtained from Strem Chemical Company. Platinum and palladium catalysts were obtained from Engelhard; Raney catalysts were purchased from W. R. Grace and Company.

Stereoselective Cyclohexene Oxidation Catalyzed by $[C_5H_5V(CO)_4]$

Reactions were run in the neat olefin or in 1:1 mixtures of cyclohexene and 1,2-dichloroethane. In the first case, oxygen were bubbled, 2.0 l/hr., through 80 ml of cyclohexene containing 0.25 gram of the catalyst with stirring at 65°C. Solutions became pale amber and remained clear during reaction. Products were vacuum transferred at 65°C and 0.01 mm from the catalyst and residue and analyzed by glpc. Spinning band distillation of the vacuum transferred reactions products yielded IV, b.p. 96°/10mm, in 99.5% glpc purity which was identified by comparison of its ir and nmr spectra with literature data (Chamberlain et al, 1970). A small amount of trans-1,2-epoxycyclohexane-3-ol was separated from the cis-isomer by preparative glpc and identified by its ir and nmr spectra (Chamberlain et al, 1970). In the second case, oxygen was bubbled, 0.5 l/hr. through 6 ml of 1,2-dichloroethane and 6 ml of cyclohexene containing 0.05 gram of the catalyst at 65°C. Analytical procedure was identical to that used in the case of reactions run in the neat olefin.

Reaction of Cyclohexenyl Hydroperoxide with 2-Cyclohexene-1-ol

Cyclohexenyl hydroperoxide, 8.0 mmoles, in 10 ml of cyclohexene prepared by autoxidation of cyclohexene according to the method of Gould and Rado (1969), was added to 0.20 mmole of $[C_5H_5V(CO)_4]$ in 0.50 ml of 2-cyclohexene-1-ol, VI, and stirred for 3 hours at 70°C. Then the reaction mixture was vacuum transferred and analyzed as in previous runs. Iodometric titration of the solution showed 3.9 mmoles of unreacted hydroperoxide while glpc analysis showed that 4.0 mmoles of IV was produced.

Metal-Catalyzed Reactions of 2-Cyclohexene-1-ol with tert-Butyl Hydroperoxide

tert-Butyl hydroperoxide, 2.5 grams (0.028 mole), was added to 0.10 gram of each of the complexes listed in Table II in 6.0 grams (0.061 mole) of 2-cyclohexene-1-ol. This mixture was stirred under nitrogen at 75°C for 30 minutes after which another 2.5 grams (0.028 mole) of tert-butyl hydroperoxide were added. The reaction mixture was stirred at 75°C for the time designated in the table after which it was immediately vacuum transferred (0.01 mm/80 C) from the catalyst and analyzed by glpc using a 6 ft x 0.125 inch column packed with 10% silicone UC-W98 on 80-100 mesh chromasorb W (Hewlett-Packard). Benzene was used as an internal standard and response factors were determined by analyses of known standards containing the pure compounds.

Preparation of cis-1,2-Epoxycyclohexane-3-ol

Pure cis-1,2-epoxycyclohexane-3-ol was obtained from spinning band distillation of a vanadylacetylacetonate catalyzed reaction ten times the size of those described above. Careful distillation yielded 24.8 grams of cis-1,2-epoxycyclohexane-3-ol having a glpc purity of 98%. Redistillation of fractions rich in the cis-epoxyol gave 7.4 grams of product bringing the total recovery of cis-1,2-epoxycyclohexane-3-ol to 32.2 grams (58% yield based on starting tert-butyl hydroperoxide). The fraction boiling at 96°C/10 mm, glpc purity = 99.5% was shown to be the cis-isomer by a comparison of its ir and nmr [τ7.9-9.1 (6H, complex m), 6.8 (2H, s) and 6.1 (1H)] spectra with literature data (Chamberlain et al, 1970). The only impurity in this sample, 0.5%, was trans-1,2-epoxycyclohexane-3-ol. The cis-isomer, when obtained in 99% purity, crystallized at room temperature as white needles, m.p. 33-35°C.

Preparation of trans-1,2-Epoxycyclohexane-3-ol

The vacuum transferred reaction mixtures from several $Mo(CO)_6$-catalyzed epoxidation reactions were combined and 1.9 grams of trans-1,2-epoxycyclohexane-3-ol, glpc purity = 99.0%, was isolated by preparative glpc using a 20 ft x 0.25 inch column packed with 5% SP-2250 silicone on chromasorb G (Varian). The product was identified by a comparison of its ir and nmr [($CDCl_3$)τ7.9-9.0 (6H, complex m), 6.9 (1H, δ, J 4.0 Hz), 6.8 (1H, fine str), 6.0 (1H, m), 7.5 (1H, s, OH, conc. var.)] spectra with literature data (Chamberlain et al, 1970).

Catalytic Hydrogenolysis of cis- and trans-1,2-Epoxycyclohexane-3-ol

A 10% solution of the epoxy alcohol in toluene, 5.0 ml, containing 0.1 g catalyst was stirred under 50 psi of hydrogen in a glass pressure vessel for the designated time. After filtration the clear, colorless product mixture was analyzed by glpc. Identification of the isomeric diols was made by comparison of glpc retention times on a 10 ft x 1/8" column (Carbowax on 45/50 M Chromasorb W, base treated) with retention times of authentic pure isomers. Isolation of pure isomers was accomplished either by preparative glpc or by fractional crystallization of the pure isomer from reaction mixtures. Identification was made by comparison of ir and nmr spectra of isolated products with those of authentic compounds.

Geometrical Isomerization of cis-1,2-Epoxycyclohexane-3-ol

cis-1,2-Epoxycyclohexane-3-ol, 44 mmoles, in 10.0 ml toluene was stirred at 110°C with the homogeneous catalysts listed in Table IV for the designated time and then analyzed by glpc. The identity of reaction products was determined by isolation of pure materials by preparative glpc and comparison of ir and nmr spectra with authentic compounds.

Regioselective Dehydrogenation of cis-1,2-Epoxycyclohexane-3-ol

cis-1,2-Epoxycyclohexane-3-ol was stirred at 200±1°C in the presence of the catalyst, 0.20 g, listed in Table V, for the designated time under nitrogen. The solution was filtered hot, then analyzed by glpc. Yield of the major products are listed in Table V. In addition to those products listed, small amounts of cyclohexane diones and hydroxycyclohexanones were found but no resorcinol was detected.

ACKNOWLEDGEMENT

The author thanks Caroline Link and Martha Hibberd for technical assistance, and Robert Warren and Ronald Bingeman for assistance in the isolation and identification of products.

REFERENCES

1. Allison, K., Johnson, P., Foster, G., and Sparke, M., Ind. Eng. Chem. Prod. Res. Develop., 5, 166 (1966).
2. Arzoumanian, H., Blanc, A., Hartig, U., and Metzger, J., Tetrahedron Letters, 1011 (1974a).
3. Arzoumanian, H., Blanc, A., Metzger, J., and Vincent, J.E., J. Organometal. Chem., 82, 261 (1974b).
4. Chamberlain, P., Roberts, M., and Whitham, G. H., J. Chem. Soc. (B), 1374 (1970).
5. Cherniphkova, F., and Mushenko, D., Khim. Prom., 48, 249 (1972).
6. Collman, J., Kubota, M., and Hosking, J., J. Amer. Chem. Soc., 89, 4811 (1967).
7. Eberhardt, G., Tadros, M., and Vaska, Chem. Commun., 290, (1972).
8. Fusi, A., Ugo, R., Fox, F., Pasini, A., and Cenini, S., J. Organometal. Chem., 26, 417 (1971).
9. Fusi, A., Ugo, R., and Sanderighi, G., J. Catal., 34, 175 (1974).
10. Gould, E., and Rado, M., J. Catal., 13, 238 (1969).
11. Hartman, B., and Rickborn, B., J. Org. Chem., 37, 4246 (1972).
12. Henbest, H., and Trocha-Grimshaw, J., J. Chem. Soc. Perkin I, 601 (1974).
13. Henbest, H., Wilson, R., Chem. and Ind., 659 (1966).
14. Henbest, H., Wilson, R., J. Chem. Soc., 1958 (1957).
15. Imai, H., Nichiguchi, T., and Fukuzumi, K., J. Org. Chem., 39, 1622 (1974).
16. Kaloustian, J., Lena, L., and Metzger, J., Bull. Soc. Chim. Fr., 4415 (1971).
17. Kaneda, K., Itoh, T., Fugiwara, Y., and Teranishi, S., Bull. Chem. Soc. Japan, 46, 3810 (1973).

18. Kurkov, V., Pasky, J., and Lavigne, J., J. Amer. Chem. Soc., 90, 4744 (1968).
19. Kwart, H., Starcher, P., and Tinsley, S., Chem. Commun., 335 (1967).
20. Lyons, J. E., Advan. Chem. Ser., 132, 64 (1974a).
21. Lyons, J. E., Tetrahedron Letters, 2737 (1974b).
22. Lyons, J. E., Homogeneous Catalysis - II, Joint Symposium of the Division of Industrial and Engineering Chemistry and the Division of Petroleum Chemistry, 166th Meeting, ACS, Chicago, Ill., August (1973).
23. Lyons, J. E., and Turner, J., J. Org. Chem., 37, 2881 (1972).
24. Metelitsa, D., Russian Chemical Reviews, 41, 807 (1972).
25. Milstein, D., Buchman, O., and Blum, J., Tetrahedron Letters, 2257 (1974).
26. Mimoun, H., Serre de Roch, I., and Sajus, L., Tetrahedron Letters, 37 (1970).
27. Mitsui, S., Senda, Y., Shimodaira, J., and Ichikawa, H., Bull. Chem. Soc. Japan, 38, 1897 (1965).
28. Regen, S. L., J. Org. Chem., 39, 260 (1974).
29. Saito, Y., Yuki Gosei Kagaku Kyokai Shi, 26, 943 (1968).
30. Sasson, Y., and Blum, J., Chem. Commun., 309 (1974).
31. Sasson, Y., Alkin, P., and Blum. J., Tetrahedron Letters, 833 (1974).
32. Sharpless, K., and Michaelson, R., J. Amer. Chem. Soc., 95, 6136 (1973).
33. Sheng, M., and Zajacek, J., J. Org. Chem., 76, 418 (1970).
34. Yasui, T., Kogyo Kagaku Zasshi, 72, 1615 (1969).
35. Zable, H., and Buchman, J., "Selective Organic Transformations", Vol. 2, B. S. Thyagarajan, Ed., Wiley-Interscience, New York, New York, 1972.

SOME ASPECTS OF NITROALKANE HYDROGENATION VIA HOMOGENEOUS CATALYSIS

JOHN F. KNIFTON

Beacon Research Laboratories, Texaco Inc.,
Beacon, New York 12508

ABSTRACT

The homogeneous catalytic hydrogenation of aliphatic nitrocompounds is described for the selective synthesis of various linear, cyclic, and substituted oximes and amines. Ruthenium complexes of tertiary Group VB donor ligands, such as $RuHCl(PPh_3)_3$, are efficient catalysts for the preparation of secondary alkyl primary amines under superatmospheric pressures of hydrogen. The effectiveness of Group IB metal salts for aliphatic oxime synthesis varies in the order: Cu(I)≈Cu(II)>Ag(I), the specific activity of each catalyst system being highly sensitive to the structure and basicity of the solvent and other ligands coordinated to the metal center. Evidence is presented for a mechanism of cyclohexanone oxime formation involving initial cuprous hydride formation by heterolytic H_2 splitting, followed by deoxygenation of the coordinated nitroalkane anion as a rate determining step. Qualitative data for the aliphatic amine synthesis are consistent with a somewhat similar mode of H_2 and substrate activation by the ruthenium catalysts.

INTRODUCTION

While a variety of homogeneous catalytic techniques presently exist for the synthesis of amines and other partially reduced species from aliphatic nitrocompounds, many of these methods have inherent limitations. Commonly used reagents for $C-NO_2$ reduction, such as Fe/HCl, Fe/CH_3COOH, Sn/HCl, oftentimes lead to Nef-type hydrolysis of the nitroalkane as a competing reaction, and the cleavage of other functional groups (Smith, 1966). Alternate catalysts may require stringent reaction conditions (e.g. $Co_2(CO)_8$, Murahashi and Horiie, 1960, and $Fe(CO)_5$, Kmiecik, 1965), have a history of erratic behavior (e.g. Zn/NH_4Cl, Neunhoeffer and Liebrich, 1938), or require essentially stoichiometric amounts of catalyst, as in the case of $LiAlH_4$. The purpose of these studies has been to develop novel homogeneous hydrogenation catalysts for the selective reduction of nitroaliphatic compounds, which show improved turnover numbers, and which are active in media that are non-acidic (thereby avoiding Nef-type hydrolysis as a competing reaction), non-aqueous (since many higher MW nitrocompounds are only marginally soluble in water) and preferably basic, thereby favoring the formation of the more reactive nitroalkane anion (Smith, 1966). Two homogeneous catalyst systems are described in this paper; both satisfy the above criteria. Group IB metal salts in alkylpolyamine solvents have been found to be excellent for the selective hydrogenation of nitroalkanes to oximes (eq 1). Here particular attention has been given to the synthesis of cyclohexanone oxime in view of its importance as a precursor of Nylon-6. Ligand-stabilized ruthenium and iron complexes such as dichlorotris(triphenylphosphine)ruthenium (II) are catalyst precursors for the hydrogenation of nitroalkanes to amines

$$\begin{matrix} R' \\ R \end{matrix}\!\!>CHNO_2 + H_2 \longrightarrow \begin{matrix} R' \\ R \end{matrix}\!\!>C = NOH + H_2O \qquad (1)$$

(eq 2) both in neutral and strongly basic media. Other nitrogen-containing functional groups may also be hydrogenated with these catalysts; they may also be used for the selective and sequential hydrogenation of nitroaromatics (Knifton and Suggitt, 1974).

$$\begin{matrix} R' \\ R \end{matrix}\!\!>CHNO_2 + 3H_2 \longrightarrow \begin{matrix} R' \\ R \end{matrix}\!\!>CHNH_2 + 2H_2O \qquad (2)$$

RESULTS

Synthesis of Aliphatic Oximes

The synthesis of alkyl oximes from nitroalkanes may be most conveniently carried out in alkylpolyamine solvents containing copper(I) salts under moderate pressures of hydrogen. Illustrated in Table I are typical syntheses from various primary and secondary nitroalkanes. While the technique most readily lends itself to the preparation of linear and cyclic aliphatic oximes, such as dodecanone oxime and cyclohexanone oxime (expt 3-5), certain substituted nitroalkanes, exemplified here by α-nitrotoluene and methyl 4-nitropentanoate, may also be selectively reduced to the corresponding oximes in modest yields (eq 3 and 4). The cyano group in α-(2-cyanoethyl)-nitro-

TABLE I. *Synthesis of Aliphatic Oximes from Nitroalkanes Catalyzed by Copper (I) Salts in Alkylpolyamine Solvents*[a]

Expt	Nitroalkane	Nitroalkane conv. (mole %)	Major Product[b]	Oxime Yield (mole %)[c]
1	1-Nitropropane	100	Propanal Oxime	61
2	2-Nitropropane	100	Acetone Oxime	75
3	Nitrocyclohexane	100	Cyclohexanone Oxime	93
4	Nitrododecanes[d]	100	Dodecanone Oximes	90
5	Nitrated n-Dodecane[e]	100	Dodecanone Oximes	80
6	α-Nitrotoluene	95	Benzaldoxime	69
7	Methyl 4-Nitropentanoate	100	Methyl 4-(hydroxyimino)-pentanoate	60
8	α-(2-Cyanoethyl)-nitrododecanes[d]	95	α-(3-Aminopropyl)-nitrosododecanes	12
9	2-Nitrocyclohexanone	100	f	--
10	3-Nitro-2-Pentanol	100	f	--
11	β-Nitrostyrene	100	g	--

[a]Experimental conditions, 2.0-10 mmole CuCl; 25-50 mmole RNO_2; 80-95°; 35 atm H_2. [b]Oximes identified by elemental analyses and comparison of melting points and spectral properties (Infrared, NMR) with those reported in the literature. [c]Based upon moles of nitroalkane charged. [d]An isomeric mixture of 2- through 6-nitrododecanes. [e]A mixture of 25.9% (v/v) nitrododecanes, 64.1% (v/v) n-dodecane and 10% other materials including dodecanones, prepared by liquid/vapor phase nitration of n-dodecane. [f]A mixture of products containing some oxime but with loss of carbonyl or hydroxyl functionality. [g]Oligomers.

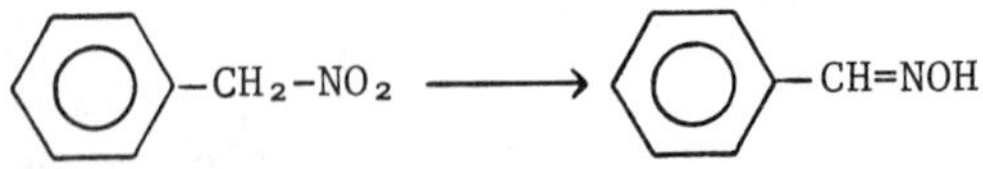

(3)

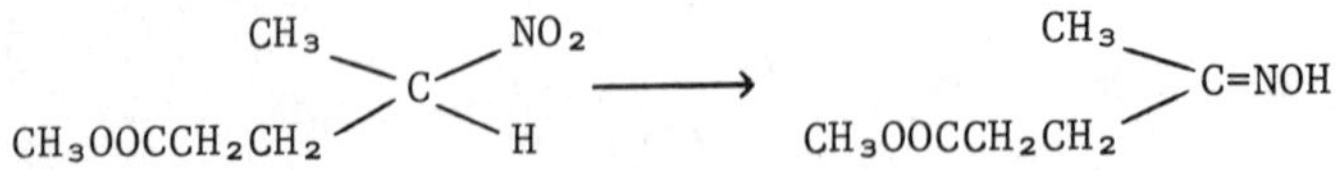

(4)

dodecanes undergoes at least partial hydrogenation to amine during concomitant reduction of the nitro group, and a mixture of products, including α-(3-aminopropyl)-nitrosododecanes in low yields (eq 5), are obtained. Other substituted nitroalkanes, notably those containing carbonyl, alcohol and olefinic linkages, suffer loss of functionality during hydrogenation in the copper(I)-amine media, due in part to solvent displacement of the functional group as in the case of 2-nitrocyclohexanone and 3-nitro-2-pentanol (expt 9 and 10) or because of competing oligomerization reactions, as in the case of β-nitrostyrene (expt 11). The aliphatic oximes may also be prepared in good yields from paraffin diluted nitroalkanes such as might be produced by paraffin nitration. This is illustrated in Table I for nitrated n-dodecane.

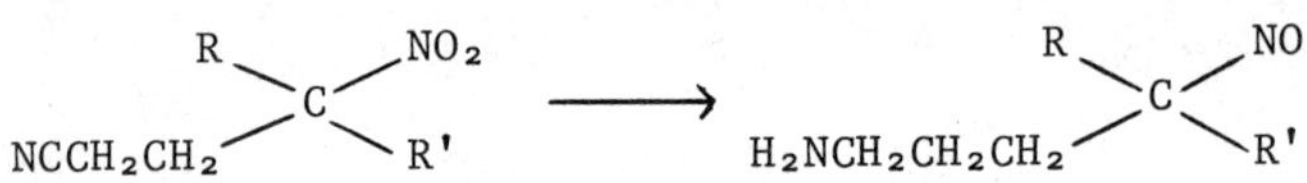

(5)

Of particular note is the truly catalytic character of this oxime synthesis. While initial nitrocyclohexane-to-copper (I) mole ratios normally range from 10 to 30, catalyst recycle has been demonstrated, and in a typical experimental series, a total of 0.4 mole of nitrocyclohexane, in four 0.1 mole batches, was hydrogenated to cyclohexanone oxime with a sample of copper (I) chloride (0.013 mole) in ethylenediamine. Both the catalytic activity, as measured by the rate of consumption of nitrocyclohexane (see Figure 1) and the yield of cyclohexanone oxime, isolated by solvent extraction, were maintained over the four cycles; the maximum productivity of the catalyst would certainly exceed this figure. The major by-product is water (eq 1) and there is no evidence for reduction of the copper salts to the metal, or hydrolysis to the oxide, except at high catalyst concentrations.

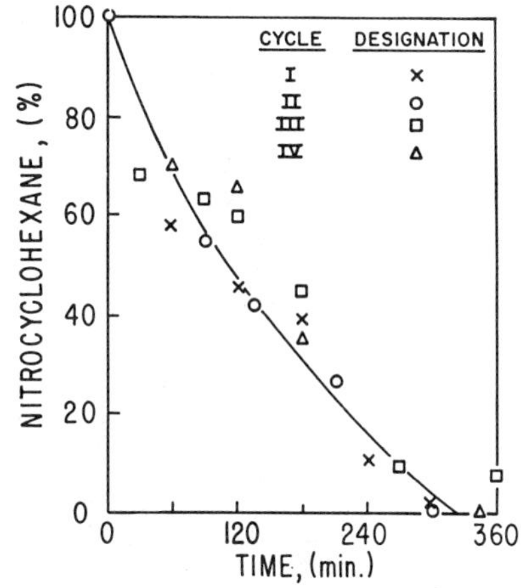

Fig. 1. Nitrocyclohexane hydrogenation, catalyst recycle study.

Catalyst Composition

The aldoxime and ketoxime syntheses shown in Table I were each carried out using copper(I) salts in alkylpolyamine as catalyst, primarily solutions of copper(I) chloride in ethylenediamine. Cyclohexanone oxime synthesis may also be demonstrated with solutions of silver(I) and copper(II) salts (see Table II). The slower rates of reduction, and low oxime yields, obtained with silver(I) acetate and nitrate (expt 15-17) are due primarily to accompanying reduction of the silver ions to the metal, which can only be avoided by operating at much lower

TABLE II. *Cyclohexanone Oxime Synthesis Catalyzed by Group IB and IIB Metal Salts*[a]

Expt	Metal Salts	Nitrocyclohexane Conversion (%)	Nitrocyclohexane Rate ($M \cdot hr^{-1}$)	Cyclohexanone Oxime Yield (mol %)[b]
12	Copper(I) Chloride	100	1.10	93
13	Copper(I) Acetate	100	0.41	80
14	Copper(II) Acetate Monohydrate	100	0.40	78
15	Silver(I) Acetate	8.9	≈0.01[c]	6.7
16	Silver(I) Nitrate	71	0.06	43
17	Silver(I) Nitrate[d]	<1	<0.01	Trace
18	Mercury(II) Chloride	None	--	--

[a]Experimental Conditions: 0.47 M [RNO_2]; 57 mM [Cu] or [Ag]; Solvent, Ethylenediamine; 95°; 50 atm H_2. [b]Cyclohexanone oxime yield based on nitrocyclohexane charged. [c]Extensive precipitation of silver metal during this run. [d]Reaction Temperature 22°.

temperatures (≈22°) where oxime formation is extremely slow.

Rates of nitrocyclohexane reduction for comparable solutions of copper(I) and copper(II) acetates in ethylenediamine are essentially equivalent (expt 12, 14). This result is somewhat surprising since copper(I) salts in basic media are reported to be significantly more active for hydrogen activation than those of copper(II) (Halpern, 1965). Furthermore, reduction of copper(II) by hydrogen is itself copper(I) catalyzed in solvents like quinoline (Calvin and Wilmarth, 1956). These latter reactions are evidently not rate determining under the conditions of nitroalkane reduction described in Table II, utilizing ethylenediamine as solvent and superatmospheric pressures of hydrogen, since reduction rates with copper(I) and copper(II) acetate are so similar. The apparent order of activity for the Group IB and IIB metal salts, for constant anion, is then:

$$\mathrm{Cu(I)} \simeq \mathrm{Cu(II)} > \mathrm{Ag(I)} \gg \mathrm{Hg(II)} \qquad (6)$$

Selective reduction of nitroalkanes to oximes is normally favored by strongly basic reaction conditions (Knifton, 1973). In this work, solvent studies were concerned with the effect of the amine solvent structure upon the rate and yield of cyclohexanone oxime formation (see Table III). The trends are similar to those noted previously for the CO-reduction of nitroalkanes by copper(I)-amine solutions (Knifton, 1973), and for the hydrogenation of silver salts in amines (Halpern and Milne, 1960). Reaction rates, as measured by the formation of cyclohexanone oxime, are maximized with highly basic alkylpolyamine

TABLE III. *Cyclohexanone Oxime Synthesis in Various Amine Solvents*[a]

Expt	Solvent Composition	Solvent pK_a[b]	Nitrocyclohexane Conversion (%)	Yield of Isolated Cyclohexanone Oxime (mole %)[c]
19	Pyridine	5.45	<5	None
20	Triethanolamine	8.02	50	Trace
21	Morpholine	8.70	35	20
22	Diethanolamine	9.00	80	22
23	Diethylenetriamine	9.94	98	50
24	Ethylenediamine	10.18	100	78
25	*n*-Hexylamine	10.4	100	14
26	3,3'-Iminobispropylamine	10.65	100	74
27	Piperidine	11.28	100	20

[a]Typical run conditions: 0.1 *M* [CuCl], 1.0 *M* [$C_6H_{11}NO_2$], 95°, 50 atm H_2. [b]Data taken from "Stability Constants of Metal-Ion Complexes," Section II: Organic Ligands, Chem. Soc. Spec. Publ. No. 17, 1964, and Supplement No. 1, Spec. Publ. No. 25, 1971. [c]Cyclohexanone oxime yield based on nitrocyclohexane charged.

solvents such as ethylenediamine and 3,3'-iminobispropylamine (expt 23, 26), although the high solubility of the cyclohexanone oxime in certain of these solvents (e.g. diethylenetriamine, expt 24) precludes ease of separation of this oxime from the crude hydrogenation product. Less effective are highly basic monoamines like piperidine (expt 27), and other liquid amines of base strength less than about 9 pK_a units. An earlier more exhaustive, study of solvent effects for the Cu(I)-CO system (Knifton, 1973), served to demonstrate that here too catalyst activity is favored by alkylpolyamine solvents, with no nitroalkane reduction being detected with amine solvent of pK_a <9.2.

Also considered, briefly, was the effect upon the hydrogenation activity of the copper catalyst of certain Lewis bases capable of π-back bonding. Copper(I) is known to form series of stable complexes with both triaryl and trialkylphosphines (Booth, 1964), the trialkylphosphines, in particular, stabilizing copper(I) against oxidation to the +2 state (Axtell and Yoke, 1973). In a series of experiments, rates of nitrocyclohexane reduction were measured using standard catalyst solutions (0.1 *M* copper(I) chloride in ethylenediamine) containing added triphenylphosphine, triphenylphosphite and tri-n-butylphosphine (P:Cu=3.3). In all three cases the added ligands resulted in slower rates of reduction and lower oxime yields. The order of increasing inhibition was:

$$P(OC_6H_5)_3 < P(C_6H_5)_3 < P(\underline{n}\text{-}C_4H_9)_3 \qquad (7)$$

with the more weakly complexing triphenylphosphite showing the smallest effect.

Synthesis of Aliphatic Amines

Various soluble ruthenium and iron complexes with π-acceptor ligands have been found active for the hydrogenation of nitroalkanes to amines. Particular attention has been given to those complexes known to catalyze homogeneous hydrogenation reactions, and/or transformations of the C-NO_2 function. In Table IV are illustrated some typical syntheses of dodecylamines from 2-through 6-nitrododecanes, carried out in oxygen-free benzene, ethanol mixture. Generally ruthenium complexes such as dichlorotris(triphenylphosphine)ruthenium(II), proved to be superior catalyst precursors in this application (Knifton, 1975). With solutions of the $RuCl_2(PPh_3)_3$ complex, advantages of this technique over existing methods of reducing nitroalkanes *via* homogeneous catalysis (see for example Smith, 1966; Murahashi and Horiie, 1960; Kmiecik, 1965; and Alper, 1972) include the good (up to 88 mole %) yields of alkyl amine obtained, with

TABLE IV. *Synthesis of Dodecylamines from Nitrododecanes Catalyzed by Ruthenium and Iron Complexes*[a,b]

Expt	Complex	Added Base	Mole Ratio $C_{12}H_{25}NO_2$:Ru,Fe:Base			Press H_2 (atm)	$C_{12}H_{25}NH_2$ Yield (mole %)[c]	Relative Rate[d]
28	$RuCl_2(PPh_3)_3$	KOH	100	1	200	90	54	
29	"	"	10	1	20	"	88	1-1.5
30	"	"	3	1	6	"	81	
31	"	"	3	1	6	34	59	
32	"	"	3	1	6	1	<5	
33	"	"	10	1	20	0[e]	<1	
34	"	C_5H_5N	3	1	20	90	33	
35	"	$(Et)_3N$	10	1	20	"	83	0.90
36	"	None	10	1		"	57	1.00
37	$RuHCl(PPh_3)_3$	"	10	1		"	60	0.95
38	$RuCl_3(AsPh_3)_2$	"	10	1		"	79	2.1
39	$RuCl_2(SbPh_3)_3$	"	10	1		"	77	2.9
40	$RuCl_2(DIPHOS)_2$[g]		10	1		"	57	1.1
41	$RuCl_2(PPh_3)_3$ $+2PPh_3$	"	10	1		"	1.7	<0.1
42	$RuCl_2(CO)_2(PPh_3)_2$	KOH	10	1	20	"	23	
43	$Ru(CO)_3Cl_2$	"	10	1	20	"	67[f]	
44	$Fe(CO)_5$	"	1	1	2	"	67[f]	
45	$Fe(CO)_3(PPh_3)_2$	"	2	1	4	"	16	

[a]A mixture of isomers 2- through 6-nitrododecanes. [b]Run conditions: 0.001-0.02 M Ru, 120°, 1-6 hr. [c]$C_{12}H_{25}N_2$ yield data refer to maximum dodecylamine yields, based upon nitrododecane charged, for reaction times up to 6 hr. The data were estimated by both ir and glpc techniques. [d]Relative rate data are based upon the maximum observed rates of nitrododecane reduction for each experiment, as determined by glpc, with exp. 9 as the base (reference) case. [e]Run under N_2 (68 atm). [f]Extensive precipitation of ruthenium or iron complex. [g]DIPHOS, $(C_6H_5)_2PCH_2CH_2P(C_6H_5)_2$.

improved catalyst turnover, without the need for an aqueous, acidic media which could result in competing Nef-type hydrolysis. No amine is detected in the absence of ruthenium or iron complex, nor does reduction proceed in the absence of hydrogen (expt 33). This is in spite of reports by Sasson and Blum (1971) that primary carbinols may act as hydrogen donors to the $RuCl_2(PPh_3)_3$ complex, and the fact that many nitroaromatics have been reduced by catalytic transfer hydrogenation (Brieger and Nestrick, 1974).

For a series of ruthenium complexes with tertiary Group VB donor ligands (expt 37-39) the rate of hydrogenation appears to increase with decreasing ligand strength (Hendrici-Olive and Olive, 1971), in the order:

$$PPh_3 < AsPh_3 < SbPh_3 \qquad (8)$$

Improved yields of dodecylamine are generally obtained under superatmospheric pressures of hydrogen (>30 atm; expt 30-32) and in the presence of added alkali or organic bases such as triethylamine (cf. expt 29, 35 and 36). Generally, as in the reduction of nitroaromatics (Knifton and Suggitt, 1974; L'Eplattenier, et al, 1970) iron complexes were found less effective than the ruthenium complexes; they also show lower stability in the alkali media.

DISCUSSION

For a wide variety of transition and post-transition metal ions in solution, activation of molecular hydrogen leads to the formation of reactive metal-hydride complexes (Harmon, et al, 1973). Generally hydride formation proceeds by one of two mechanisms involving homolytic or heterolytic cleavage of the H_2 molecule. Copper(I) and silver(I) ions are known to activate molecular hydrogen by both paths (e.g. eq 9 and 10). Generally, however, the path involving heterolytic cleavage (eq 9)

$$Cu^+ + H_2 \longrightarrow CuH + H^+ \quad (9)$$

$$2Cu^+ + H_2 \longrightarrow 2CuH^+ \quad (10)$$

will be favored wherever the metal ion is surrounded by basic ligands, or solvents of high polarity, due in part to a) stabilization of the released proton, and b) suppression of the reverse reaction. This effect can be seen quantitatively in the activation energy data for heterolytic splitting of H_2 by silver(I) in the presence of various complexing agents of increasing base strength (Table V).

Group IB hydride species are generally labile, and in the absence of a suitable substrate, will tend to dissociate to the metal, or undergo the reverse reaction to regenerate hydrogen. In the presence of certain oxidizing agents, however, catalytic reduction by hydrogen has been observed. Examples include the reduction of dichromate and permanganate by hydrogen with aqueous solutions of silver(I) salts (Webster and Halpern, 1957),

TABLE V. *Ligand Effects Upon the Activation Energies for Reaction:*

$$Ag^+ + H_2 \longrightarrow AgH + H^+$$

Silver Salt	Solvent	E(kcal mol^{-1})[a]
Silver Acetate	Water	24
Silver Heptanoate	Heptanoic Acid	19
Silver Acetate	Pyridine	13-16

[a]Data taken from Halpern, 1959.

and the reduction of p-benzoquinone (Calvin, 1938) and carboxylic acid (Stouthamer and Vlugter, 1965) by hydrogen and solutions of copper(I). For the reduction of aliphatic nitro compounds to the corresponding oximes, the involvement of intermediate copper(I) and silver(I) hydride complexes in the alkylpolyamine solvents would be consistent with these studies, and related work by Wright and Weller (1954) and by Halpern (1965) on H_2 activation by Group IB ions in quinoline and other amine bases. In the case of the oxime synthesis, the hydride source need not necessarily come from molecular hydrogen. Certain other hydride sources are also effective, although only under non-catalytic, near-stoichiometric, conditions (see Table VI). Sodium borohydride, which does not normally reduce nitroalkanes (Gold and Klager, 1963), and hydrazine, both yield significant quantities of cyclohexanone oxime in the presence of ethylenediamine solutions of copper(I) (expt 46-47). Lithium aluminum hydride, which is reported to reduce nitroalkanes to the corresponding amines in ethereal solution (Nystrom and Brown, 1948), and ketoximes in basic media, is not effective. Control experiments under nitrogen (expt 49), in the absence of either molecular hydrogen, or an alternate hydride source, also failed to show the formation of oxime.

The importance of intermediate hydride species may be inferred also from available rate and isotopic exchange studies (Knifton, 1974). For the reduction of nitrocyclohexane catalyzed by solutions of copper(I) chloride in ethylenediamine, the rate of cyclohexanone oxime formation is sensitive to applied hydrogen pressure only below about 5 atm. Under higher H_2 pressures, the rate expression is of the form,

$$\frac{-d[RNO_2]}{dt} = k\frac{[Cu][RNO_2]}{1+k\ [RNO_2]} \qquad (11)$$

TABLE VI. *Nitrocyclohexane Reduction by Various Reducing Agents*[a]

Expt	Reducing Agent	$[RNO_2]/[H]$	Reaction Time (min)	Nitrocyclohexane Conversion (%)	Cyclohexanone Oxime Yield (mole %)[b]
46	$NaBH_4$	0.42	210	59	17
47	N_2H_4	0.31	180	57	49
48	$LiAlH_4$	0.31	180	1[c]	1
49	None[d]	--	150	10.3	1

[a]All experiments run under 1 atm. nitrogen, 57 mmole [CuCl]; 0.47 mole $[RNO_2]$; 85°. [b]Yield based upon nitrocyclohexane charged. [c]Extensive precipitation of solids during this run. [d]A control experiment with no added reducing agent.

for which the effective constant, k, is 17 M^{-1}, hr^{-1}. Deuterium studies indicate a kinetic isotope ratio of at least 1.2. This ratio is comparable to those reported for other cuprous catalyzed hydrogenations (see Chalk and Halpern, 1959). No deuterium could be detected in the cyclohexanone oxime after work-up; apparently, the primary function of the CuH intermediate (eq 9) is to effect deoxygenation of the nitro group, rather than hydride addition.

Preferred reaction conditions for the paraffin amine synthesis (50-150 atm H_2, 90-130°, excess alkali, see Table IV), should also favor the formation of labile ruthenium-hydrido species. Solutions of the $RuCl_2(PPh_3)_3$ complex in benzene, ethanol and other solvents have found extensive application in hydrogenation catalysis, particularly for the reduction of 1-olefins (Hallman, et al, 1968), and the enhancement of activity by base additives is well documented. Nishimura, et al, (1973) attribute this improved activity to the base promoted formation of intermediate hydrochlorotris(triphenylphosphine)ruthenium(II) as depicted by eq 12. Even in the absence of alkali, the alkanol cosolvent, or in this case the amine product, may serve as promoters for this hydride formation step. The importance of

$$RuCl_2(PPh_3)_3 + H_2 + Base \rightleftharpoons RuHCl(PPh_3)_3 + Base{\cdot}HCl \qquad (12)$$

the $RuHCl(PPh_3)_3$ complex in the amine synthesis is consistent with the observed similar hydrogenation rates for $RuHCl(PPh_3)_3$ and $RuCl_2(PPh_3)_3$ (expt 36 and 37), induction periods prior to hydrogenation with $RuCl_2(PPh_3)_3$, and the spectra of recovered catalyst samples (ν (Ru-H) 2020 cm^{-1}). Basic reaction conditions should also favor deprotonation of the nitroalkane to its anionic form, by shifting the equilibrium of eq 13 further to the right. Generally, the selective reduction of nitroalkanes

$$RR'CHNO_2 + B: \rightleftharpoons (RR'CNO_2)^- + BH^+ \qquad (13)$$

to alkyl amines or oximes proceeds most readily via the prior formation of the nitroalkane anion (Knifton, 1973, 1974). While the pK values for short chain primary and secondary nitroalkanes in aqueous media are of the order 8-10, in highly basic solvent media the nitroalkane molecule will be extensively deprotonated to the anionic form (Smith, 1966). This has been confirmed spectroscopically (Knifton, 1973). In the alkyl amine synthesis, the addition of certain other Lewis bases, such as pyridine, leads to catalyst deactivation (expt 34, Table IV), as does the presence of the strongly coordinating CO molecule (expt 42 and 43).

The critical role played by the solvent in reaction (1),

particularly regarding its basicity and chelation properties, is evident from the data in Table III. The function of the preferred alkylpolyamine solvents, like ethylenediamine, may be at least three-fold. Firstly, the more basic solvents will favor both the formation of CuH by stabilizing the released hydrogen ion (eq 9), and as better σ-donors to the copper(I), the electron transfer steps involved in subsequent RNO_2 reduction to oxime. Weller and Mills (1953), for example, have reported the activation of molecular hydrogen by cuprous salts in organic bases to vary significantly with the base strength of the solvent (see also Table V for silver(I) complexes). The fact that in this work, however, no nitrocyclohexane reduction was detected with pyridine as solvent, even though activation of H_2 by copper(I) salts in pyridine is well established (Wright, et al, 1955), implies that solvent basicity is a critical factor in some step other than the initial metal hydride formation.

A second function of the amine solvent will be to favor the formation of the nitroalkane anion (eq 13) as discussed above. A third function will be to stabilize the cuprous ion against disproportionation to the metal, although it is evident from the generally negative effect upon catalyst activity of adding strongly coordinating phosphines capable of π back-bonding that, as with numerous other homogeneous catalysts, there is a rather delicate balance here between the stability, nucleophilicity and catalytic activity of the copper complexes.

Hydrogenation Mechanisms

A rationale for the nitroalkane reduction to oxime, consistent with the available data, is presented in Scheme 1. The initial reaction, heterolytic splitting of H_2 by solvated copper(I) to form a hydride complex, is followed by nucleophilic displacement, or addition, to the coordination sphere of the copper complex (Axtell, et al, 1973), by the nitroalkane anion, to give the labile intermediate I. Partial deoxygenation of the coordinated nitroalkane anion by hydride attack through the quasi-cyclic transition state II would yield the product oxime without the need to involve the nitroso tautomer (Smith, 1966). Consistent with the isotopic and kinetic data, at moderate hydrogen pressures, deoxygenation of the nitroalkane anion is regarded as a slow, rate determining step.

Structure I in Scheme 1 is shown involving metal-oxygen rather than metal-carbon bonding by analogy with known nitroalkane complexes (Lee, 1972). Subsequent deoxygenation of the coordinated nitroalkane has been invoked earlier in studies of nitrocompounds reduced by iron and ruthenium carbonyls (Alper, 1972), although there reductions were believed to proceed via complete deoxygenation of the $-NO_2$ group, with the formation of

SCHEME 1[a]

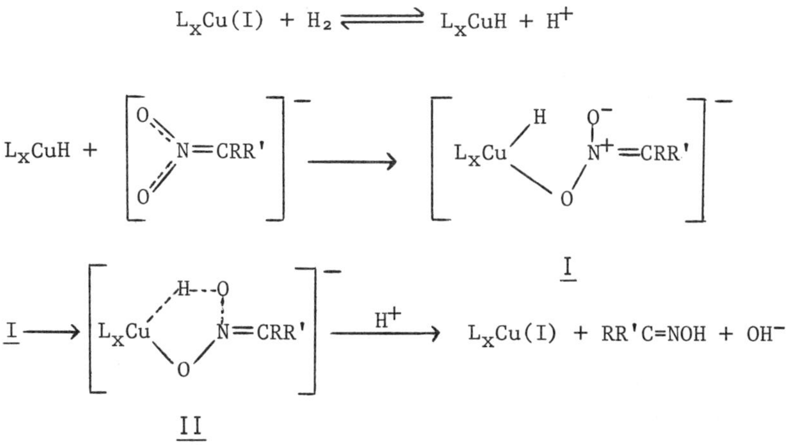

[a]Where L_x refers to coordinated solvent.

---

nitrene-like intermediates stabilized by bonding to the metal. The lack of evidence for alkylamine by-products and dimeric derivatives in this work, particularly azo and azoxy compounds, mitigates against the involvement of nitrenes here and is consistent with only partial deoxygenation of the nitro anion.

The proposed mechanism for nitroalkane hydrogenation to amine (Scheme 2) also involves initial heterolytic splitting of molecular hydrogen, followed by dissociation to the <u>trans</u>-hydrido chlorobis(triphenylphosphine)ruthenium(II) complex, <u>III</u> (James, 1973). This would be consistent with similar hydrogenation rates for $RuHCl(PPh_3)_3$ and $RuCl_2(PPh_3)_3$ (expt 36 and 37), the observed promotional effect of added alkali, inhibition by excess triphenylphosphine (expt 41), and the increasing rate with decreasing ligand strength in the order of eq 8. Subsequent coordination of the nitroalkane anion with <u>III</u> could proceed through halide or solvent displacement, according to the degree of ionic dissociation of the ruthenium complex (James and Markham, 1974), and involve the formation of five coordinate ruthenium species. Certainly recovered catalyst samples often show strong maxima at 1580 cm^{-1}, assignable to $\nu(NO_2)$ vibrations of the coordinated $RR'CNO_2^-$ anion (Lee, 1972), together with some evidence for the presence of ruthenium carbonyl species ($\nu(C{\equiv}O)$ 1950 cm^{-1}) as a result of ethanol decarbonylation (Vaska, 1960). Deoxygenation of the coordinated nitroanion to give <u>IV</u> will be followed by the stepwise oxidative addition of

SCHEME 2

$$RuCl_2(PPh_3)_3 + H_2 + base \rightleftharpoons RuHCl(PPh_3)_3 + base \cdot HCl$$

$$RuHCl(PPh_3)_3 \rightleftharpoons \underset{\underline{III}}{RuHCl(PPh_3)_2} + PPh_3$$

$$RuHCl(PPh_3)_2 + (RR'CNO_2)^- \longrightarrow \underset{\underline{IV}}{RuCl(PPh_3)_2(RR'CNO)} + OH^-$$

$$RuCl(PPh_3)_2(RR'CNO) + 3H_2 \xrightarrow{k} RuHCl(PPh_3)_2 + RR'CHNH_2 + H_2O$$

---

hydrogen (James, 1973) and the elimination of product amine and elements of water. The dependence of the rate upon applied H_2 pressure, and substrate concentration, makes it likely that the final equation of Scheme 2 contains one or more slow, rate determining, steps, but more detailed kinetic data would be needed before commenting further on the nature of these steps. Since no alkyl amine is detected in the absence of molecular hydrogen (expt 33), it is unlikely that the ethanol cosolvent is an important hydride source here. A third potential hydrogen source comes from the ortho-metallation of the $RuHCl(PPh_3)_2$ (eq 14).

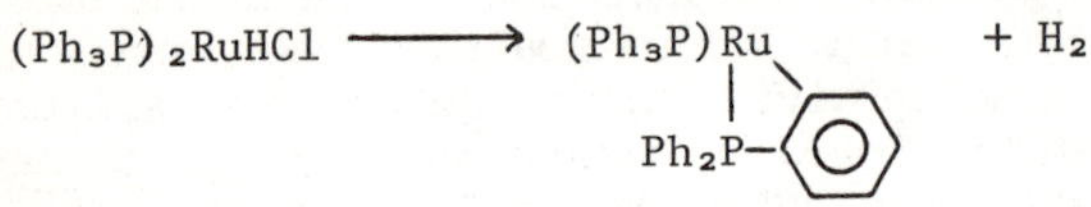
(14)

Ortho-metallation has been applied to the selective deuteration of $HRuCl(PPh_3)_2$ (Parshall, et al, 1969), and the stoichiometric reduction of alkenes (James, et al, 1974), but is normally slow compared to hydrogenations catalyzed by $RuHCl(PPh_3)_3$ (see Hallman, et al, 1968). Other alternatives to Scheme 2 include the possible formation of polyhydrido species, like $H_2Ru(PPh_3)_3$, prepared by Eliades, et al, (1970), by the hydrogenation of $RuCl_2(PPh_3)_3$ in the presence of triethylamine (eq 15).

$$RuHCl(PPh_3)_3 + H_2 + Base \rightleftharpoons RuH_2(PPh_3)_3 + Base \cdot HCl \qquad (15)$$

A more detailed examination of C-NO_2 hydrogenation to amine by solubilized ruthenium complexes has been found possible with nitroaromatic substrates (Knifton and Suggitt, 1974), where the technique has been extended to both selective and sequential hydrogenation catalysis.

ACKNOWLEDGEMENT

The author thanks Texaco Inc. for permission to publish this paper, and Messrs. T. S. Strothers and C. A. Dondero for experimental assistance.

REFERENCES

1. Alper, H. Inorg. Chem., 11, 976 (1972).
2. Axtell, D. D., Yoke, J. T., Inorg. Chem., 12, 1265 (1973).
3. Axtell, D. D., Good, B. W., Porterfield, W. W., Yoke, J. T., J. Amer. Chem. Soc., 95, 4555 (1973).
4. Booth, G., "Advances in Inorg. Chem. and Radiochem." (Emelius, H. J. and Sharpe, A. G., Eds), Vol. 6, p 47, Academic Press, New York, 1964.
5. Brieger, G., Nestrick, T. J., Chem. Rev., 74, 567 (1974).
6. Calvin, M., Trans. Faraday Soc., 34, 1181 (1938).
7. Calvin, M., Wilmarth, W. K., J. Amer. Chem. Soc., 78, 1301 (1956).
8. Chalk, A. J., Halpern, J., J. Amer. Chem. Soc., 81, 5846 (1959).
9. Eliades, T. I., Harris, R. O., Zia, M. C., Chem. Commun., 1709 (1970).
10. Gold, M. H., Klager, K., Tetrahedron, 19, Suppl. 1, 77 (1963).
11. Hallman, P. S., McGarvey, B. R., Wilkinson, G., J. Chem. Soc., A, 3143 (1968).
12. Halpern, J., J. Phys. Chem., 63, 398 (1960).
13. Halpern, J., Milne, J. B., Proc. Int. Congr. Catal. 2nd, 1, 445 (1960).
14. Halpern, J., Ann. Rev. Phys. Chem., 16, 103 (1965).
15. Harmon, R. E., Gupta, S. K., Brown, D. J., Chem Rev., 73, 21 (1973).
16. Hendrici-Olive, G., Olive, S., Angew. Chem. Int. Ed. Engl., 10, 105 (1971).
17. James, B. R., "Homogeneous Hydrogenation," J. Wiley, New York, 1973, p 90.
18. James, B. R., Markham, L. D., Inorg. Chem., 13, 97 (1974).
19. James, B. R., Markham, L. D., Wang, D. K. W., J. Chem. Soc. Chem. Comm., 439 (1974).
20. Neunhoeffer, O., and Liebich, H. G., Ber., 71, 2247 (1938).
21. Knifton, J. F., J. Org. Chem., 38, 3296 (1973).

22. Knifton, J. F., Suggitt, R. M., U. S. Patent 3,832,401 (1974).
23. Knifton, J. F., J. Catal., 33, 289 (1974).
24. Knifton, J. F., J. Org. Chem., 40, 519 (1975).
25. Kmiecik, J. E., J. Org. Chem., 30, 2014 (1965).
26. Lee, A. G., Spectrochimica Acta, 28A, 133 (1972).
27. L'Eplattenier, F., Matthys, P., Calderazzo, F., Inorg. Chem., 9, 342 (1970).
28. Murahashi, S., Horiie, S., Bull. Chem. Soc. Jap., 33, 78 (1960).
29. Nishimura, S., Ichino, T., Akimoto, A., Tsuneda, K., Bull. Chem. Soc. Jap., 46, 279 (1973).
30. Nystrom, R. F., Brown, W. G., J. Amer. Chem. Soc., 70, 3738 (1948).
31. Parshall, G. W., Knoth, W. H., Schunn, R. A., J. Amer. Chem. Soc., 91, 4990 (1969).
32. Sasson, Y., Blum, J., Tetrahedron Lett., 2167 (1971).
33. Smith, P. A. S., "Open Chain Nitrogen Compounds," Vol. II, W. A. Benjamin, New York, 1966, Chapter 14.
34. Stouthamer, B., Vlugter, J. C., J. Amer. Oil Chem. Soc., 42, 646 (1965).
35. Vaska, L., Z. Naturforsch, B, 15, 56 (1960).
36. Weller, S., Mills, G. A., J. Amer. Chem. Soc., 75, 769 (1953).
37. Webster, A. H., Halpern, J., J. Phys. Chem., 61, 1239 (1957).
38. Webster, A. H., Halpern, J., Trans. Faraday Soc., 53, 51 (1957).
39. Wright, L., Weller, S., J. Amer. Chem. Soc., 76, 3345 (1954).
40. Wright, L., Weller, S., Mills, G. A., J. Phys. Chem., 59, 1060 (1955).

USE OF HYDRAZINE-RUTHENIUM ON CARBON FOR SELECTIVE SYNTHESIS OF NITRO-SUBSTITUTED o-PHENYLENEDIAMINES

JOHN L. MIESEL, GEORGE O. P. O'DOHERTY, AND JOHN M. OWEN

Lilly Research Laboratories
Division of Eli Lilly and Company, Greenfield, In 46140

INTRODUCTION

Among the precursors of heterocycles, o-phenylenediamines are especially useful as starting materials. They may be used in the synthesis of several heterocyclic systems of varying ring size including benzimidazoles, quinoxalines, and 1,5-benzodiazapines. Often it is desired to have a nitro group at a specific position in the product either to observe the effect on some biological activity or to use the nitro group for further synthetic transformations. In order to obtain a single isomer, the synthesis of a specific nitro-o-phenylenediamine is often required.

Nitro-substituted o-phenylenediamines have been synthesized by selective reduction of polynitroanilines or polynitrobenzofurazan-N-oxides as shown in Figure 1. Historically the method of choice has been the reduction of polynitroanilines, such as the 2,4-dinitroaniline shown, with sulfide ion, especially in alkaline medium, to give the o-phenylenediamine. Schröter (1957) has presented a brief review of the scope of this reaction. The utility of this reduction is limited at times by poor yields, problems with removing sulfur or sulfur-containing by-products, and odor and toxicity arising from the use of hydrogen sulfide or other sulfides--especially on a large scale. Recently, Lyle and LaMattina (1974) reported a controlled catalytic reduction of certain 2,6-dinitroanilines. Reduction of the anilines shown in Figure 1 at low pressure over palladium on carbon in 1,2-dimethoxyethane-chloroform to the theoretical uptake of hydrogen gives good yields of the o-phenylenediamine. Extension of this reduction beyond these examples has not been reported. As another approach, Boyer and Schoen (1956) reduced 3,5-dinitro-1,2-dinitrosobenzene with hydriodic acid to give 3,5-dinitro-o-phenylenediamine.

One further single example of such a selective reduction was reported by Pitré and Lorenzotti (1965). They synthesized

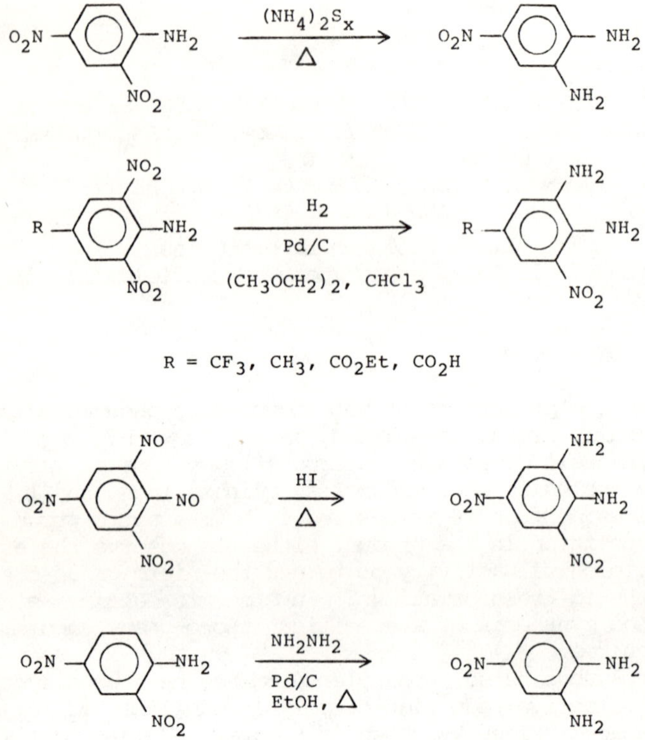

Figure 1. Syntheses of nitro-substituted o-phenylenediamines.

4-nitro-o-phenylenediamine by reaction of 2,4-dinitroaniline with 1.5 equivalents of hydrazine (the calculated amount) in the presence of 5% palladium on carbon in refluxing ethanol. Further studies of their procedure in our laboratories have shown that it is an exceptionally useful synthesis of nitro-substituted o-phenylenediamines. With certain adaptions and variations, which will be discussed, we have found this synthesis is quite broad in scope, usually devoid of experimental difficulties, and adaptable to quantities from 1 g to 1 kg.

RESULTS AND DISCUSSION

Reduction of 2,6-Dinitroanilines

The range of 4-substituted-2,6-dinitroanilines subjected

to reduction by hydrazine-ruthenium on carbon is shown in Figure 2.

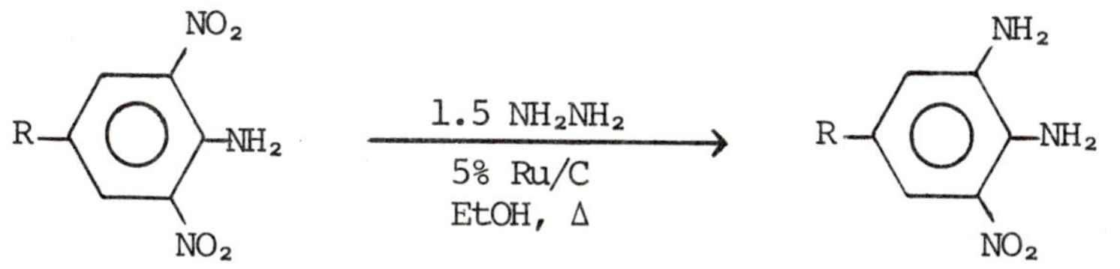

Reaction succesful when R is:			Reaction unsuccessful when R is:
CH_3	CF_3	$SO_2N(CH_3)_2$	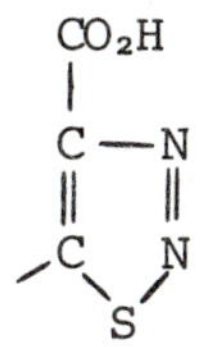
C_2H_5	CHF_2	Cl	
$C(CH_3)_3$	$CO_2C_2H_5$	I	
C_6H_5	CN	SCH_3	
	SO_2NH_2	OCH_3	

Figure 2. 2,6-Dinitroanilines reduced by NH_2NH_2-Ru/C

Detailed experimental procedures for reactions on both a laboratory scale and a prep lab scale are given in the Experimental Section; however, some general comments can also be made regarding starting materials and reaction conditions. Commercial hydrazine has been used in all these reactions either as anhydrous hydrazine or as any of the readily available hydrate solutions. Five percent ruthenium on carbon catalyst has been normally used, but the use of other catalysts will be discussed later in the paper. Commercial denatured ethanol has been the only solvent in any of our reactions, although Furst et al, (1965), in a review, reports the use of other solvents for non-specific hydrazine-catalyst reductions. Some variation in reaction conditions may be needed for best results in individual examples. The starting aniline may either be in solution or in a solution-suspension, but some solubility in ethanol is essential. It is best for the reaction mixture to be above 60° before the addition of the hydrazine; this avoids side reactions in a case such as the 4-methoxy compound. The quantity of hydrazine used may vary from the theoretically calculated amount to a 5-10% excess to give the cleanest product. In kilogram-scale reactions, 5-10% excess hydrazine is always necessary as the tremendous volumes of nitrogen evolved in these large-scale reductions apparently sweep out some of the added hydrazine. In most reactions excellent results are achieved by adding the hydrazine solution in a small amount of ethanol in one portion. But in the case of the strong electron-donating 4-substituents, methoxy and thiomethyl, this procedure leads to the formation of small quantities of the triamine, <10% by nmr of the crude reaction mixture, along with unreacted aniline. Dropwise addition of the ethanolic hydrazine to the hot aniline-catalyst solution gives a slightly cleaner product in these two cases.

In most cases work-up of the diamine product is extremely simple. Filtration of the solution to remove catalyst and evaporation of the filtrate generally gives the diamine product in better than 90% purity. The slight impurity is often only a trace of unreacted aniline. For most synthetic purposes the crude product can be used directly, but a single recrystallization will give analytically pure product. This reduction procedure gave neither hydrogenolysis of halogen nor was it affected by the catalyst-deactivating group - SCH_3. Groups potentially liable to attack by hydrazine, difluoromethyl and carboethoxy, were not transformed in the reaction probably due to the speed of the reduction.

The two failures of this reduction of 2,6-dinitroanilines should be discussed. The 4-thiadiazolyl derivative is insoluble in boiling ethanol and gives only starting aniline from the normal reduction procedure. Reduction of the 4-carboxylic acid could have failed either because of extreme insolubility or because of formation of the benzoic acid salt of the hydrazine.

Reduction of 2,4-Dinitroanilines

Reduction of 2,4-dinitroanilines with hydrazine-ruthenium on carbon is not as facile as the reduction of 2,6-dinitroanilines. The clean reduction of 2,4-dinitroaniline itself was the reaction originally reported by Pitré and Lorenzotti. The substituted anilines shown in Figure 3 were cleanly reduced to the corresponding o-phenylenediamines using the usual conditions. A typical example is given in the Experimental Section.

Reduction of 5-chloro-2,4-dinitroaniline under these conditions is unsuccessful. The major product of the reaction is the phenylhydrazine arising from displacement of the active halogen, isolated as its acetone addition product (Figure 4). Even in this reaction a very small amount (>2%) of the diamine is obtained; this is determined by the synthesis of the expected benzimidazole from a crude reaction fraction.

A more serious probelm is found in the reduction of a series of 6-alkyl-2,4-dinitroanilines. In all cases, the crude reaction product is a mixture of diamine, triamine, and unreacted aniline (Figure 5). These mixtures do not separate easi-

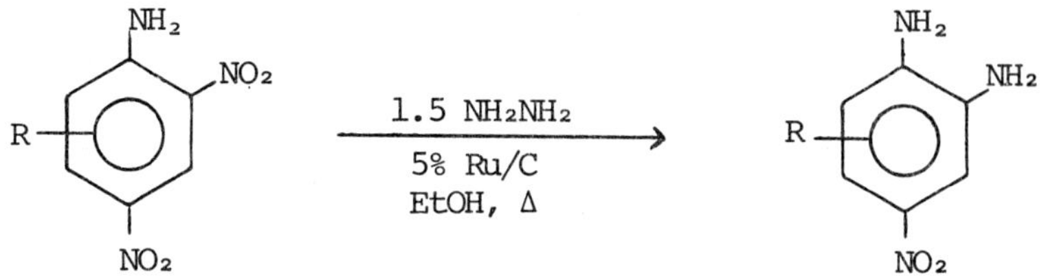

R = 5-NH_2, 5-CF_3, 6-CF_3, 6-Cl

Figure 3. 2,4-Dinitroanilines reduced by NH_2NH_2-Ru/C

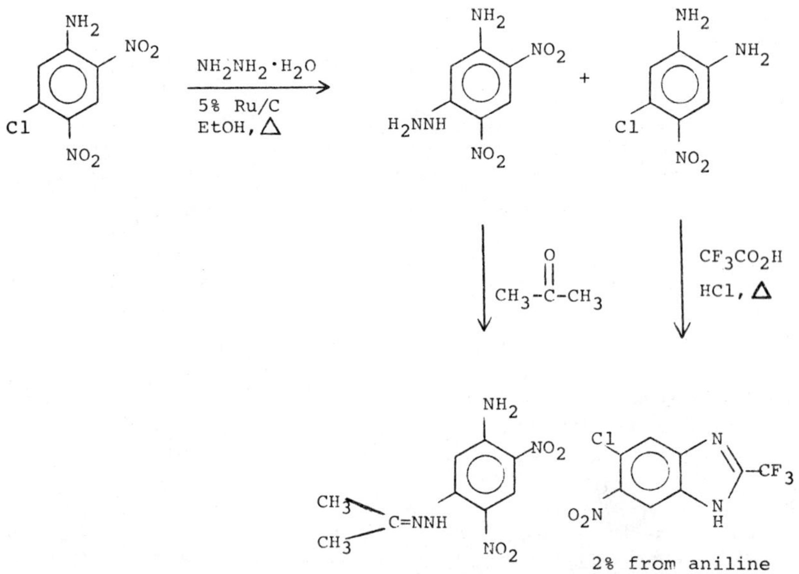

Figure 4.

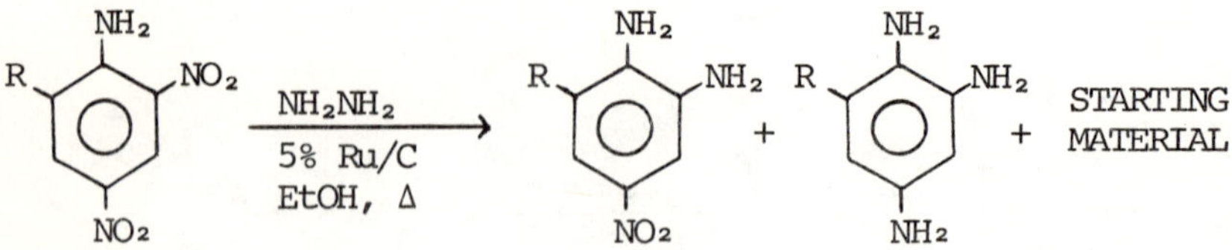

$R = C_2H_5$, $\underline{i}$-C_3H_7, $\underline{sec}$-C_4H_9, $\underline{tert}$-C_4H_9, $\underline{sec}$-C_5H_{11}

Figure 5.

ly, and the presence of substantial quantities of triamine interferes with further reactions of the crude product. As the alkyl group becomes more bulky, the proportion of triamine increases. When R is tert-butyl, even the best reaction conditions, slow dropwise addition of hydrazine and large amount of catalyst, gave only 64% diamine after very careful column chromatography. Even then the diamine was contaminated with a trace of triamine. Ammonium polysulfide reduction of the sec-amyl compound gives a mixture of diamine, triamine, and aniline not significantly different from the hydrazine reduction.

Reduction of Miscellaneous Nitroanilines

The only 2,3-dinitroaniline reduced in our studies is the anisidine shown in Figure 6. Reduction using the hydrazine-ruthenium procedure is totally unsatisfactory. The reaction product apparently is a mixture of starting material and triamine. Reduction of the two trinitroanilines in Figure 6 is more successful. The reduction of a suspension of picramide under our usual conditions gives a crude product which is essentially pure o-phenylenediamine by nmr. The tert-butyl compound is a more complex case. TLC analysis of the crude reaction mixture indicates two major products. One product can be shown to be the diamine by synthesis of a benzimidazole derivative (Figure 6). The benzimidazole derived from the other possible o-phenylenediamine is available from an unambiguous route and was not isolated from the reaction. The identity of the other major reduction product is not known.

Reduction of N-Substituted Anilines

Although the products were less useful as synthetic intermediates, we were interested in the effect of N-substitution on the selectivity of this reduction procedure. The examples in Figure 7 show some of the systems studied. In two examples

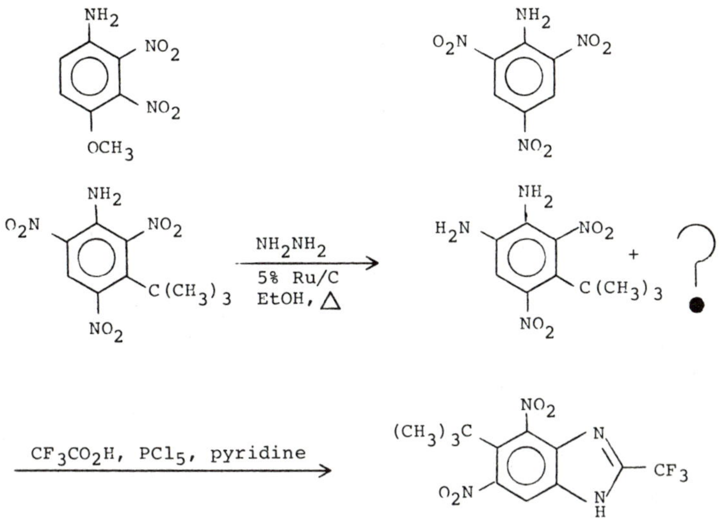

Figure 6. Miscellaneous Anilines Studied

simple alkyl substitution on the amino group does not affect the selectivity of diamine formation. Even in the example where the amino nitrogen forms part of an imidazole ring, the ortho-nitro group is selectively reduced to give 7-amino-5-nitro-2-trifluoromethylbenzimidazole. However, nitroanilides cyclize to N-hydroxybenzimidazoles by capture of the intermediate hydroxylamine. Reductive cyclization of o-nitroanilides was reported by O'Doherty and Fuhr (1973) at the last of these conferences.

Given these results, the attempted selective reduction of some dinitrophenylhydrazine derivatives was of some interest (Figure 8). Reaction of a simple alkylated phenylhydrazine with hydrazine-ruthenium under our standard conditions does not give selective reduction but an unworkable mixture of products. However, examples of acylated and phosphorylated nitrophenylhydrazines do give selective monoreduction. In the case of the acyl compound, care is needed to avoid cyclization of the intermediate amino product.

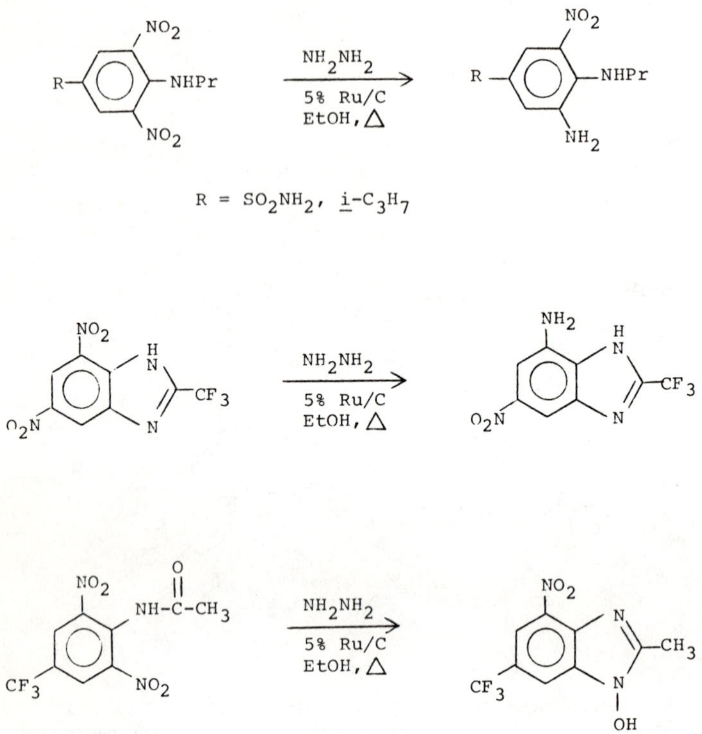

Figure 7. Reduction of N-substituted Anilines

Reduction of Nitro-1,2-dinitrosobenzenes

It was previously mentioned that hydrazine-ruthenium or ammonium polysulfide reduction of 2-alkyl-4,6-dinitroanilines gives mixtures of aniline, *o*-phenylenediamine, and triamine. The other possible literature method for effecting this reaction was hydriodic acid reduction of the corresponding 1,2-dinitrosobenzenes (benzofurazan-N-oxides). The needed dinitroso compounds were prepared by thermolysis of the dinitrophenylazides. Hydriodic acid reduction of the *tert*-butyl-nitro compound in Figure 9 was totally unsuccessful. However, a simple adaptation of the hydrazine-ruthenium reduction procedure surprisingly gives synthetically useful yields of reasonably pure products. An exact experimental procedure for a typical

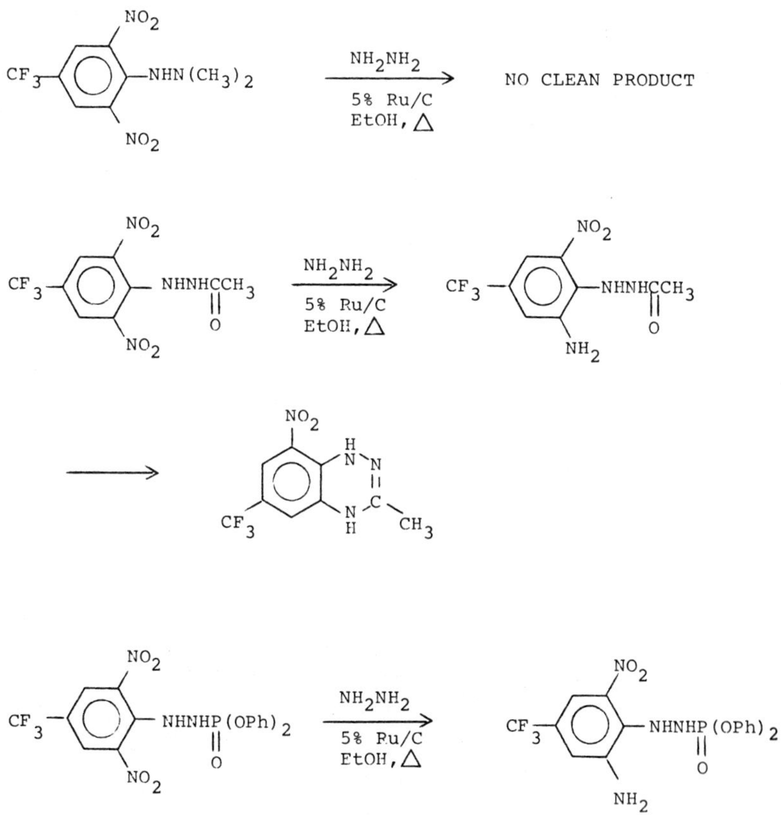

Figure 8. Reduction of Dinitrophenylhydrazine Derivatives

example is given in the Experimental Section. Better results from the reduction are apparently related to a higher proportion of catalyst to substrate and slow dropwise addition of a dilute hydrazine solution. Several 2-alkyl-4-nitro-o-phenylenediamines were successfully prepared by this procedure (Figure 9).

3-Nitro-5-(1,2,3-thiadiazol-4-yl)-o-phenylenediamine cannot be prepared by hydrazine-ruthenium reduction of the aniline due to the insolubility of the aniline. The corresponding

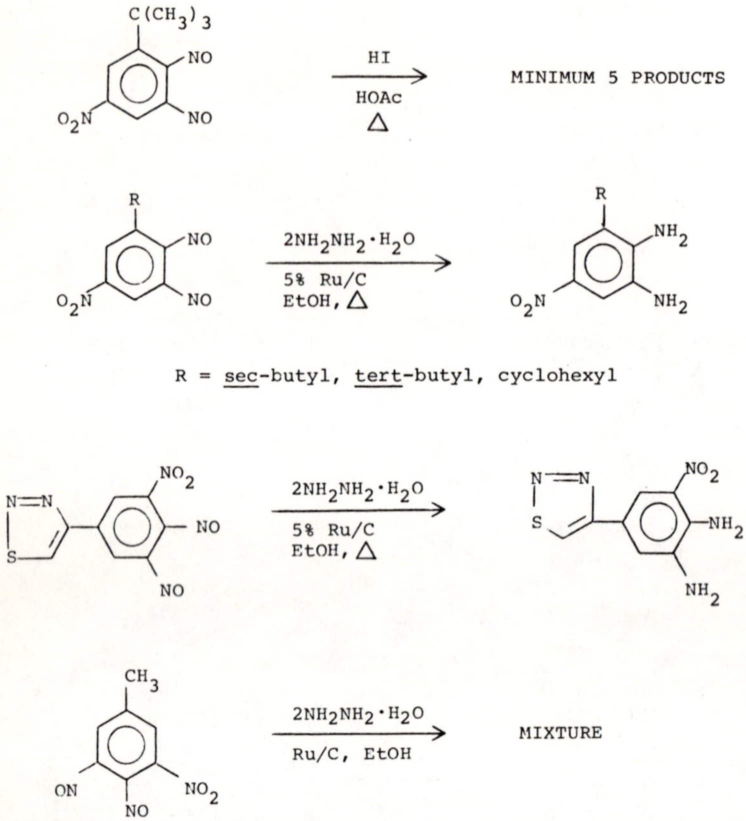

Figure 9. Reduction of 1,2-Dinitrosobenzenes

dinitroso compound is much more soluble and its reduction gives the diamine sufficiently pure to use as an intermediate. However, this reduction may not be general. 3-Nitro-5-methyl-o-phenylenediamine is easily prepared by hydrazine-ruthenium reduction of the dinitroaniline, but the reduction of the dinitroso compound is rather unselective.

Examination of Other Catalysts

Most of this work has been done using 5% ruthenium on carbon with general success. However, in two examples other catalysts were examined for specificity. The first of these cases was the reduction of 4-amino-3,5-dinitrobenzotrifluoride. The reductions were run under identical conditions varying only the catalyst. The nmr's of the crude products from evaporation of the filtered reaction mixture were compared with the spectra of aniline, diamine, and triamine (Figure 10).

	Relative Proportions Of			
Catalyst	Aniline	Diamine	Triamine	Other
5% Ru/C	4	96	-	-
5% Pd/C	6	88	6	-
5% Pt/C	8	82	3	7
5% Rh/C	-	47	6	47
5% Ru/Alumina	-	60	10	30
Raney Ni	mixture too complex to analyze			

Figure 10. Reduction of 4-Amino-3,5-dinitrobenzotrifluoride with Hydrazine-catalyst

In this case although 5% ruthenium on carbon is clearly the most selective catalyst, 5% palladium could be used. Substitution of an alumina support for the carbon greatly decreases the selectivity of the reduction. A study of the rather unselective reduction of 6-tert-butyl-2,4-dinitroaniline showed no qualitative advantage of palladium over ruthenium and Raney Ni was again clearly the most unselective catalyst.

CONCLUSIONS

(1) Reduction of dinitro and trinitroanilines with hydrazine-ruthenium on carbon is a widely useful selective synthesis of substituted nitro-o-phenylenediamines. Examples of the scope of the reaction and some of its limitations are presented.

(2) Some substituted-nitro-1,2-dinitrosobenzenes may be successfully reduced by hydrazine-ruthenium on carbon to the o-phenylenediamines.

(3) Ruthenium on carbon is the best catalyst for these selective reductions within the small group of catalysts tested.

EXPERIMENTAL SECTION

5-Cyano-3-nitro-*o*-phenylenediamine

To a stirred mixture of 10.0 g of 4-cyano-3,5-dinitroaniline (0.048 mole) and 1.0 g of 5% ruthenium on carbon (Engelhard Industries) in 250 ml of 2B ethanol at 65° was added in one portion 2.5 g of anhydrous hydrazine (0.078 mole, 10% excess) in a small amount of ethanol. An immediate exotherm and evolution of gas was observed. After the evolution of gas had slowed, the reaction was refluxed for one hour. The hot solution was filtered. The solid precipitate and catalyst were refluxed in 100 ml of acetone and filtered. The combined filtrates were evaporated to dryness. The solid residue was recrystallized from ethanol to give 7.5 g of red needles, mp 211-211.5° (88% yield). Anal. Calcd. for $C_7H_6N_4O_2$: C, 47.19; H, 3.39; N, 31.45. Found: C, 46.95; H, 3.56; N, 31.20.

3-Nitro-5-trifluoromethyl-*o*-phenylenediamine

One kilogram of 2,6-dinitro-4-trifluoromethylaniline (3.99 mole) and 25 g of 5% ruthenium on carbon in 12 l of ethanol were stirred in a 22-l five-neck round-bottom flask equipped with an overhead stirrer, two large-bore condensers, a thermometer, and an addition funnel. This mixture was heated to 55-60° and the heating bath drained. To this was added in a rapid dropwise fashion 370 g of 85% hydrazine hydrate (6.29 moles, 5% excess). The reaction temperature was allowed to rise to reflux. When the exotherm had ceased, the reaction was refluxed for one hour. The hot solution was filtered through a pad of Hyflo-Super Cel which was then washed with hot ethanol. The filtrates were concentrated under vacuum and chilled. The precipitate was filtered, washed, and dried to give 657 g of crude product. The crude product was recrystallized from 2 l of methanol by the addition of 2 l of water and chilling to give 600 g (68% yield) of red solid, mp 125° [Lit. mp 123.5-125°; Lyle and LaMattina (1974)].

3-Chloro-5-nitro-*o*-phenylenediamine

To a stirred mixture of 8.7 g of 2-chloro-4,6-dinitroaniline (0.04 mole) and 1.0 g of 5% ruthenium on carbon in 400 ml of ethanol at 65-70° was added in one portion 3.0 g of 99-100% hydrazine hydrate (0.06 mole) in 30 ml of ethanol. The reaction mixture was refluxed for one hour and then filtered while hot. The filtrate was evaporated under vacuum to give a crude product. The crude product was recrystallized from boiling toluene with filtration of the hot solution to give 4.8 g of red-purple needles, mp 203-09° (64% yield). Anal. Calcd. for $C_6H_6ClN_3O_2$: C, 38.42; H, 3.22; N, 22.40; Cl, 18.90. Found: C, 38.44; H. 3.02; N, 22.43; Cl, 18.60.

3-sec-Butyl-5-nitro-o-phenylenediamine

To a stirred mixture of 9.9 g of 3-sec-butyl-5-nitro-1,2-dinitrosobenzene (0.04 mole) and 2.0 g of 5% ruthenium on carbon in 250 ml of ethanol at 65-70° was added in a slow dropwise fashion 4.0 g of 99% hydrazine hydrate (0.08 mole) in 100 ml of EtOH. After addition was complete, the reaction mixture was refluxed for one hour and then filtered while hot. The filtrate was evaporated under vacuum to give 7.8 g of red solid. The crude solid was recrystallized from boiling toluene to give 4.2 g of deep-red needles, mp 63-66° (48% yield). Anal.Calcd. for $C_{10}H_{15}N_3O_2$: C, 57.40; H, 7.23; N, 20.08. Found: C, 57.69; H, 6.92; N, 19.78.

REFERENCES

1. Boyer, J. H., and Schoen, W., J. Am. Chem. Soc., 78, 423 (1956).
2. Doherty, G. O., and Fuhr, K. H., Ann. N. Y. Acad. Sci., 214, 221 (1973).
3. Furst, A., Berlo, R. E., and Hooton, S., Chem. Rev., 65, 51 (1965).
4. Lyle, R. E., and LaMattina, J. L., Synthesis, 726 (1974).
5. Pitré, D., and Lorenzotti, E., Chimia, 19, 462 (1965).
6. Schröter, R., "Amine durch Reduktion," Methoden der Organishcen Chemie, 11/1, p474 ff, ed. Eugen Müller, Georg Thieme Verlag, Stuttgart, 1957.

SELECTIVE HYDROGENATIONS CATALYZED BY DICHLORODICARBONYLBIS(TRIPHENYLPHOSPHINE)RUTHENIUM(II)

DARRYL R. FAHEY

Research and Development, Phillips Petroleum Company
Bartlesville, Oklahoma 74004

ABSTRACT

The titled complex catalyzes partial hydrogenations of multi-unsaturated organic compounds, often with a high degree of selectivity to the monounsaturate. Polyenes, α,β-unsaturated ketones and esters, nitroaromatics, and aldehydes are hydrogenated in a wide variety of solvents between 110-200° and under 1-1200 psig of hydrogen. There seems to be no advantage in using $RuClH(CO)_2(PPh_3)_2$ as the catalyst over the more easily synthesized $RuCl_2(CO)_2(PPh_3)_2$. Polyene hydrogenation products are produced as an equilibrium mixture of alkenes throughout the reaction. Double-bond migration in 1-alkenes is very rapid and initially yields 2-alkenes with cis to trans ratios of 2-3:1; eventually the equilibrium mixture of isomers is obtained. Hydrogenation rates of several olefins were measured and analyzed in terms of the rate equation:

$$\text{Rate} = \underline{k}'\underline{K}_2[H_2][\text{Cat}]_0[S]/(1 + \underline{K}_2[S] + [PR_3]/\underline{K}_1)$$

where $\underline{k}'$, $\underline{K}_1$, and $\underline{K}_2$ are rate and equilibrium constants, and $[H_2]$, $[\text{Cat}]_0$, $[S]$, and $[PR_3]$ are hydrogen, initial catalyst, olefin, and free t-phosphine concentrations.

INTRODUCTION

In 1968 we became involved in a search for catalysts capable of selectively hydrogenating cyclic polyenes to cyclic monoenes. Most of our interest was in the development of a practical synthesis of cyclododecene from 1,5,9-cyclododecatriene. Cyclododecene is an industrially valuable chemical intermediate for the preparation of nylon polymers. It is of special interest to Phillips Petroleum Company as a precursor to 1,13-tetradecadiene via the olefin metathesis reaction with ethylene (eq. 1). Phillips currently is operating a multi-purpose olefin metathesis pilot plant that produces a number of specialty olefin products, one of which is 1,13-tetradecadiene, a useful crosslinking agent in α-olefin polymers.

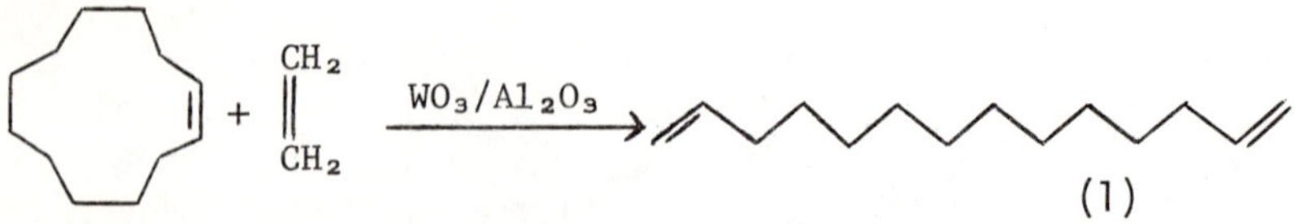

(1)

In many applications of cyclic monoenes only small amounts of residual polyenes can be tolerated in the product. Since the residual polyenes cannot always be easily separated from the monoene product by distillation (this is true for cyclododecene), the extent of conversion to which the polyenes must be taken is of necessity very high. Initially, we screened known catalysts that were described in the literature as being capable of performing certain specific selective hydrogenations. None of these catalysts displayed both the high selectivity and productivity we desired. After testing a large number of additional coordination complexes, we found that ruthenium complexes were capable of catalyzing selective hydrogenations. Our goal was finally reached on our 198th hydrogenation: at ~100% conversion of cyclododecatriene and cyclododecadiene, a >98% yield of cyclododecene was realized. The catalyst was $RuCl_2(CO)_2(PPh_3)_2$, and the hydrogenation was carried out in the presence of added PPh_3 at 140° and 150-200 psig of H_2 (Fahey, 1973a, 1974). Under these conditions, the ruthenium complex is very stable and highly productive. The complex is easily synthesized, is not readily poisoned, can be handled in air, and can be recycled without a loss of activity. Other multifunctional unsaturated organic molecules besides cyclododecatriene have been hydrogenated by $RuCl_2(CO)_2(PPh_3)_2$, and various additional selective hydrogenations have been discovered. This report surveys the kinds of selectivity that may be achieved in hydrogenations and describes some chemistry of the $RuCl_2(CO)_2(PPh_3)_2$ complex. Where relevant, work by other scientists is also discussed.

RESULTS AND DISCUSSION

Catalyst

Stephenson and Wilkinson (1966) were the first to prepare $RuCl_2(CO)_2(PPh_3)_2$. This was accomplished by bubbling CO into a refluxing ethanolic solution of $RuCl_3$ for five hours followed by reaction with excess PPh_3. They reported a 45% yield of colorless crystalline product. Using their procedure, we normally obtained yields between 40 and 60%. In some instances, we isolated a pale yellow $RuCl_2(CO)(PPh_3)_3$ complex instead.

This complex could in turn be converted to the dicarbonyl derivative by refluxing its solution in 2-methoxyethanol, containing a small amount of $CHCl_3$, for two hours as CO was bubbled into the solution. On cooling, $RuCl_2(CO)_2(PPh_3)_2$ precipitated from solution in an overall 56% yield from $RuCl_3$. Probably the most economical synthetic technique is the in situ preparation of the complex prior to a hydrogenation reaction. As a 0.02 molar ethanolic solution of $RuCl_3$ containing excess PPh_3 is stirred and heated under an atmosphere of H_2 and CO (excess CO preferred), the solution gradually loses its brown color as the colorless $RuCl_2(CO)_2(PPh_3)_2$ complex is formed. After the solution is completely colorless, the solution is cooled and the gases are vented. The olefin may then be added and hydrogenation can be carried out conventionally. After hydrogenation, when the solution has cooled, $RuCl_2(CO)_2(PPh_3)_2$ precipitates from solution in 86-100% yields. Attempts to synthesize large quantities of the complex by greatly increasing the concentration of ruthenium in the solution sometimes led to an unidentified black solid instead of $RuCl_2(CO)_2(PPh_3)_2$.

Analytically pure high-melting colorless crystals of $RuCl_2(CO)_2(PPh_3)_2$ are obtained by recrystallization of the complex from hot 2-methoxyethanol containing a small amount of chloroform. A single stereoisomer is isolated which is presumably the thermodynamically most stable. Portions of its infrared and proton decoupled ^{13}C nmr spectra are shown in Figure 1. The presence of two C=O and two Ru-Cl stretching vibrations in the infrared spectrum indicates that the CO groups are cis and that the Cl groups are cis. In the ^{13}C nmr spectrum, the substituted, ortho, and meta carbon resonances all occur as "apparent" 1:2:1 triplets as a result of strong P-P coupling, a feature characteristic of trans $R_3P-M-PR_3$ complexes. The complex therefore has the configuration shown in eq. 2.

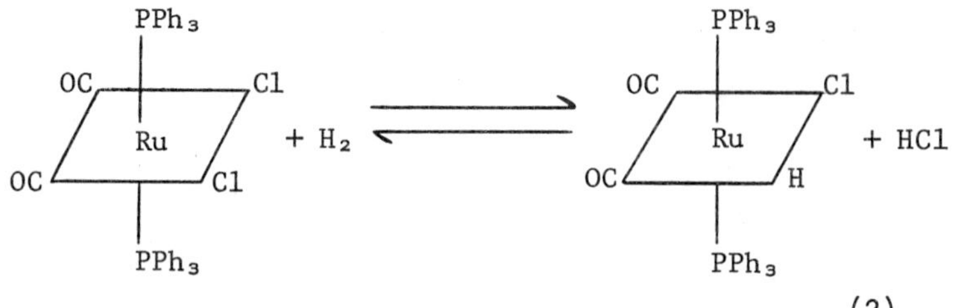

(2)

Although $RuCl_2(CO)_2(PPh_3)_2$ can be isolated unchanged after hydrogenations, it is most likely in equilibrium with a hydridoruthenium chloride complex under the reaction conditions.

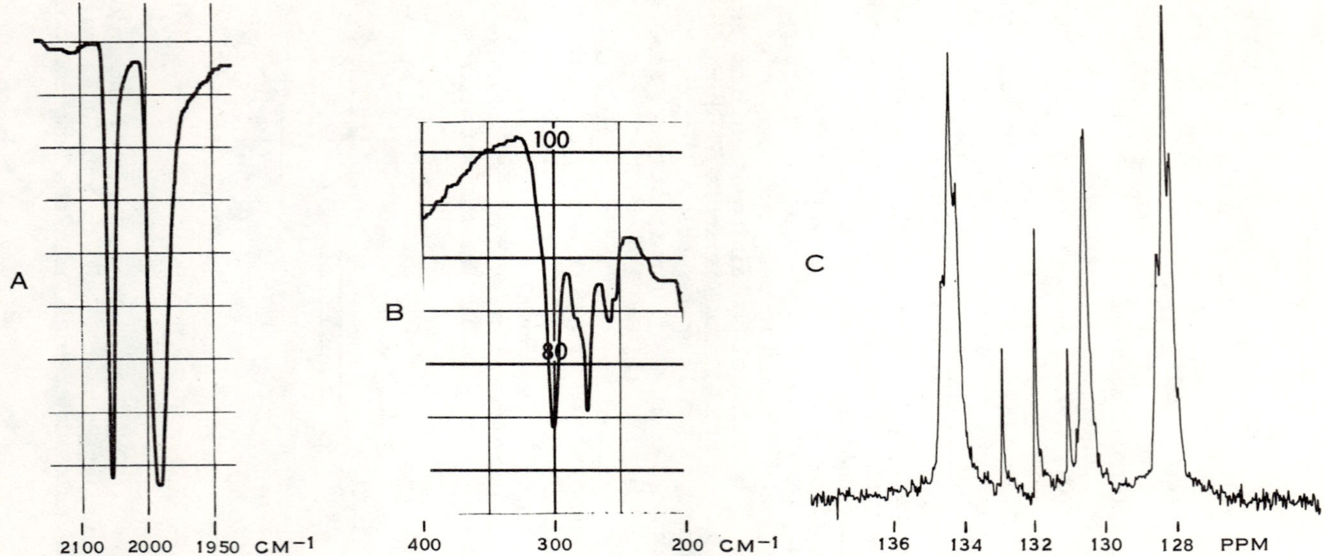

Fig. 1. *Selected portions of infrared and ^{13}C nmr spectra of $RuCl_2(CO)_2(PPh_3)_2$: A, infrared C=O stretching region (CH_2Cl_2); B, infrared Ru-Cl stretching region (CsI); C, 25 MHz proton-decoupled ^{13}C nmr of aryl carbons ($CDCl_3$). Peak assignmnents are: δ 128.2 meta carbons, δ 130.5 para carbons, δ 131.9 substituted carbons, and δ 134.4 ppm ortho carbons.*

It would be very desirable and informative to also study the catalytic properties of $RuClH(CO)_2(PPh_3)_2$. With this in mind, we made several attempts to synthesize the hydridoruthenium chloride complex from the dichloride. Treatment of $RuCl_2(CO)_2$-$(PPh_3)_2$ with $LiAlH_4$, $NaBH_4$, or N_2H_4 produced colorless to pale yellow solids whose infrared spectra were rich in absorptions between 1850 to 2060 cm^{-1}. Although most of the absorptions were carbonyl peaks, some likely were Ru-H stretching vibrations. Extensive recrystallization of the solid products sometimes yielded the starting dichloride, but we were unable to isolate the desired $RuClH(CO)_2(PPh_3)_2$ complex. The catalytic behaviour of these products resembled that of the parent dichloride complex, with the exception that some of the products deposited ruthenium metal during hydrogenations. We also attempted the synthesis of organoruthenium complexes as possible models of the reaction intermediates. It was anticipated that hydrogenation of the organoruthenium complex might also yield the desired hydridoruthenium chloride complex. However, reactions of CH_3Li and C_6H_5MgBr with $RuCl_2(CO)_2(PPh_3)_2$ both produced intractable product mixtures. Another route to the hydridoruthenium chloride complex that was tried was the reaction of $RuCl_2(CO)_2(PPh_3)_2$ with hydrogen in the presence of $HNEt_2$. At moderate temperatures, no reaction occurred; at 140° reaction took place but the mixture turned black, and a black solid, possibly elemental ruthenium, was deposited. When an olefin was present, it was hydrogenated under these conditions. Our final attempts to synthesize the hydridoruthenium chloride involved the treatment of $Ru(CO)_3(PPh_3)_2$ with HCl. Even with deficiencies of HCl, the only component that we isolated from the product mixture was $RuCl_2(CO)_2(PPh_3)_2$. Thus, the pure $RuClH(CO)_2(PPh_3)_2$ complex eluded us. For routine hydrogenations, it appears that the $RuCl_2(CO)_2(PPh_3)_2$ complex is the most convenient form in which the catalyst can be handled.

After we had completed these attempts to prepare the hydride, James and Markham (1971) reported the isolation of a $RuClH(CO)_2(PPh_3)_2$ complex by carbonylation of $RuClH(PPh_3)_3$ (eq. 3). The hydridoruthenium chloride was formed in a mixture with the dicarbonyl dichloride, but it could be isolated in 10% yield by careful fractional crystallization. The complex was assigned a trans H-Ru-Cl configuration on the basis of spectroscopic studies (James et al, 1973). The configurational stability of the complex was apparently not studied.

Why have we concentrated our efforts on the dicarbonyl complex? Primarily because this complex is the most stable catalyst we have encountered that will selectively hydrogenate 1,5,9-cyclododecatriene. It is not the most selective hydrogenation catalyst. If the dicarbonyl complex is good, wouldn't

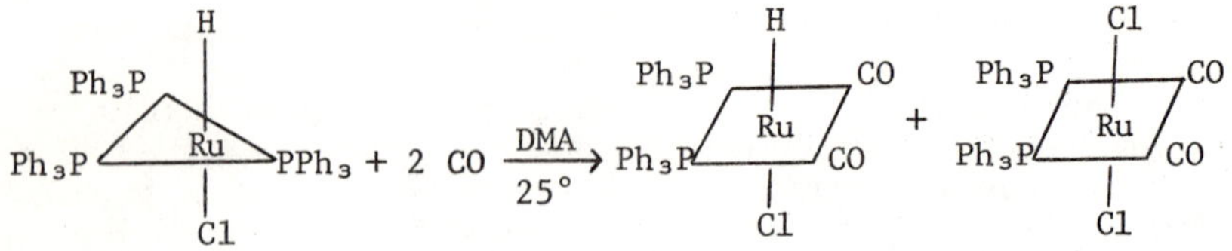

(3)

the tricarbonyl complex be better? Selectivities in 1,5,9-cyclododecatriene hydrogenation by mono-, di-, and tricarbonyl ruthenium complexes are compared in Table I. In the presence of added PPh_3, the dicarbonyl complex provides maximum conversion of the polyenes without excessive overhydrogenation of the monoene.

TABLE I. *Ruthenium Carbonyls as Catalysts for Selective Hydrogenation of 1,5,9-Cyclododecatriene in Ethanol*[a]

Complex	Molarity	Temp. °C	Time Hr.	Yield, % CDA	CDE	CDD + CDT
$RuClH(CO)(PPh_3)_3$[b]	0.017	130-140	1.5	24	76	0
$RuCl_2(CO)_2(PPh_3)_2$	0.006	140-150	2.3	0.1	98.8	1.1
$Ru(CO)_3(PPh_3)_2$+HCl[c]	0.006	140-150	3.5	0.1	92.5	7.4

[a]CDT was 0.55 M and added PPh_3 was 0.068 M. [b]Complex was prepared in situ and was isolated and characterized after the hydrogenation was completed. [c]Hydrogen uptake was not perceptible after 2.2 hours. $RuCl_2(CO)_2(PPh_3)_2$ was isolated from the product solution.

Hydrogenations

For convenience, most reactions were performed in aerosol compatibility bottles heated in a controlled-temperature oil bath. A dip tube extending into the reaction solution allowed samples to be withdrawn during hydrogenations. This kind of system is desirable since $RuCl_2(CO)_2(PPh_3)_2$ will function as a catalyst only at elevated temperatures and preferably at superatmospheric hydrogen pressures. The minimum threshold temperature at which hydrogenation occurs at a significant rate is 110-120° for reactive conjugated dienes and nitroaromatics, 135-140° for non-conjugated dienes and terminal olefins, and 150-160° for α,β-unsaturated carbonyl compounds, aldehydes, and some non-terminal olefins. Hydrogenations can be conducted at temperatures up to 200°, above which the catalyst slowly decomposes (Fahey, 1974). The catalytic properties of $RuClH(CO)_2(PPh_3)_2$ were briefly examined by James et al, (1973) under an atmosphere of CO and H_2 (1:1), and it is instructive to mention their results here. They observed very slow hydrogenation of methyl vinyl ketone at 80°, while 1-octene and 1-hexene were essentially unreactive at 70 and 100°, respectively. Thus, even the preformed hydridoruthenium chloride complex requires high temperatures for catalytic activity. We had previously observed that the presence of a variety of reducing agents in hydrogenations catalyzed by $RuCl_2(CO)_2(PPh_3)_2$ failed to lower the threshold hydrogenation temperature (Fahey, 1973a). Most of our studies have been performed under 100-200 psig of hydrogen, but hydrogen uptake can occur at any positive hydrogen pressure. In the absence of hydrogen, neither olefin hydrogenation nor isomerization occurs in benzene solution. If the available hydrogen is completely consumed during a hydrogenation, the catalyst ceases all activity.

Hydrogenations can be run in most organic solvents including hydrocarbons, ethers, alcohols, esters, and amides, or in no solvent at all. Dimethyl sulfoxide, sulfolane, and acetonitrile all retard olefin hydrogenations. For some olefins, N,N-dimethylformamide as solvent possesses some interesting solubility characteristics that can facilitate separation of the product from the catalyst. An example of this is mentioned later.

In many hydrogenations, high selectivities are achieved only in the presence of added PPh_3 or certain other electron-pair donors. This is clearly demonstrated by comparing the composition of reaction mixtures during the hydrogenation of 1,5,9-cyclododecatriene in the absence and in the presence of added PPh_3 (Fahey, 1973a). Cyclododecane is rapidly formed in the hydrogenation without added PPh_3, but its formation is severely inhibited in the hydrogenation containing added PPh_3. This effect is dependent on the concentration of PPh_3. Since our primary interest in the catalyst is its capability with regard to selective hydrogenations, most of our studies were carried out in the presence of added PPh_3. The minimum concentration necessary for a highly selective hydrogenation of 1,5,9-cyclododecatriene is 0.0356 molar (Fahey, 1973b,c), and

TABLE II. *Selectivity of Polyene to Monoene Hydrogenation at 99-100% Polyene Conversion*[a]

Polyene	PPh_3, M	$RuCl_2(CO)_2$-$(PPh_3)_2$, M	Time Hr	Yield, % Polyenes	Yield, % Alkenes	Yield, % Alkanes	Selectivity[b]
1,4-Cyclohexadiene	0.037	0.0031	1.7	0.5[c]	99.5	0.1	1.00
1,3-Cycloheptadiene	0.036	0.0031	10.3	0.3	76.6	23.0	0.77
1,5-Cyclooctadiene[d]	0.078	0.0031	3.2	0.8	93.4[e]	6.1	0.94
1,5,9-Cyclododecatriene[d]	0.036	0.0025	12.9	0.6	97.3	1.9	0.98
Norbornadiene[d]	0.078	0.0031	1.0	0.4	80.2	19.4[f]	0.81
1,3-Pentadiene[d]	0.036	0.00031	0.5	0.0	99.3	0.7	0.99
1,4-Pentadiene	0.067	0.0058	1.0	0.0	97.0	3.0	0.97
3,3-Dimethyl-1,4-Pentadiene[d]	0.078	0.0015	6.0	0.3	37.5	62.3	0.37
4-Vinylcyclohexene	0.036	0.0031	64	1	85	14	0.86

[a]Reactions were run in benzene or toluene solutions at 140° under 150-200 psig. Polyene concentrations were 0.25-0.59 M. [b]Defined in text. [c]Polyene is benzene. [d]Fahey (1973c). [e]Alkenes are cyclooctene (93.0%) and bicyclo[3.3.0]oct-2-ene (0.4%). [f]Alkanes are norbornane (2.4%) and nortricyclene (17.0%).

many of the hydrogenations to be described used this concentration.

In a consecutive reaction of the type polyene ⟶ monoene ⟶ alkane, selectivity is normally defined as [monoene]/([monoene] + [alkane]). Under this definition, selectivities are dependent on the extent of polyene conversion, and as polyene conversion increases, the selectivity decreases. Therefore comparisons of selectivities in hydrogenations are preferably made at the same polyene conversions. Selectivity values may be of more theoretical than practical value since the maximum obtainable yield of monoene is not necessarily revealed.

Table II contains the results of several hydrogenations carried to at least 99% conversion of the polyenes. In most cases, the selectivity is very high. When more than one alkene product is possible, an equilibrium mixture of isomers is produced (e.g., the pentenes derived from the two pentadienes are an equilibrium mixture of 1-pentene, cis-2-pentene, and trans-2-pentene). By-products were formed during the hydrogenations of 1,4-cyclohexadiene (disproportionation to benzene and cyclohexene) and 1,5-cyclooctadiene and norbornadiene (ring closures to bicyclo[3.3.0]oct-2-ene and nortricyclene, respectively). Benzene formation from the cyclohexadiene could have occurred spontaneously, but the ring closure reactions are most certainly metal-catalyzed processes. Double-bond migrations in the polyene were observed during hydrogenations of 1,4-cyclohexadiene, 1,5-cyclooctadiene, 1,4-pentadiene, and 4-vinylcyclohexene, so that most of the hydrogenation actually occurred via an isomer of the starting polyene. In examples with the lower selectivity values, high monoene yields can be realized at lower polyene conversions. In Figure 2, the mole % of alkene is 85% at 95% conversion of 1,3-cycloheptadiene. Similarly, Figure 3 shows that there is 67% alkene at 84% conversion of 3,3-dimethyl-1,4-pentadiene. In a truly non-selective hydrogenation (i.e., one in which hydrogenation rates of the alkene and the polyene are equal), the concentration of the alkene would not exceed 40% during the reaction. An interesting contrast is provided in the hydrogenation of 3,3-dimethyl-1,4-pentadiene by PtO_2, where the monoene concentration never reached 5% during the reaction. In this case, the hydrogenation shows a high selectivity to the alkane throughout the reaction. 5-vinyl-1,3-cyclohexadiene was partially hydrogenated to see if isomerization to ethylbenzene would predominate over a hydrogenation reaction, but only a few percent of ethylbenzene formed and hydrogen uptake proceeded normally.

Other kinds of selective hydrogenations have also been explored. α,β-Unsaturated esters and ketones can be hydrogenated to the saturated derivative without disturbing the func-

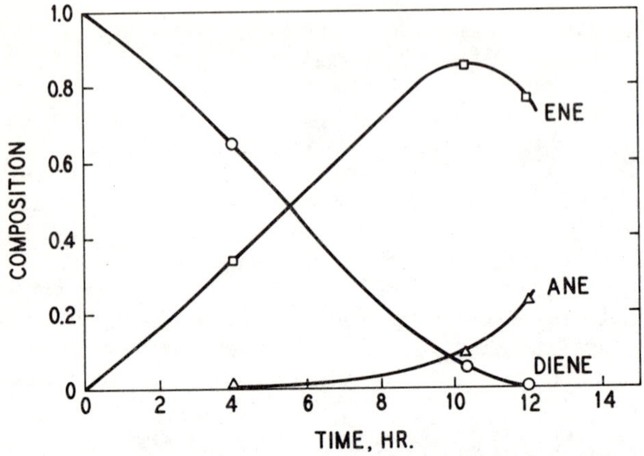

Fig. 2. Composition of reaction mixture during hydrogenation of 1,3-cycloheptadiene catalyzed by $RuCl_2(CO)_2(PPh_3)_2$. Reaction conditions are listed in Table II.

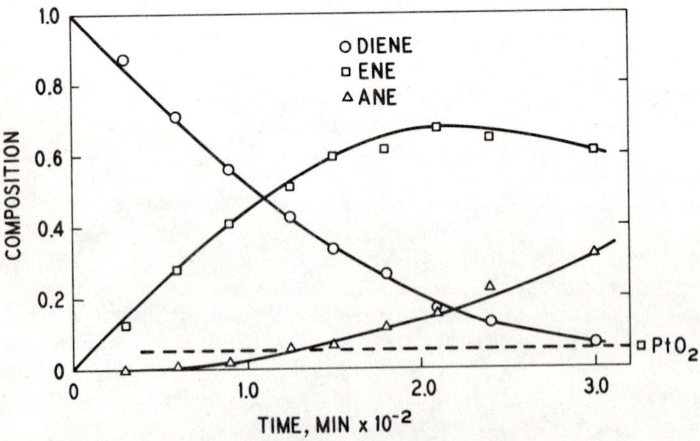

Fig. 3. Composition of reaction mixture during hydrogenation of 3,3-dimethyl-1,4-pentadiene catalyzed by $RuCl_2$-$(CO)_2(PPh_3)_2$. Reaction conditions are listed in Table II.

tionality, as shown in Table III. When maleic anhydride was hydrogenated, succinic anhydride was the major product, but compounds containing ester and carboxylic acid groups were also formed.

TABLE III. *Hydrogenation of Olefins Bearing Functional Unsaturation*[a]

Olefin	Time, Hrs.	Products (%)
Ethyl acrylate	14.5	Ethyl acrylate (0.4), Ethyl propanoate (99.6), Ethanol (0)
Cyclopent-2-enone[b]	12.8	Cyclopentanol (1.1), Cyclopentenone (0.7), Cyclopentanone (98.2)
Maleic anhydride	7.0	Succinic anhydride (>50)

[a]Solutions were 0.0031 M $RuCl_2(CO)_2(PPh_3)_2$ and 0.3-0.6 M olefin in benzene. The reaction was performed at 150° under 160-190 psig H_2. [b]Solution contained 0.037 M PPh_3.

Certain functional groups can be selectively hydrogenated by $RuCl_2(CO)_2(PPh_3)_2$. Knifton and Suggit (1973) found that 1-nitronaphthalene was selectively hydrogenated to the amine in the presence of 2-nitromesitylene (eq. 4). This selectivity must be attributed to the difficulty in hydrogenating the sterically hindered mesityl nitro group. Electronic factors presumably govern the selective hydrogenation of p-dinitrobenzene to p-nitroaniline in the presence of nitrobenzene (eq. 5) (Knifton and Suggit, 1973). Electron-withdrawing substituents appear to activate the nitro group towards hydrogenation.

Hydrogenation of aldehydes has also been catalyzed by a $[RuCl_2(CO)_2PPh_3]_2$ complex. Frediani et al, (1973) observed that the hydrogenation of cyclohexylformaldehyde gave a 77% yield (97% selectivity) of cyclohexylmethanol at 79% aldehyde conversion in 0.5 hrs at 150°. They also noted that this complex is a poor hydroformylation catalyst. Since $Co_2(CO)_8$ is an active hydroformylation catalyst but a poor hydrogenation

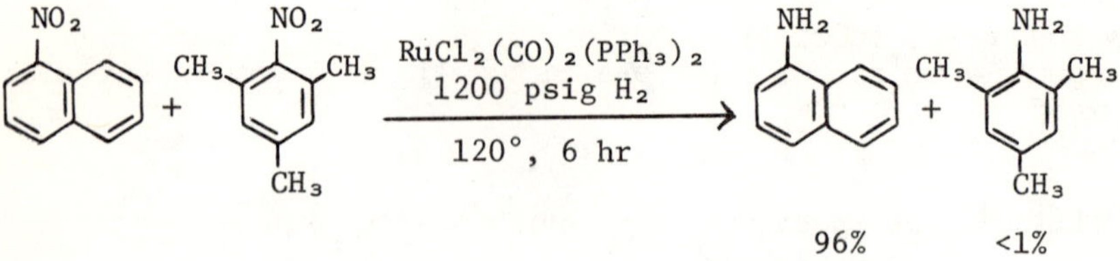

(4)

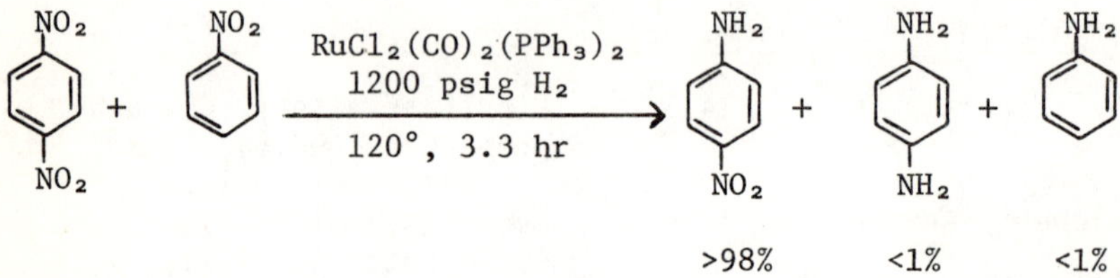

(5)

catalyst, it is usually difficult to obtain alcohols directly from the alkenes. Frediani et al, (1973) teamed $Co_2(CO)_8$ with $[RuCl_2(CO)_2PPh_3]_2$ and successfully produced alcohols from olefins in one step (Table IV). This reaction takes advantage of the greater reactivity of $Co_2(CO)_8$ over $[RuCl_2(CO)_2PPh_3]_2$ towards alkenes, and the ability of $[RuCl_2(CO)_2PPh_3]_2$ to rapidly hydrogenate aldehydes.

TABLE IV. *Hydroformylation/hydrogenation of Olefins by* $Co_2(CO)_8/[RuCl_2(CO)_2PPh_3]_2$[a,b]

Olefin	Time, hr.	Yield, % Alkane	Alcohol	Aldehyde
Cyclohexene	5	5	92	trace
1-Pentene	4	4	86	7
1-Octene	4	5	83	10
4-Methyl-1-pentene	5	6	83.5	10

[a]Co/Ru is 3/1; 40-50 atm of CO; 100 atm of H_2; 100% conversion of olefin; temp. = 150°. [b]Taken from Frediani et al, (1973).

Stereoselectivity

The hydrogenation of a 1:2 mixture of cis- and trans-1,3-pentadiene yielded the thermodynamic equilibrium mixture of pentenes: 1-pentene (5%), cis-2-pentene (29%), and trans-2-pentene (66%). Approximately the same mixture of pentenes was formed even during the initial stages of the hydrogenation. cis-1,3-Pentadiene is consumed slightly faster than the trans isomer since the cis/trans ratio of unreacted pentadienes was 1:4 at 98% conversion. Thus the reaction does not yield a hydrogenated product stereospecifically but does show a slight rate preference in hydrogenating cis vs trans dienes. The same mixture of pentenes is also eventually formed in the hydrogenation of 1,4-pentadiene and in the isomerization of 1-pentene; but, for these reactions, the initial products differ.

Isomerization of 1-pentene to 2-pentene occurs with a high preference for the cis isomer over the trans. At 15% isomerization of 1-pentene, the cis to trans ratio of 2-pentenes is 3:1. At equilibrium the cis to trans ratio is 0.44:1. A similar preference for formation of the cis-2-ene isomer was observed in an isomerization-hydrogenation experiment with 1,4-pentadiene (eq. 6). Under nitrogen, no reaction occurred. The pressure in the system was then increased by 10 psig with a dose of hydrogen which was rapidly absorbed.

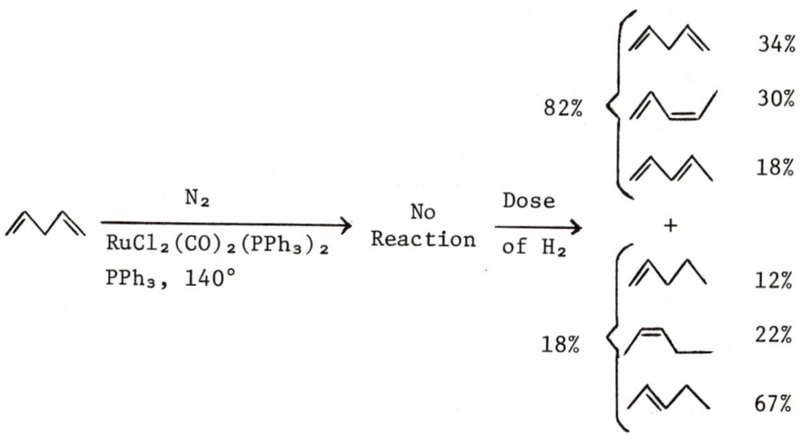

(6)

The product mixture shown in eq. 6 was formed, and it did not change further. Again the stereochemical preference for isomerization of the terminal alkene to the cis-2-ene isomer is exhibited. In eq. 6 the 2-pentenes are formed in the near-equilibrium ratio, while the relatively large amount of 1-pentene can be attributed to a direct hydrogenation of 1,4-pentadiene.

In a practical situation, any stereochemical preferences shown by the catalyst are eventually erased by the isomerization action of the catalyst.

Kinetics and Mechanism

Several features must be accommodated in any proposed mechanism. The relative hydrogenation rates of several alkenes and alkadienes, in the presence of added PPh_3, were found to decrease in the order: conjugated dienes > nonconjugated terminal dienes > nonconjugated internal dienes > terminal alkenes > internal alkenes (Fahey, 1973b,c). Double-bond isomerization is very rapid and is normally faster than hydrogenation. Added PPh_3 slows hydrogenations to different degrees.

Homogeneous hydrogenations catalyzed by $RuClH(PPh_3)_3$ (James and Markham, 1973), $RuH(OCOCF_3)(PPh_3)_3$ (Rose et al, 1969), $RhH(CO)(PPh_3)_3$ (O'Connor and Wilkinson, 1968), and $CoH(CO)_2(PBu_3)_2$ (Ferrari et al, 1972) all obey the rate law shown in eq. 7. The terms $\underline{K}_1$, $\underline{K}_2$, and k' correspond to the equilibrium and rate constants for the processes shown in eqs. 8-10, and $[H_2]$, $[Cat]_o$, [S], and $[PR_3]$ are hydrogen, initial catalyst, olefin, and free t-phosphine concentrations. We have also analyzed our rate studies with the $RuCl_2(CO)_2(PPh_3)_2$ catalyst using eq. 7, and values determined for $\underline{K}_1$, $\underline{K}_2$, and $\underline{k}'$ are presented in Table V.

$$\text{Rate} = \frac{\underline{k}'\underline{K}_2[H_2][Cat]_o[S]}{1 + \underline{K}_2[S] + [PR_3]/[\underline{K}_1]} \quad (7)$$

$$MHL_n(PPh_3) \underset{}{\overset{\underline{K}_1}{\rightleftharpoons}} MHL_n + PPh_3 \quad (8)$$

$$\text{Alkene} + MHL_n \overset{\underline{K}_2}{\rightleftharpoons} M(\text{alkyl})L_n \quad (9)$$

$$M(\text{alkyl})L_n + H_2 \xrightarrow{\underline{k}'} \text{Alkane} + MHL_n \quad (10)$$

For comparison, similar data for hydrogenations catalyzed by $RuClH(PPh_3)_3$ and $CoH(CO)_2(PBu_3)_2$ are included in the table. These data do, in fact, meet our requirements. Relative hydrogenation rates for olefins are simply based on the values of

TABLE V. *Rate and Equilibrium Constants for Catalytic Hydrogenations*

Substrate	K_2, $mol^{-1}l$	k', $mol^{-1}l\,sec^{-1}$	Relative Rate
	for $RuCl_2(CO)_2(PPh_3)_2$ in benzene at 140°; K_1 = 0.025 mol l^{-1}		
1,3-Pentadiene	~6.0	~6.2	20,000
3,3-Dimethyl-1,4-pentadiene	4.0	2.6	320
1,5,9-Cyclododecatriene	2.0	0.77	61
Styrene	3.6	0.38	41
Cyclododecene	~0.5	~0.09	1
	for $RuClH(PPh_3)_3$ in DMA at 35°;[a] K_1 = 0.03 mol l^{-1}		
1-Alkenes	1.1-2.6	520-870	1,500
4-Methoxystyrene	0.7	410	
Cyclooctene	0.6	280	
Cyclohexene	0.2	18	
Maleic Acid	260	4	1
	for $CoH(CO)_2(PBu_3)_2$ in hexane at 130°;[c] K_1 = 0.0016 mol l^{-1}		
Cyclohexene	0.17	0.058	

[a]Data taken from James and Markham (1973). [b]More recent scrutiny of the $RuClH(PPh_3)_3$ system by Professor James has led to the conclusion that K_1 is very small ($<10^{-5}$ mol l^{-1}). This reevaluation does not affect the relative rates and equilibrium constants. Professor James also feels that eqs. 8 and 9 are best combined (at least for the $RuClH(PPh_3)_3$ system to:

$$RuClH(PPh_3)_3 + \text{alkene} \underset{}{\overset{K_2}{\rightleftharpoons}} RuCl(\text{alkyl})(PPh_3)_2 + PPh_3.$$

(B.R. James, personal communication, June 1975). [c]Data taken from Ferrari et al (1972).

$\underline{K}_2$ and $\underline{k}'$. The importance of double-bond isomerization is a consequence of the small values for $\underline{k}'$. Rate dependencies on added PPh_3 concentration are related to the relative magnitudes of $\underline{K}_1^{-1}$ vs $\underline{K}_2$. The larger $\underline{K}_1^{-1}$ is with respect to $\underline{K}_2$, the greater will be the rate retarding influence of added PPh_3.

The rate and equilibrium constants for $RuCl_2(CO)_2(PPh_3)_2$ and $RuClH(PPh_3)_3$ follow similar trends. For conjugated dienes, however, a difference exists. Compared to 1-alkenes, conjugated dienes are hydrogenated relatively slowly by $RuClH(PPh_3)_3$ but rapidly by $RuCl_2(CO)_2(PPh_3)_2$. Both catalysts certainly must have large $\underline{K}_2$ values for conjugated dienes, hence $\underline{k}'$ must be very small for the $RuClH(PPh_3)_3$ system. The data in Table V also indicate that, for cyclic olefins, the rate of hydrogenation would follow $RuClH(PPh_3)_3 > RuCl_2(CO)_2(PPh_3)_2 > CoH(CO)_2$-$(PBu_3)_2$.

Preparative Scale Hydrogenation of 1,5,9-Cyclododecatriene

An important test of a catalyst comes when a small-scale reaction is scaled up to a preparative reaction. This was studied at catalyst-to-substrate mole ratios of 1:1200 and 1:4800 which are economically more attractive than the 1:100 ratio generally used in other studies. Since the desired cyclododecene product had to be relatively free from polyenes, the hydrogenations were taken to high conversions. Table VI shows that poorer selectivities and longer reaction times resulted. The products from these reactions were separated from

TABLE VI. *Preparative Scale Hydrogenations*[a]

CDT, mmol	$RuCl_2(CO)_2(PPh_3)_2$, mmol	PPh_3, mmol	Reaction Time, hr.	Yield, %		
				Polyenes	CDE	CDA
930	0.2	11	24	1.7	81.7	16.6
930	0.8	23	20	2.9	91.7	5.4
930	0.8	23	24	1.5	91.0	7.5

[a]Reactions were performed in a 300-ml autoclave in 60 ml of benzene at 150° and 600 psig of H_2.

the catalyst and solvent by distillation and were then subjected to the ethylene cleavage reaction in a scale-model olefin metathesis reactor. Satisfactory yields of 1,13-tetradecadiene were obtained.

Recycling of the catalyst was demonstrated in several hydrogenations. Probably the most informative example was when seven 50-g samples of 1,5,9-cyclododecatriene were hydrogenated in benzene solution by recycling a single 0.050 g sample of $RuCl_2(CO)_2(PPh_3)_2$ (Fahey, 1973a). No diminution of catalytic activity was observed throughout the study. A total of 32,100 moles of cyclododecatriene per mole of catalyst was hydrogenated without catalyst decay.

A useful technique that can simplify product isolation and catalyst recycle is the use of DMF as solvent. In general, polyenes are much more soluble in DMF than alkenes and alkanes so that, when cyclododecatriene is hydrogenated, the monoene product forms a separate phase. The DMF-catalyst phase can be drawn off the bottom of the vessel and recycled. For example, 15 g of CDT was hydrogenated by 0.05 g of $RuCl_2(CO)_2(PPh_3)_2$ in 20 ml of DMF. After cooling the solution, the top phase was decanted giving an 82% yield of cyclododecene (actual analysis: CDA, 1.7%; CDE, 96.7%; CDD+CDT, 1.5%; trace of DMF). The DMF-catalyst phase (presumably containing some residual CDD and CDT) was then recycled. A product distillation step may still be required, but catalyst separation is simplified.

CONCLUSIONS

The complex $RuCl_2(CO)_2(PPh_3)_2$ catalyzes the partial hydrogenations of multiunsaturated organic compounds, often with a high degree of selectivity. The catalyst is easily synthesized, convenient to use, and very stable. Its most important application is for the selective hydrogenation of 1,5,9-cyclododecatriene to cyclododecene.

ACKNOWLEDGMENTS

Valuable discussions with Drs. E. A. Zuech and J. E. Mahan are gratefully acknowledged. I also thank Dr. W. B. Hughes for permission to quote his hydrogenation studies (Table VI), Professor B. R. James for a copy of L. D. Markam's thesis, and G. R. Birdsong and B. Loffer for helpful experimental assistance.

ABBREVIATIONS USED

CDA	cyclododecane
CDD	cyclododecadiene
CDE	cyclododecene
CDT	cyclododecatriene
DMA	N,N-dimethylacetamide
DMF	N,N-dimethylformamide
L	ligand
n	unspecified integer
R	alkyl or aryl group
Ph	phenyl

REFERENCES

1. Fahey, D. R., J. Org. Chem., 38, 80 (1973a).
2. Fahey, D. R., Abstract 59, Sixth International Conference on Organometallic Chemistry, Amherst, Mass., Aug. 1973b.
3. Fahey, D. R., J. Org. Chem., 38, 3343 (1973c).
4. Fahey, D. R. (to Phillips Petroleum Co.), U.S. Patent 3,804,914 (April 16, 1974).
5. Ferrari, G. F., Andreetta, A., Pregaglia, G. F. and Ugo, R., J. Organometal. Chem., 43, 213 (1972).
6. Frediani, P., Bianchi, M. and Piacenti, F., Chem. Ind. (Milan), 55, 543 (1973).
7. James, B. R. and Markham, L. D., Inorg. Nucl. Chem. Letters, 7, 373 (1971).
8. James, B. R. and Markham, L. D., Abstract 58, Sixth International Conference on Organometallic Chemistry, Amherst, Mass., Aug. 1973; Markham, L. D., Ph.D. Thesis, U. of British Columbia (1973).
9. James, B. R., Markham, L. D., Hui, B. C. and Rempel, G.L., J. Chem. Soc. Dalton, 2247 (1973).
10. Knifton, J. F. and Suggit, R. M. (to Texaco Development Co.), British Patent 1,340,458 (Dec. 12, 1973).
11. O'Connor, C. and Wilkinson, G., J. Chem. Soc. (A), 2665 (1968).
12. Rose, D., Gilbert, J. D., Richardson, R. P. and Wilkinson, G., J. Chem. Soc. (A), 2610 (1969).
13. Stephenson, T. A. and Wilkinson, G., J. Inorg. Nucl. Chem., 28, 945 (1966).

POLYMER HYDROGENATIONS WITH SOLUBLE LITHIUM/COBALT AND ALUMINUM/COBALT CATALYSTS

JOHN CARL FALK
Roy C. Ingersoll Research Center
Borg-Warner Corporation
Des Plaines, Illinois 60018

ABSTRACT

Soluble polymer hydrogenation catalysts can be prepared by reacting an organoaluminum or organolithium compound with a transition metal salt of 2-ethylhexanoic acid. The catalyst activity is a function of lithium (or aluminum) - transition metal ratio, catalyst concentration, temperature, and hydrogen pressure. With the proper choice of catalyst composition and reaction conditions, the hydrogenation catalysts can discriminate between different types of polymeric unsaturation. Polymer hydrogenation may be used to prepare a number of novel polymer structures which are difficult and, in many cases, impossible to prepared by alternate routes.

INTRODUCTION

The early hydrogenation of polymers was limited to natural polymers and was, for the most part, of a destructive nature. For example, Berthelot in 1869 hydrogenated natural rubber with hydriodic acid isolating liquid products which he was unable to characterize.

Since Berthelot's experiment, the literature on polymer hydrogenation has been voluminous. In the early 1920's, when the concept of a macromolecule containing a large number of repeating units was not universally accepted, two structures of rubber were thought possible: (1) Either it was a low molecular weight unsaturated hydrocarbon, associated into colloidal particles, or (2) the molecules of rubber were long chains with many isoprene molecules connected by primary valences. According to C. Harries (1919, 1923) who, along with E. Fisher (1913), was an exponent of the low molecular weight colloid theory, the distinction between the two possibilities could be made by hydrogenation, "Hydrorubber could probably be distilled in vacuo without decomposition. Thus, its structure could be easily proved". Staudinger and Fritschi (1922) hydrogenated natural rubber at 100 atmospheres hydrogen at 270°C. The hydrorubber, which they isolated, had a molecular formula C_5H_{10}; it could not be distilled; it was a colloidal solution and had properties similar to natural rubber. This proved to be a major step in the acceptance of the concept of macromolecules as long molecules containing a number of repeating units for which Staudinger eventually received a Nobel Prize.

More recently, Parker (1948) reported the hydrogenation of polyacrylonitrile to polyallylamine. The polyacrylonitrile latex was hydrogenated with Raney nickel in the presence of ammonium hydroxide at 135-145°C and 750 psi hydrogen.

A poly(ethylene-vinyl chloride) copolymer may be formed by the hydrogenation of poly(vinyl chloride) with a Pd on charcoal catalyst (5%) with 900 psi hydrogen at 100 C in dimethylacetamide (Gluesenkamp, 1958). After 17 hours, 20% of the chlorine atoms were replaced with hydrogen atoms.

Isotactic poly(2-methoxy-1-vinylcyclohexane) was prepared by hydrogenation of poly(o-methoxystyrene) with a nickel catalyst at 180°C (Natta, 1962).

Hydrogen is present in the Ziegler-Natta catalyzed polymerization of propylene to control molecular weight and regenerate catalyst. The reaction may be viewed as hydrogenation of an active catalytic site on a polymer (Vandenberg, 1962 and Natta, 1959).

As a final example, Steinhofer (1965) hydrogenated polystyrene to poly(vinylcyclohexane) at 2,000 psi hydrogen at 165°C with a supported nickel catalyst.

A hydrogenated polymer was introduced into the marketplace in the early 50's by Phillips Petroleum Co. Hydropol®, made by hydrogenating emulsion polybutadiene with a nickel-kieselguhr catalyst at 150°C for eight hours, had excellent low temperature properties, good chemical resistance, and good electrical properties. It was used by the Department of Defense as a cable jacket in Alaska (Jones, 1953).

The foregoing polymer hydrogenations have all been carried out with heterogeneous metal catalysts. The remainder of the discussion will concern polymer hydrogenations using recently developed catalysts which have been variously called "soluble", "homogeneous", or "coordination" catalysts.

DISCUSSION

Heterogeneous catalysts for polymer hydrogenations have a number of characteristics. Reaction occurs on solid surfaces.

The surface species are difficult to characterize. In many cases, high temperatures and high hydrogen pressures are needed. High catalyst concentrations are required and reaction times are long. It is difficult to filter the heterogeneous catalyst from an organic solution of the hydrogenated polymer since at practical concentrations for polymer hydrogenation solution viscosities are high. Removal of the heterogeneous catalyst from a polymer latex, such as emulsion polybutadiene, is even more difficult.

By comparison, homogeneous catalysts offer many advantages. The solution species are often characterizable. Reaction temperature and hydrogen pressure are low. Catalyst concentration is low combined with rapid hydrogenation. Removal of catalyst from the reaction mixture is easy. The lithium-cobalt and aluminum-cobalt catalysts which I am about to discuss can be removed by washing the hydrogenated polymer in an organic solvent with dilute acid.

One of the earlier reports of hydrogenation with a soluble catalyst is the work of Sloan, et al (1963). Their catalyst was prepared by adding a trialkylaluminum such as triethylaluminum to a transition metal salt such as cobalt acetylacetonate. This catalyst was used to hydrogenate a number of olefins, two of which are listed in Table I. The hydrogenations were carried at 51-55 psi hydrogen in n-heptane or toluene with 0.5% catalyst. 1-Hexene was quantitatively hydrogenated in 2 hours at 30°C with a catalyst having an aluminum/cobalt ratio of 6.0. The hydrogenation of cyclohexene was more rapid with complete conversion occurring after 20 minutes at the same temperature with a catalyst having an aluminum/cobalt ratio of 4.0. In their study, Sloan and co-workers also examined n-butyllithium as the organometallic part of the catalyst. The lithium/cobalt catalyst had a lithium/cobalt ratio of 12.6 and was much less active than aluminum/cobalt catalysts. After 22 hours at 50°C, only 69% of the cyclohexane was hydrogenated. Approximately 1/70th of the time at a lower temperature was required to hydrogenate cyclohexene with an aluminum/cobalt catalyst.

They evaluated acetylacetonates other than cobalt and found that hydrogenation catalysts made from the cobalt (II) salt were more active than catalysts made from the iron (III) salt which was more reactive than catalyst made from chromium (III) salt.

The following reaction mechanism was proposed for these hydrogenation reactions.

TABLE I. *Olefin Hydrogenation*

Olefin	Al/Co	T(°C)	Time (Hours)	Conversion(%)
1-Hexene	6.0	30	2	100
Cyclohexene	4.0	30	0.33	100
Cyclohexene	12.6(Li/Co)	50	22	69

$$R_3Al + MX_n \rightarrow R_2AlX + RMX_{n-1} \quad \text{(A)}$$

$$RMX_{n-1} + H_2 \rightarrow RH + HMX_{n-1} \quad \text{(B)}$$

$$C = C + HMX_{n-1} \rightleftarrows HC - C - MX_{n-1} \quad \text{(C)}$$

$$HC - C - MX_{n-1} + H_2 \rightarrow HC - CH + HMX_{n-1} \quad \text{(B)}$$

In the first step of the reaction, the trialkylaluminum reacts with the transition metal salt to form an organo transition metal salt (A). The organo transition metal salt then reacts with hydrogen to form a transition metal salt hydride (B) which then adds across the olefin double bond to form a second organo transition metal salt (C). In the final step of the reaction organo transition metal salt (C) reacts with hydrogen to form the hydrogenated olefin product and regenerate the active catalyst species (B) which may then react with more olefin continuing this cycle until hydrogenation is complete.

Although this mechanism has generated some controversy and may or may not be the true mechanism, and the evidence for it has been meager, it is a starting point for a discussion of the hydrogenation of polymers with both soluble lithium/cobalt and aluminum/cobalt catalysts. According to Sloan's mechanism, the function of the organoaluminum is to supply a carbanion to the cobalt salt which is ultimately replaced by a hydride to form the active catalytic species. If so, then an organolithium at some ratio of organolithium to transition metal salt, should be good source of carbanions. The lithium/cobalt catalyst studied by Sloan may not have been the optimum composition.

The data in Table II (Falk, 1971) show that with the proper lithium/cobalt ratio, the catalytic activity is similar to aluminum/cobalt systems. The lithium/cobalt catalyst was made by adding a solution of cobalt (II) 2-ethylhexanoate to a solution of n-butyllithium over a period of 90 minutes. The catalyst was used immediately. The hydrogenations were carried out at 50°C, at 50 psi hydrogen, with 0.6% (mole) catalyst. The concentration of olefin in cyclohexane, the hydrogenation medium, was 3% (weight/volume). The ratio of lithium to cobalt was held constant at 4.0.

Using these reaction conditions, both 1-hexene and cyclohexene were quantitatively hydrogenated in 20 minutes compared to Sloan's report of 1-hexene's quantitative hydrogenation with an aluminum/cobalt catalyst having an Al/Co ratio of 6 in 120 minutes, and cyclohexene's quantitative hydrogenation with an aluminum/cobalt catalyst (Al/Co = 4) in 20 minutes. Thus, hydrogenation catalysts made from n-butyllithium and cobalt (II) 2-ethylhexanoate are as active as aluminum/cobalt catalysts. The lower activity reported by Sloan was due to the choice of lithium/cobalt ratio, 13.6, which resulted in 69% hydrogenation of cyclohexene in 1320 minutes while hydrogenation is quantitative in 20 minutes if the catalyst has a lithium/cobalt ratio of 4.0

Sterically hindered olefins require more drastic hydrogenation reaction conditions suggesting the catalyst possesses a degree of selectivity. 2-Methylpentene- 2 required 60 minutes for 92% conversion and 2,3-dimethylbutene - 2 required 90 minutes for only 37% conversion.

TABLE II. *Olefin Hydrogenation*

Olefin	Li/Co	Time (Min.)	Conversion (%)
1-Hexene	4.0	20	100
Cyclohexene	4.0	20	100
2-Methylpentene-2	4.0	60	92
2,3-Dimethylbutene-2	4.0	90	37

Lithium/cobalt hydrogenation catalysts can be made from organolithiums other than n-butyllithium (Falk, 1971), Table III. Ethyllithium, s-butyllithium, and phenyllithium can be used to make catalysts as active as n-butyllithium based catalysts. Cyclooctene hydrogenation was quantitative in 10 minutes. The cyclopentyllithium based catalyst was strangely less active. 90 minutes reaction time was required for 63% hydrogenation of cyclooctene and may result from impurities in the organo lithium. The hydrogenations reported in Table III were done at 50 psi, 50°C with 0.3% catalyst. In all cases the lithium/cobalt ratio was held constant at 4.0.

TABLE III. *Olefin Hydrogenation with Li/Co Catalysts as a Function of Li Source*

Type	Time (Min.)	Conversion (%)
Ethyllithium	10	100
s-Butyllithium	10	100
Phenyllithium	10	100
Cyclopentyllithium	90	63

Polymeric organo lithiums may also be used to make lithium/cobalt hydrogenation catalysts. Living anionic polybutadiene may be prepared by initiating the polymerization of butadiene in an inert solvent like cyclohexane with an alkyllithium. Addition of cobalt (II) 2-ethylhexanoate forms an active hydrogenation catalyst as indicated by the data in Table IV. Cyclooctene was hydrogenated with this catalyst at 0.3% concentration, 50 psi hydrogen at 50°C to cyclooctane with 51% conversion after 30 minutes. Active hydrogenation catalysts may also be made from living polyisoprene and polystyrene. These catalysts were prepared exactly as the living polybutadiene/cobalt (II) 2-ethylhexanoate catalyst. Hydrogenation conditions were the same. With both catalysts hydrogenation of cyclooctene was quantitative in 30 minutes. The difference in reactivity noted between living polybutadiene/cobalt (II) 2-ethylhexanoate catalyst and analogous catalysts based on living polyisoprene and living polystyrene is difficult to explain and may be an artifact of this set of experiments. Recently (Falk, 1973), polydienes have been transmetalated in high yield (>95%) with s-butyllithium and an equivalent amount of tetramethylethylenediamine.

TABLE IV. *Alternate Polymeric Sources of R-Li*

Type	% Conversion to Cyclooctane*
———$(CH_2CH{=}CHCH_2)$——— θ Li ⊕	51
ΔΔΔΔ$(CH_2CH{=}C(CH_3)CH_2)$ΔΔΔ θ Li ⊕	100
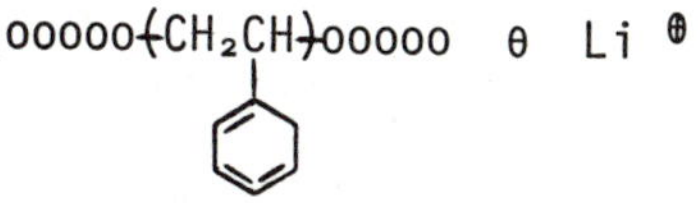	100
———————— PBD θ θ θ θ Li ⊕ Li ⊕ Li ⊕ Li ⊕	34

*30 Minute Reaction Time

During the transmetalation no polydiene degradation occurs if the reaction is run at room temperature for 2 hours. 100,000 molecular weight anionic polybutadiene was metalated to 10 sites per chain in this manner. This metalated polymer was used as the organolithium moiety to prepare a lithium/cobalt hydrogenation catalyst. Cyclooctene was hydrogenated in 34% conversion to cyclooctane in 30 minutes using this catalyst.

During the hydrogenation of cyclooctene with catalysts based on living polydienes or polylithiated polydienes, the polydiene moieties were also hydrogenated.

Hydrogenation catalysts made by reacting organo aluminums or lithiums with cobalt(II) salts are referred to in the literature as "soluble", "coordination", or "homogeneous" catalysts. They may be either/all or neither since little work has been reported on the nature of the catalysts species. Solutions of organo aluminums or lithiums are transparent with a slight yellow cast. The cobalt (II) 2-ethylhexanoate solution is deep blue. When the solutions are mixed together to form the hydrogenation catalyst solution, the color is black. In one experiment, a lithium/cobalt catalyst was filtered under nitrogen thru a 0.45μ millipore filter. A jelly like substance was filtered off. Figure 1 is a scanning electron micrograph photograph of this gel at a 5,000-fold magnification. No particulate structure is noted and the nature of the filtered material is not known. It may or may not be part of the active hydrogenation catalyst. Light scattering indicates the catalyst is at least, in part, colloidal in nature with 3,500-3,900Å size particles. Further work on the characterization of the catalyst is needed.

The remainder of the paper will discuss the hydrogenation of polybutadiene, polyisoprene and polystyrene which were prepared by anionic polymerization of the appropriate monomer in cyclohexane (Hsieh, 1965). The living polymers were quenched with an amount of isopropanol equivalent to the living ends and then diluted with sieve dried cyclohexane to the concentration desired for hydrogenation. The microstructure of the polybutadiene was 43% of cis 1,4, 49% trans 1,4, and 8% 1,2 while the polyisoprene was predominantly cis 1,4.

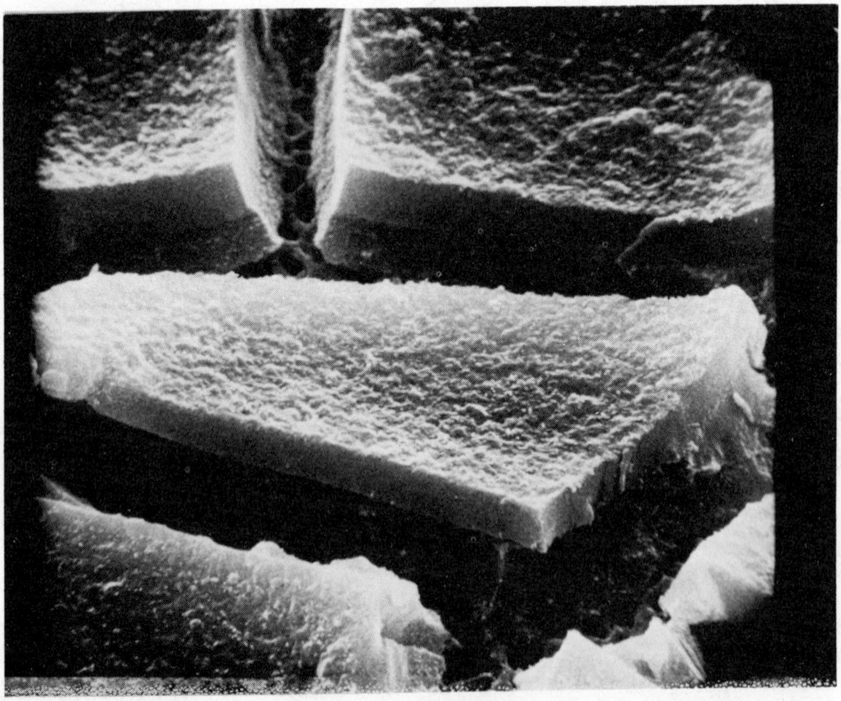

Fig. 1. Scanning Electron Micrograph of Li/Co Filtered Catalyst 5,000X

The hydrogenation of 100,000 molecular weight polybutadiene is described in Table V and how it is affected by the lithium/cobalt ratio of the catalyst. The hydrogenations were carried out at 50°C, 50 psi hydrogen, with 0.3 mole percent catalyst. The concentration of poly-1,4-butadiene in cyclohexane was 2-3% (w/vol).

TABLE V. *Hydrogenation of Polybutadiene*

$$\left(CH_2CH{=}CHCH_2\right) \xrightarrow{H_2} \left(CH_2CH_2CH_2CH_2\right)_n$$

Li/Co	Time (Min.)	Conversion (%)
0.95	60	0
3.0	20	100
5.0	30	100
9.9	90	0

Hydrogenation activity results from careful choice of lithium/cobalt ratio. At low lithium/cobalt ratios (0.95) no hydrogenation occurs. As the lithium/cobalt ratio is increased, catalyst activity increases, and at a lithium/cobalt ratio of 3.0 hydrogenation is complete in 20 minutes. Higher lithium/cobalt ratios, up to 5.0, also give active hydrogenation catalysts. At lithium/cobalt ratios greater than 5.0, catalyst activity diminishes and ceases at 9.9.

The polydienes cited in Table V and throughout the remainder of the paper are terminated with a 9% block of polystyrene to facilitate the determination of the percent conversion by infrared spectroscopy. The starting material is used as a reference against which the hydrogenated samples are compared. The block of polystyrene is common to all samples and acts as an internal standard. Thus, the amount of polydiene unsaturation may be determined, throughout the course of the reaction, from absorptions in the infrared characteristic of polystyrene and the polydiene with an accuracy of ±1-2%.

The hydrogenation of a polymer such as polybutadiene could involve random saturation of the double bonds, or the catalyst could be bound to the same polymer chain until hydrogenation is complete before moving on to the next chain. Anderson (1974) has shown that the saturation in partially hydrogenated polybutadiene containing 63% 1,2 polymerized butadiene was uniformly distributed. He reacted the partially hydrogenated polybutadiene with 2,3-dinitrobenzenesulfenyl chloride which adds across olefin unsaturation to form an UV active chromophor. The derivatized polybutadiene was then analyzed by a gel permeation chromatograph (GPC) equipped with both refractive index and ultraviolet detectors. The refractive index detector measures the macromolecular sizes in the usual manner and the ultraviolet detector measures the amount of unsaturation therein. One of Anderson's GPC curves is reproduced below (Fig. 3). He concluded that this polymer, a partially hydrogenated polybutadiene, has a relatively uniform distribution of unsaturation. The bell shaped curve is the refractive index curve and the horizontal line is the ultraviolet curve. The deviation in distribution of unsaturation at the low GPC count values is typical of the variations introduced at the molecular weight extremes due to experimental error.

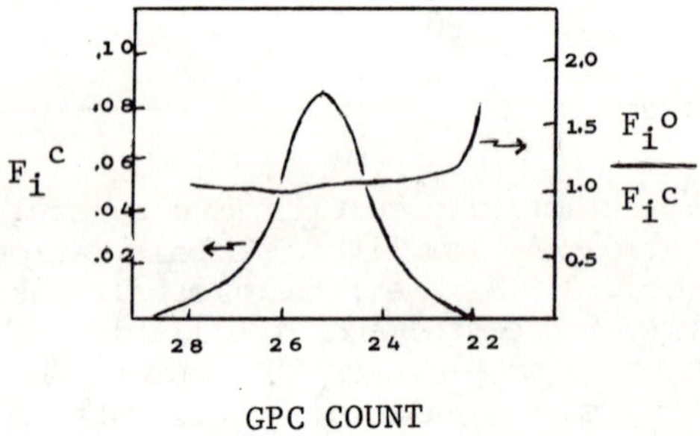

Fig. 3. The distribution of unsaturation in sample 2, a partially hydrogenated polybutadiene.

The work of Anderson confirms unpublished work in our laboratories on unsaturation distribution. Preparative GPC was used to isolate several molecular weight fractions of partially hydrogenated polybutadiene. Infrared analysis indicated that the amount of unsaturation was constant for the fraction collected.

The rate of hydrogenation of poly-1,4-butadiene decreases with a decrease in catalyst concentration at a constant temperature of 50°C and 50 psi hydrogen, Table VI (Falk, 1972). The catalyst was made from n-butyllithium and cobalt (II) 2-ethylhexanoate at a lithium/cobalt ratio of 5.0. At 0.3 mole % catalyst poly-1,4-butadiene is rapidly hydrogenated. Similar ac-

TABLE VI. *Hydrogenation of Polybutadiene*

	Catalyst (%)	Temperature (°C)	Hydrogen Pressure (psi)	Time (Min.)	Conversion (%)
CATALYST LEVEL	0.30	50	50	20	98
	0.20	50	50	20	95
	0.15	50	50	30	83
	0.10	50	50	20	50
TEMPERATURE	0.30	9	50	20	63
	0.30	25	50	20	73
	0.30	50	50	20	98
	0.30	75	50	20	97
	0.30	85	50	20	91
HYDROGEN PRESSURE	0.30	50	5	60	100
	0.30	50	25	20	99
	0.30	50	50	20	98

tivity is noted at 0.2 mole % catalyst. At 0.15 and 0.10 mole % catalyst a noticeable loss of activity is observed, 83% hydrogenation and 50% hydrogenation, respectively, after thirty and twenty minutes.

The activity of these catalysts is a function of temperature, Table VI. At a constant catalyst concentration of 0.3 mole % and hydrogen pressure of 50 psi, a minimum temperature of 50°C is required for rapid hydrogenation of poly-1,4-butadiene. At higher temperatures, 75 and 85°C, similar facile hydrogenation occurs. Temperatures lower than 50°C, e.g., 25 and 9°C, permit only partial hydrogenation of poly-1,4-butadiene.

Hydrogen pressure, as expected, effects the rate of hydrogenation of poly-1,4-butadiene. The catalyst concentration (0.3 mole %), temperature (50°C) and lithium/cobalt ratio (5.0) were held constant and the hydrogen pressure was varied, 50 psi, 25 psi, and 5 psi, Table VI. At pressures of 25 and 50 psi, hydrogenation occurs rapidly, while at 5 psi a longer reaction time is required for complete hydrogenation.

These coordination catalysts may be used to hydrogenate poly-1,4-isoprene (Falk, 1972). Table VII shows the effect of catalyst concentration on the hydrogenation of poly-1,4-isoprene. The polyisoprene was made by anionic polymerization and, although predominantly in the <u>cis</u> 1,4 configuration, contains a small amount of 1,2-polymer. The small amount of poly-1,2-isoprene is hydrogenated at the same rate or slightly slower than poly-1,4-isoprene. In this study, the catalyst was made from n-butyllithium and cobalt (II) 2-ethylhexanoate. The hydrogenations were carried out at 50°C with a hydrogen pressure of 50 psi. At 0.3 mole % catalyst concentration and lithium/cobalt ratios of 5.0 and 3.0, little hydrogenation occurs, 0, and 39% conversion. At these catalyst concentrations, poly-1,4-butadiene is quantitatively hydrogenated. The lack of hydrogenation of poly-1,4-isoprene is unexpected and may reflect loss of catalyst by oxidation, by reaction of catalyst with other impurities in the system, or by the innate lower reactivity of polyisoprene to hydrogenation. At higher catalyst concentrations, 1.2 and 3.0% (with a lithium/cobalt ratio of 3.0) hydrogenation is rapid, 77 and 100% conversion occurring respectively.

TABLE VII. *Hydrogenation of Polyisoprene*

Li/Co	Catalyst Concentration (%)	Time (Min.)	Conversion (%)
5.0	0.3	90	0
3.0	0.3	10	39
3.0	1.2	10	77
3.0	3.0	20	100

At a higher catalyst concentration, 3.0%, a catalyst concentration at which polyisoprene hydrogenation is quantitative, the rate is dependent on lithium/cobalt ratio, Table VIII (Falk, 1972). At ratio of 1.74 or lower, negligible hydrogenation occurs. As the lithium/cobalt ratio is increased hydrogenation rate increases until, at 2.5, conversion is quantitative in 20 minutes. Between lithium/cobalt ratios 2.5 and 5.0, hydrogenation is quantitative and rapid. Activity decreases above 5.0 until at a lithium/cobalt ratio of 6.5 no hydrogenation occurs after 90 minutes.

More drastic conditions are required to hydrogenate polystyrene to polyvinylcyclohexane. This is not unexpected in

TABLE VIII. *Hydrogenation of Polyisoprene*

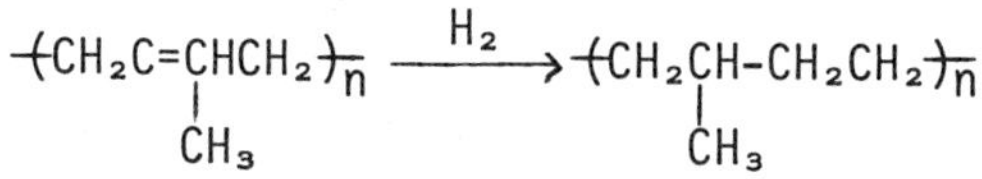

Li/Co	Time (Min.)	Conversion (%)
1.74	90	<10
2.5	20	100
5.0	10	100
6.5	90	0

view of the aromaticity of polystyrene unsaturation. The hydrogenation of polystyrene as a function of hydrogen pressure and temperature is reported in Table IX (Falk, 1972). The catalyst, made from n-butyllithium and cobalt (II) 2-ethylhexanoate with a lithium/cobalt ratio of 3.3, was evaluated at a concentration of 0.2 mole % based on styrene units. At this catalyst concentration, high pressures (a minimum of 2,000 psi) and temperatures (a minimum of 250°C) are required to achieve a reasonable conversion of polystyrene to polyvinylcyclohexane in short reaction times.

TABLE IX. *Hydrogenation of Polystyrene*

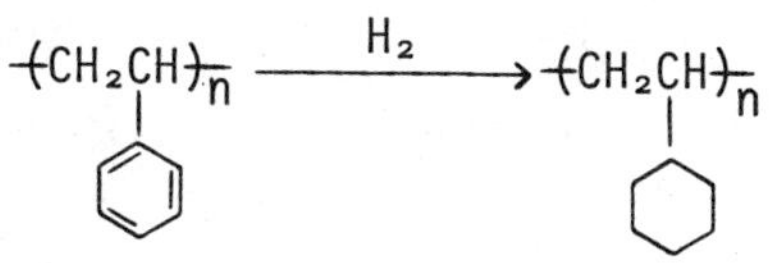

	Hydrogen Pressure (psi)	Temperature (°C)	Time (Hrs.)	Conversion (%)
HYDROGEN PRESSURE	500	300	4	9
	1000	300	2	97
	2000	300	2	97
	3000	300	2	100
	4400	300	3	100
TEMPERATURE	4400	100	3	8
	4400	250	3	81
	4400	300	3	100

The previously described work, showing the unsaturation in diene polymers to be more readily hydrogenated than unsaturation in polystyrene, is not unexpected in view of the aromaticity of the polystyrene unsaturation. More subtle differences exist in the trisubstituted ethylenic unsaturation in poly-1,4-isoprene and the disubstituted ethylenic unsaturation in poly-1,4-butadiene. Previous results reported in this paper have implied a difference in reactivity of these two structures in hydrogenation reaction.

The data of Table X (Falk, 1972) shows that hydrogenation selectivity results from careful choice of aluminum/transition metal ratios in catalysts made by reacting triethylaluminum and cobalt (II) 2-ethylhexanoate. In this study the hydrogenations were carried out at 50°C, 50 psi hydrogen, with 0.3 mole % catalyst. 100,000 molecular weight poly-1,4-butadiene and poly-1,4-isoprene were the rubbers hydrogenated.

At low triethylaluminum/cobalt (II) 2-ethylhexanoate ratios, 3.25-3.35, hydrogenation of both poly-1,4-butadiene and poly-1,4-isoprene occurs easily. At a higher aluminum/cobalt ratio, 3.40, the catalyst system hydrogenates poly-1,4-butadiene unsaturation but not poly-1,4-isoprene unsaturation.

TABLE X. *Soluble Selective Hydrogenation Catalysts Et_3Al/CoX*

Al/Co	Time (Min.)	Conversion (%) Poly-1,4-butadiene	Poly-1,4-isoprene
3.25	10	100	68
3.30	10	100	28
3.35	10	100	29
3.40	10	100	--
3.40	60	100	0

Alkyl or aryllithium compounds may be used in place of triethylaluminum in the preparation of selective hydrogenation catalysts, Table XI. The activity of these catalysts, and hence their selectivity, is a function of the molar ratio of lithium to cobalt analogous to the aluminum/cobalt catalysts. With a catalyst made from n-butyllithium and cobalt (II) 2-ethylhexanoate having a lithium/cobalt ratio of 5.0, poly-1,4-butadiene is hydrogenated with a conversion of >99% after 30 minutes while after 90 minutes under these conditions, polyisoprene is not hydrogenated.

Many other alkyllithiums may be used to prepare selective hydrogenation catalysts. Examples are s-butyllithium, ethyllithium, cyclopentyllithium, and phenyllithium. With these catalysts, the selectivity noted in Table XI is not absolute. However, improved selectivity should result with other lithium/cobalt ratios.

TABLE XI. *Selective Hydrogenation of Poly-1,4-butadiene With Soluble Hydrogenation Catalysts*[a]

Substrate[b]	Li Source	Li/Co	Time, Min.	Conversion(%)
PBD	n-Butyllithium	5.0	30	>99
PI	n-Butyllithium	5.0	90	0
PBD	Cyclopentyllithium	4.0	90	78
PI	Cyclopentyllithium	4.0	90	0
PBD	Phenyllithium	4.0	20	97
PI	Phenyllithium	4.0	90	37
PBD	sec-Butyllithium	5.0	10	95
PI	sec-Butyllithium	5.0	90	72
PBD	Ethyllithium	5.0	20	100
PI	Ethyllithium	5.0	90	27

[a]50°C; hydrogen pressure 50 psi; 0.3 mole % catalyst, 100,00 MW PBD and PI, 17% (w/w) solids in cyclohexane. [b]PBD = poly-1,4-butadiene; PI = poly-1,4-isoprene

Hydrogenation selectivity is also possible with block and random 1,4-copolymers of butadiene and isoprene. Table XII (Falk, 1971). Butadiene present either as blocks or randomly in copolymers with isoprene is hydrogenated completely, while the isoprene unsaturation remains. These hydrogenations were also carried out at 50°C, 50 psi hydrogen, at 0.3 mole % catalyst.

Hydrogenation selectivity is not restricted by molecular weight as evidenced by the data of Table XII. Whether the molecular weight of the copylymer is 1,000 or 100,000, poly-1,4-butadiene moieties are hydrogenated preferentially.

Transition metal 2-ethylhexanoate salts other than cobalt (II) give hydrogenation catalysts possessing high selectivity in polydiene hydrogenations, Table XII. Nickel (II) 2-ethylhexanoate may be used as a cocatalyst in the selective hydrogenation of poly-1,4-butadiene moieties in butadiene-isoprene copolymers. The activity of this catalyst is similar to its cobalt (II) 2-ethylhexanoate analog.

TABLE XII. *Hydrogenation of 1-4-butadiene-1,4-isoprene-1,4-butadiene Block and Random Copolymers*[a]

				Conversion (%)	
Entry	Al/Co	MW X 10^{-3}	Time, (Min.)	Butadiene Segments	Isoprene Segments
1	3.45	100	60	100	0
2[b]	3.45	100	60	100	0
3	3.45	1	20	100	0
4[c]	3.30	100	60	100	0

[a]35% polybutadiene (w/w). [b]Random butadiene, isoprene copolymers. [c]Nickel (II) 2-ethylhexanoate, triethylaluminum cocatalyst.

To summarize the effect of polymer structure on the rate of hydrogenation, reaction conditions can be chosen so that polybutadiene can be hydrogenated faster than polyisoprene which is much more reactive towards hydrogen than polystyrene.

Hydrogenation may be used to prepare a number of novel polymer structures which are difficult and, in many cases, impossible to prepare by alternate routes, Table XIII. An alternating ethylene-propylene copolymer may be prepared by hydrogenating anionic poly-1,4-isoprene (reaction 1). Linear, narrow molecular weight distribution, unbranched polyethylene may be prepared by the hydrogenation of poly-1,4-butadiene (reaction 2). Narrow molecular weight distribution polybutene-1 may be prepared by the hydrogenation of poly-1,2-butadiene (reaction 3). Sequential block copolymers or random copolymers of ethylene and butene-1 may be prepared by hydrogenating the corresponding 1,4-butadiene-1,2-butadiene copolymer (reaction 4). Selective hydrogenation of the butadiene moiety in isoprene-butadiene random or block copolymers afford copolymers of isoprene and ethylene (reaction 5). Polyvinylcyclohexane may be prepared by the hydrogenation of polystyrene (reaction 6).

A graft polymer of polystyrene onto polybutadiene may be prepared by metalating the polybutadiene backbone with a complex of s-butyllithium and N,N,N',N'-tetramethylethylenediamine followed by the use of this polylithio species to initiate styrene polymerization. Hydrogenation of the polybutadiene moiety results in a graft polymer of polystyrene on polyethylene (reaction 7) and complete hydrogenation of the graft copolymer affords a graft polymer of polyvinylcyclohexane on polyethylene (reaction 8).

TABLE XIII

NOVEL POLYMER STRUCTURES BY HYDROGENATION

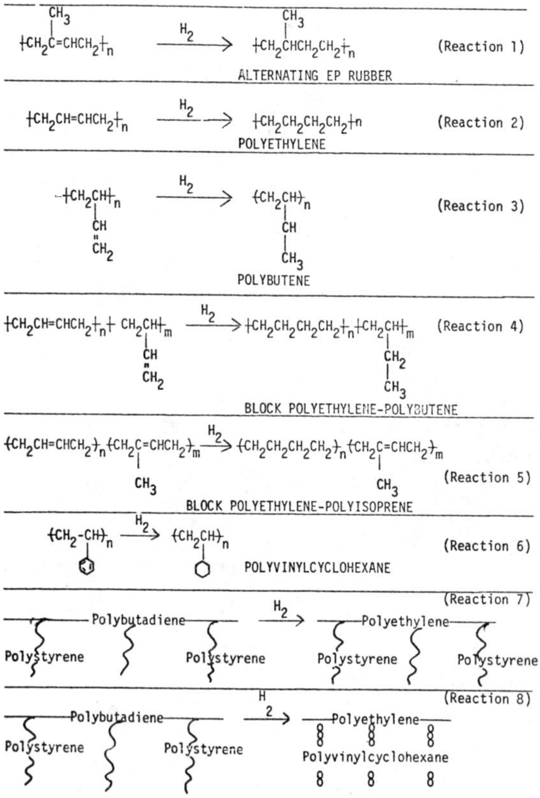

In addition to providing synthetic routes to unique polymer structures, hydrogenation may be employed to make commercially useful polymers.

Some recently developed thermoplastic elastomers owe their properties to an ABA block structure. These copolymers consist of rigid thermoplastic A blocks and soft, rubbery B segments and exhibit behavior similar to that of crosslinked rubber structures. The commercially available family of Kraton's® are examples of thermoplastic elastomers which contain blocks of polystyrene and polybutadiene arranged in an ABA sequence.

Hydrogenation may be used to prepare thermoplastic elastomers similar to Kraton®, Table XIV (Falk, 1972). An ethylene-isoprene random copolymer prepared by hydrogenation of the butadiene moieties of a 1,4-butadiene-1,4-isoprene copolymer had low tensile strength, Entry 1. The polymer contained 66% ethylene and is abbreviated 66EI. Ordering the ethylene and isoprene moieties into blocks in such a way that two polyethylene blocks flank the polyisoprene block results in a thermoplastic elastomer which has a low tensile yield, 840 psi, a high ultimate tensile strength, 2100 psi, and a high percent elongation, 690, Entry 2. These polymers are consistent with the ASTM definition of elastomers (ASTM, 1956): "A substance that can be stretched at room temperature to at least twice its original length and, after having been stretched and the stress removed, returns with force to approximately its original length in a short time." A thermoplastic elastomer also results if the polyisoprene segment contains isoprene in the 3,4 configuration, Entry 3. The physical properties of these experimental thermoplastic elastomers are similar to one of the Kraton® family, a 50% styrene-butadiene ABA block copolymer, Entry 4.

TABLE XIV. *Ethylene-Isoprene Copolymers*

Entry	Type	Yield (psi)	Ultimate (psi)	Elongation (%)	Substrate MW
1	66EI(random)	260	410	580	60,000
2	75EIE	840	2100	690	120,000
3	$66EI_{3,4}E$	760	2380	650	240,000
4	50SBS	800	2300	700	200,000

Hydrogenated graft copolymers of polystyrene on polybutadiene possess interesting properties, Table XV (Falk, 1973). A polyvinylcyclohexane graft (70%) on polyethylene is a tough, rigid material with an ultimate tensile strength of 3,900 psi coupled with a percent elongation of 7. If the backbone is a rubbery ethylene-butene-1 copolymer, the polyvinylcyclohexane graft thereon (34%) is a thermoplastic elastomer, Entry 2. Both examples of Table XV have seven polyvinylcyclohexane chains grafted to each backbone molecule.

TABLE XV. *Polyvinylcyclohexane Grafts*[a] *on Polyethylene*

Entry	PVCHX (%)	PE (MW x 10^{-3})	Tensile Strength Yield (psi)	Tensile Strength Ultimate (psi)	Elongation (%)
1	70	145	---	3900	7
2	34[b]	60	740	2400	500

[a]7 graft sites. [b]63% butene-1.

At least one commercially available polymer is prepared by hydrogenation. Shell's recently introduced Kraton GX6500® is touted as a weatherable thermoplastic elastomer based on a hydrogenated styrene-butadiene copolymer.

CONCLUSION

Catalysts have been described for the hydrogenation of low molecular weight olefins, polydienes, and polystyrene. The reaction conditions are mild; conversion is high; and no polymer degradation occurs. These catalysts are made by reacting an organoaluminum or organolithium compound with a transition metal salt of 2-ethylhexanoic acid. The catalyst activity is a function of lithium (or aluminum)-transition metal ratio, catalyst concentration, temperature, and hydrogen pressure. With the proper choice of catalyst composition and reaction conditions, the hydrogenation catalysts can discriminate between different types of polymeric unsaturation. The order of activity in the hydrogenations studied with these catalysts is poly-1,4-butadiene>poly-1,4-isoprene>>polystyrene. The hydrogenation technique described herein may be used to prepare a number of novel polymer structures which are difficult and, in many cases, impossible to prepare by alternate routes.

ACKNOWLEDGEMENTS

The author expresses his appreciation to Drs. D. F. Hoeg, J. F. Pendleton and R. J. Schlott for their encouragement and many suggestions. The author is also indebted to the Borg-Warner Corporation for permission to publish this work.

REFERENCES

1. Anderson, J. N., J. Appl. Poly. Sci., 18, 2819 (1974)
2. Berthelol, P. E. M., Bull. Soc. Chim. France, 11, 33 (1869).
3. Falk, J. C., and Schlott, R. J., Macromolecules, 4, 152 (1971).
4. Falk, J. C., J. Org. Chem., 36, 1445 (1971).
5. Falk, J. C., J. Polym. Sci. A-1, 9, 2617 (1971).
6. Falk, J. C., and Schlott, R. J., Angew. Makromol. Chem., 21, 17 (1972).
7. Falk, J. C., Makromol. Chem., 160, 291 (1972).
8. Falk, J. C., Schlott, R. J., Hoeg, D. F., and Pendleton, R. F., Rubber Chem. Technol., 46, 1044 (1973).
9. Falk, J. C., Schlott, R. J., and Hoeg, D. F., J. Macromol. Sci-Chem., A7, 1647 (1973).
10. Falk, J. C., Hoeg, D. F., Schlott, R. J., and Pendleton, J. F., J. Macromol. Sci-Chem., A7, 1669 (1973).
11. Fisher, E., Chem. Ber., 46, 3288 (1913).
12. Gluesenkamp, E. W., and Weesner, W. E., U. S. Patent 2,844,573 (1958).
13. Harries, C. D., Chem. Ber., 56, 1048 (1923); Chem. Abstr., 17, 2806 (1923).
14. Harries, C. D., "Untersuchungen über die naturlichen und künstlichen Kautschukarten", Julius Springer, Berlin, 1919, p 48.
15. Hsieh, H. L., J. Polym. Sci. A, 3, 153 (1965).
16. Jones, R. V., Moberly, C. W., and Reynolds, W. B., Ind. Eng. Chem., 45, 1117 (1953).
17. Natta, G., Die Makromol. Chem., 58, 217 (1962).
18. Natta, G., Mazzanti, G., Longi, P., and Bernardini, F., Chim. Ind., 41, 519 (1959).
19. Parker, J. H., U. S. Patent 2,456,428 (1948).
20. Sloan, M. F., Matlack, A. S., and Breslow, D. S., J. Amer. Chem. Soc., 85, 4014 (1963).
21. Staudinger, H., and Fritschi, J., Helv. Chim. Acta, 5, 785 (1922); Chem. Abstr., 17, 2974 (1923).
22. Steinhofer, A., Polster, R., and Friederich, H., Can. Patent 718,089 (1965).
23. Vandenberg, E. J., U. S. Patent 3,051,690 (1962).
24. Amer. Soc. Test. Mater., Spec. Tech. Publ., No. 184, 42 (1956).

THE HYDROGENOLYSIS OF POLYALKYLCYCLOPROPANES

ROBERT L. AUGUSTINE and BABUBHAI A. PATEL

Department of Chemistry, Seton Hall University
South Orange, New Jersey 07079

The hydrogenolysis of cyclopropane rings has been the subject of a number of publications concerned with both mechanistic studies and synthetic applications of the reaction (Newham, 1963; Augustine, 1965; Frei, et al, 1966; Rylander, 1967; Fraisse-Jullien, et al, 1968; Jorgenson, 1968; Laing and Sykes, 1968; Majerski and Schleyer, 1968; Mil'vitskaya, et al, 1968; Schultz, 1971; Roth, 1972; Pines and Noguerisa, 1972; Augustine and Rearden, Jr., 1974). It has been shown that monoalkyl and 1,1-dialkyl cyclopropanes are cleaved at the bond opposite the substituted carbon while in other polyalkylcyclopropanes (Newham, 1963; Augustine, 1965; Rylander, 1967) the bond which is broken is that between the two least substituted carbons. On the other hand, hydrogenolysis of vinyl, carbonyl, or phenyl substituted cyclopropanes generally occurs at a bond adjacent to the unsaturated substituent.

The mechanistic pathway proposed for the hydrogenolysis of cyclopropane and monoalkylcyclopropanes (Scheme 1) (Bond and Newham, 1960) is analogous to the Horiuti-Polanyi (Horiuti and Polanyi, 1934) mechanism of olefin hydrogenation. According to this proposal the formation of an olefinic intermediate could occur by desorption of 2 but under the conditions usually used for cyclopropane hydrogenolysis a high hydrogen availability to the catalyst is present and thus step 3 should be favored. The extensive deuterium exchange which has been observed during vapor phase deuterolysis (Bond and Newham, 1960; Newham, 1963) has been described as resulting from exchange of either the product alkane (Bond, 1962) or the monoadsorbed species 1 (Bond and Newham, 1960), probably occurring through the diadsorbed species, 2. Deuterolyses in solution are not, however, accompanied by such extensive exchange (Vide infra).

From a consideration of this mechanism it can be assumed that in the absence of any olefin formation the ring opening should lead to the cis addition of hydrogen to the 1 and 3 carbon atoms of the ring as has been observed in the hydrogenol-

SCHEME 1

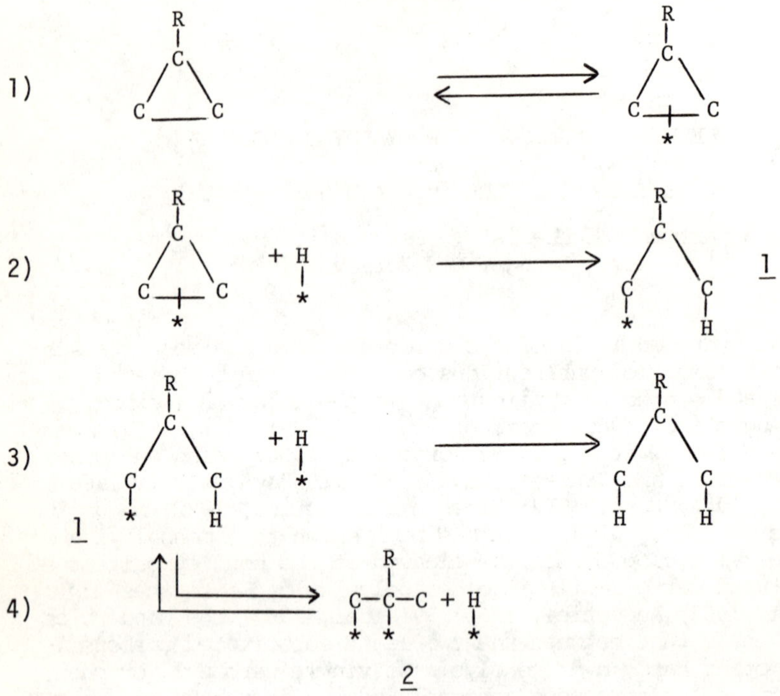

ysis of 3-acetoxynortricyclene (3) (Akhtar and Jackson, 1972). Some stereoselectivity in the hydrogenolysis of the bicyclic ester, 4 (Jorgenson, 1968), and cyclopropyl ketone, 5 (Fraisse-Jullien, et al, 1968), has also been reported but cleavage of tricyclo [4.4.1.0] undecane (6) gave a 50:50 ratio of the cis and trans 9-methyldecalins (Majerski and Schleyer, 1968).

It has also been found that the hydrogenolysis of lumitestosterone acetate (7) gives almost exclusive 5α product (cf 8) formation (Augustine and Rearden, 1974) rather than the expected 5β material, 9, which would be expected if the mechanism shown in Scheme 1 were operational. In fact, this marked absence of any appreciable 5β saturated product in the hydrogenolysis of 7 is in direct contrast to the stereoselectivity reported for the hydrogenolysis of 10β-methyl-cyclopropyl-keto steroids such as 10 (Frei, et al, 1966). This difficulty in the formation of the 5β, 10α ring fusion product is most probably a result of the fact that in this ring system the B ring of the steroid is forced into a boat conformation, thus, pro-

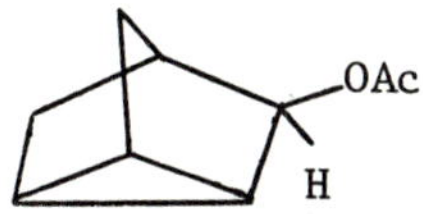

3

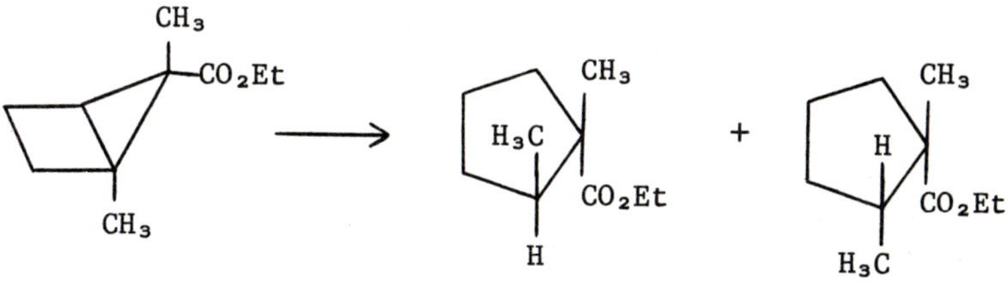

4 70% 30%

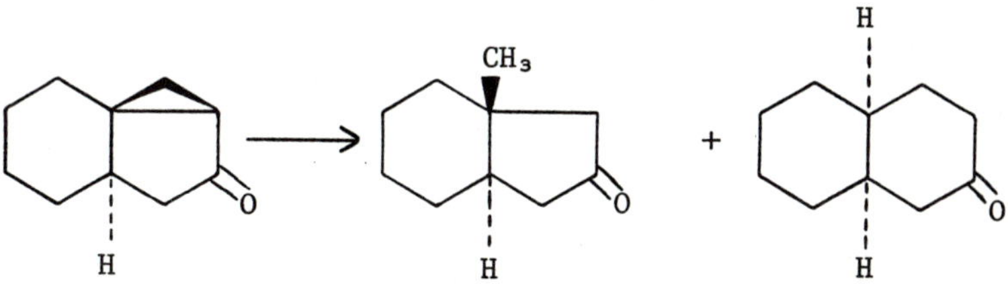

5 20% 80%

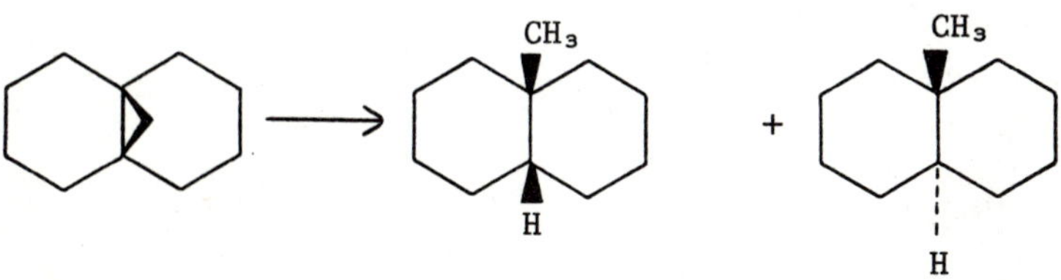

6 50% 50%

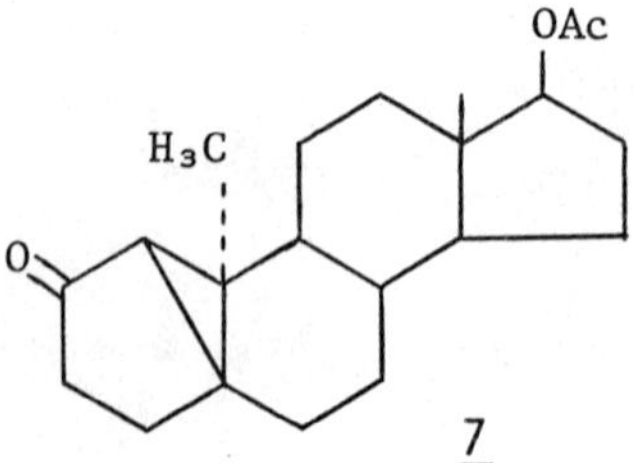

7

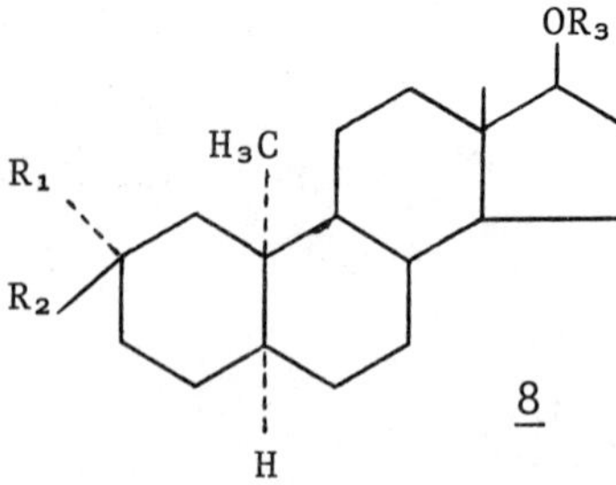

8

R_1=OH, R_2=H, R_3=Ac
R_1=OH, R_2=R_3=H
R_1=H, R_2=OH, R_3=Ac
R_1=H, R_2=OH, R_3=H

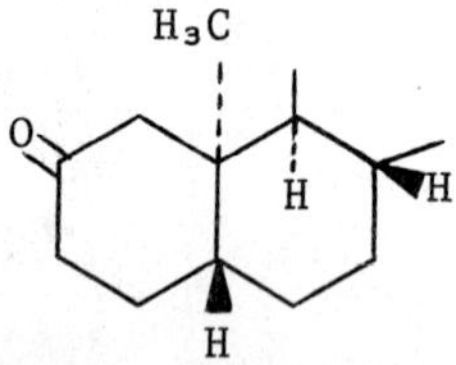

9

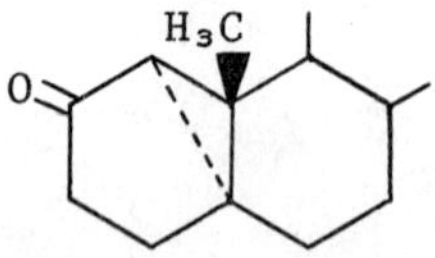

10

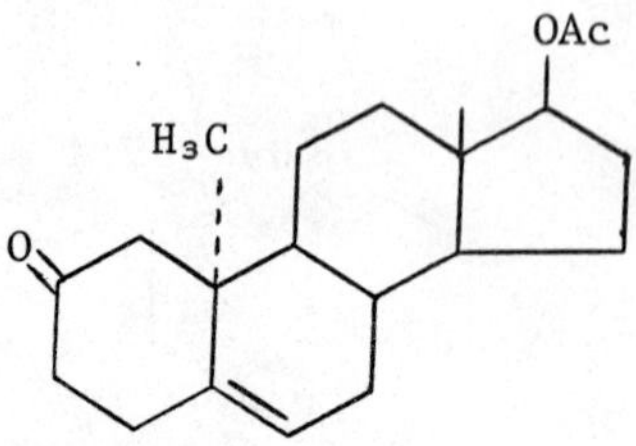

11a

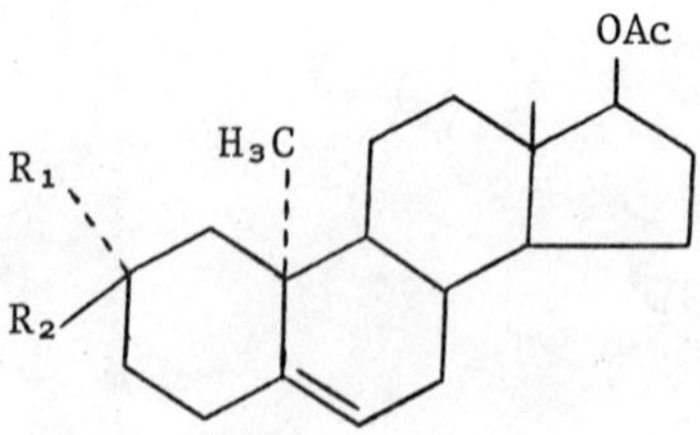

11b

R_1=OH, R_2=H
R_1=H, R_2=OH

viding an added energy barrier to the reaction. In fact the hydrogenation of either the Δ^5 ketone, 11a, or alcohols, 11b, under a variety of conditions also gives exclusive 5α product formation even though a consideration of molecular models does not clearly show why α-adsorbtion on the catalyst should be so markedly preferred in these systems. The facile acid catalysed isomerization of lumitestosterone to the Δ^5-olefin, 11a, and the almost complete 5α product formation from both 11a and 11b strongly indicate the intermediacy of an olefinic species in this hydrogenolysis. However, it was found that stirring 7 under the conditions used for hydrogenolysis, but in the absence of hydrogen, gave only recovered starting material. It was, therefore, apparent that if isomerization were occurring, it must be induced during the hydrogenation process. To account for olefin formation one must consider several possibilities. If the 1β,5β bond in 7 is cleaved as shown in Scheme 1 and if the formation of a β-half hydrogenated state as C_1 or C_5 would be anticipated, the formation of a C_1-half hydrogenated state would result in β hydrogenation at C_5, something which is not observed. A β-half hydrogenated state at C_5, 12, could then go on to form the C_5-C_6 diadsorbed species which would then be desorbed to give the Δ^5 compound. If this were the case, however, the B ring in the adsorbed species would already have at least a semblance of a boat conformation and the energy barrier to boat formation would be essentially passed. It is difficult to see then, why, under the reaction conditions used, 1500 psi of hydrogen, 150°C, simple hydrogen transfer to C_5 would not take place to give at least a moderate amount of the 5β product.

In an alternative mechanism for the cleavage of 7 neither the 5β,6β-diadsorbed species nor a 5β-half-hydrogenated state is involved. The molecular model of 7 shows that the C_6-β-hydrogen bond is directly perpendicular to the 1β,5β-bond of the cyclopropane ring, the bond which is most accessible to the catalyst surface. This relationship on a catalyst surface is shown in the diagram, 13. Extraction of the C_6-hydrogen bond by the catalyst with concomitant double bond formation between C_5 and C_6 and cleavage of the C_1-C_6 bond would give the C_1-half-hydrogenated state 14, which could then be saturated to give the Δ^5 compound which then could be readsorbed on the catalyst from the more favored α side to give the observed 5α products. A similar olefin formation step can be proposed to explain the lack of stereoselectivity in the hydrogenolysis of 6. Here, too, molecular models show a perpendicular relationship between the C_1-equatorial hydrogen and the cyclopropane bond being broken, a relationship the same as that shown in 13. In this case the octalin, 15, would be the olefinic intermediate.

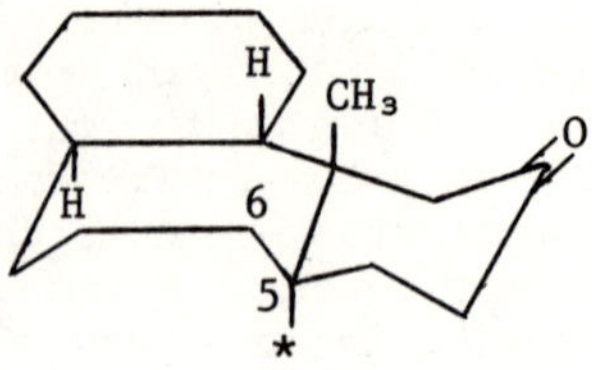

12

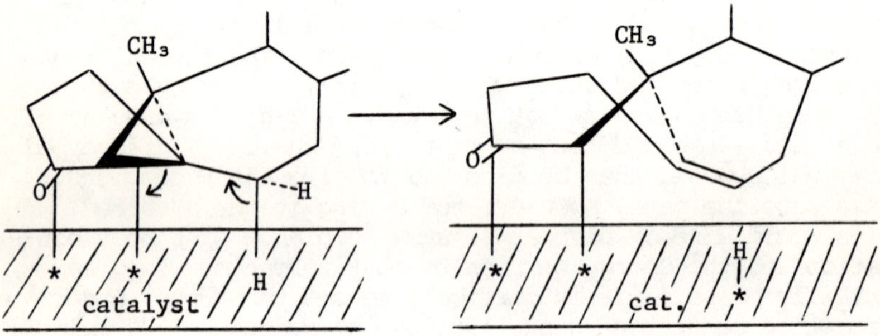

13

14

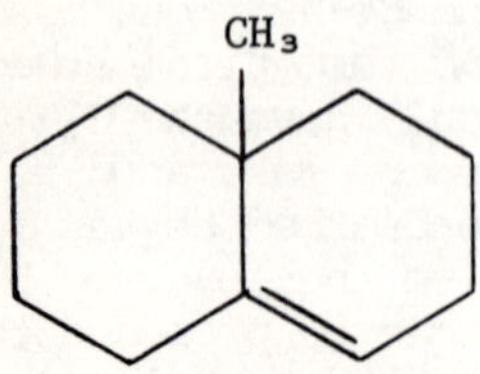

15

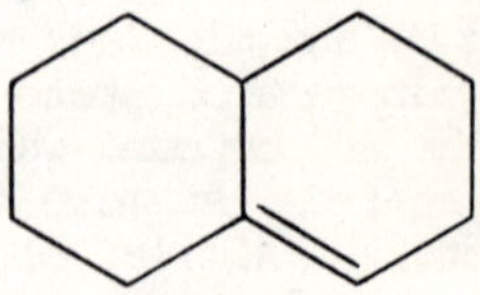

16

SCHEME 2

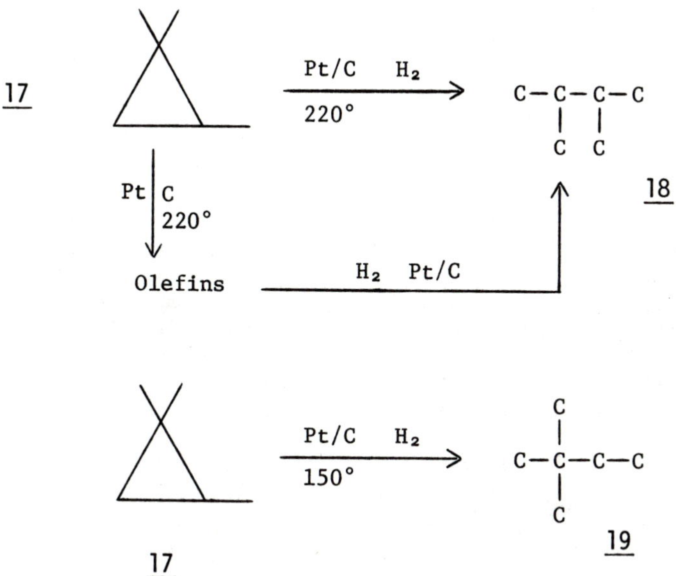

SCHEME 3

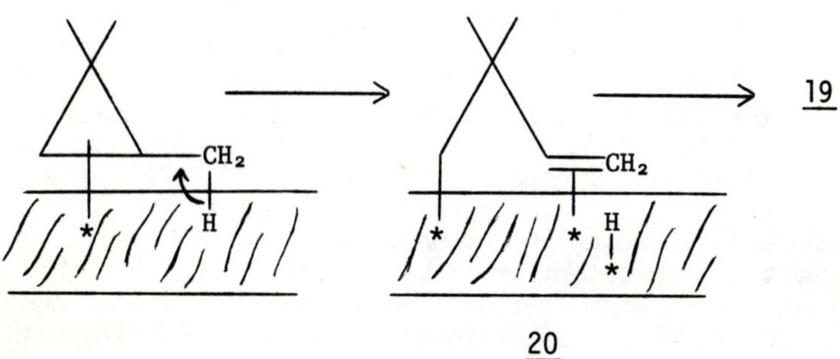

While product stereochemistry obtained on hydrogenation of 15 has not been established, a 60:40 mixture of the cis and trans decalins is formed on hydrogenation of the parent, $\Delta^{1,9}$-octalin, (16), over platinum in acetic acid (Smith and Burwell, Jr., 1962). It should be emphasized that what is proposed is not a thermal rearrangement process which sometimes appears to be taking place during the high temperature hydrogenolysis of polyalkylcyclopropanes (Lukina, et al, 1957, 1958) but, rather is an alternative pathway which can be followed even at low temperatures. The thermal stability of 7 mentioned above gives some credence to this assumption. Further support can be had by considering the reported (Lukina, et al; Prudhomme and Gault, 1966) hydrogenolysis data for 1,1,2-trimethylcyclopropane (17) as shown in Scheme 2. With this compound high temperature hydrogenolysis gives primarily 2,3-dimethylbutane (18). Passage of 17 over the hydrogenation catalyst at high temperatures gives a mixture of olefins which on hydrogenation gives essentially the same product mixture obtained from direct hydrogenolysis (Lukina, et al, 1957, 1958). In contrast, neohexane (19) is the almost exclusive product obtained from the cleavage of 17 at lower temperatures (Prudhomme and Gault, 1966). Since the 2-3 bond in 17 would be expected to be the most readily adsorbed on the catalyst the formation of 19 can be easily rationalized using the proposed reaction pathway as shown in Scheme 3. Further, deuteration of 17 gives 19 in which 60% of the molecules had between 3 and 6 deuterium atoms incorporated (Prudhomme and Gault, 1966), as should be expected from an olefinic precursor. Deuteration of 20 should give at least a D_3 compound. Attempted deuterium exchange on 19 under the reaction conditions used for the hydrogenolysis of 17 gave only about 15% of the D_3 material (Prudhomme and Gault, 1966).

The hydrogenolysis (Ghatak, et al, 1973) of 21 results in exclusive formation of the 9α-gibbane, 22, a result also in contradiction to the mechanism shown in Scheme 1 and in seeming agreement with the olefin intermediate proposal. This agreement, however, is not apparently valid since hydrogenation of 23, the olefinic intermediate expected from 21, gives a mixture of both the 9α and 9β products (Ghatak, et al, 1973).

More recently the hydrogenolysis of trans 1,2-dimethyl-1,2-diphenylcyclopropane (24) has been shown to give both the dl-(25) and meso (26) products (Kieboom, et al, 1974). The formation of 25 (suprafacial ring opening) was the result when the addition of hydrogen to carbons 1 and 2 occurred with either inversion or retention at both centers. On the other hand 26 was formed when inversion occurred at one center and retention at the other (antarafacial ring opening). These

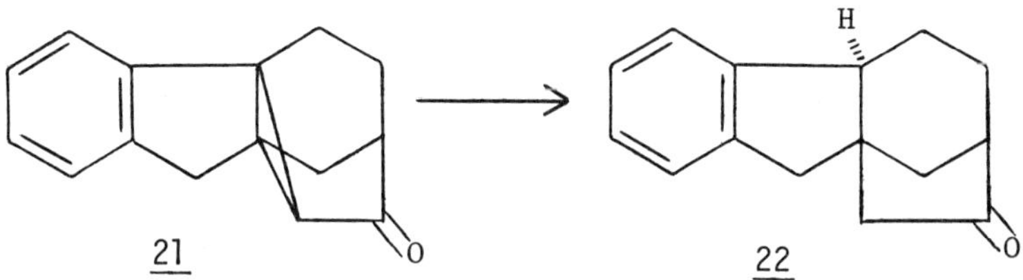

21 22

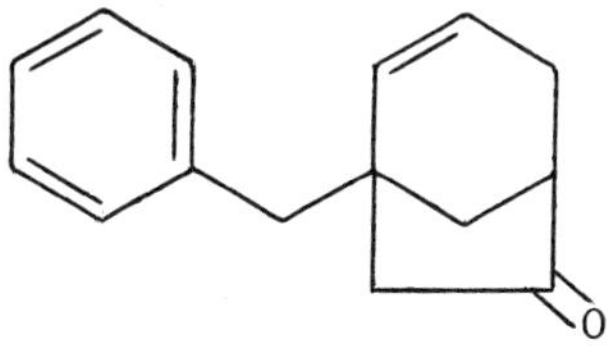

23

conventions are quite useful in describing the stereochemical outcome of cyclopropane ring cleavage. However, in this case, as with the inversion of configuration observed on hydrogenolysis of benzyl alcohols (Mitsui, et al, 1961, 1963, 1965), the process by which this inversion occurs is not clearly specified.

It was obvious, then, that more work was needed to resolve this question but that the work done should try to correlate the findings of the vapor phase deuterolyses with the reported stereochemical data obtained using solution reactions. In an attempt to do this the deuterolysis of norcarane (27) was run in d_4-acetic acid using a platinum oxide catalyst at room temperature and two atmospheres pressure. These mild conditions (Majerski and Schleyer, 1968; Hendricksen and Boeckman, Jr., 1971) were chosen to minimize the side reactions which could occur along with the cleavage of 27. Several points were found which are important to the understanding of the solution phase hydrogenolyses of cyclopropanes: 1) Ring opening did not occur in the absence of hydrogen or deuterium; 2) Recovered, unreacted norcarane contained no deuterium; 3) No deuterium was incorporated into methylcyclohexane (28) when it was exposed to deuterium under these reaction conditions; and 4) The deuterium distribution in the methylcyclohexane obtained on deuterolysis of norcarane had a maximum at d_1.

This predominant formation of a monodeutero product is apparently the direct result of a dilution of the surface deuterium pool by hydrogen obtained from an exchange process.

25

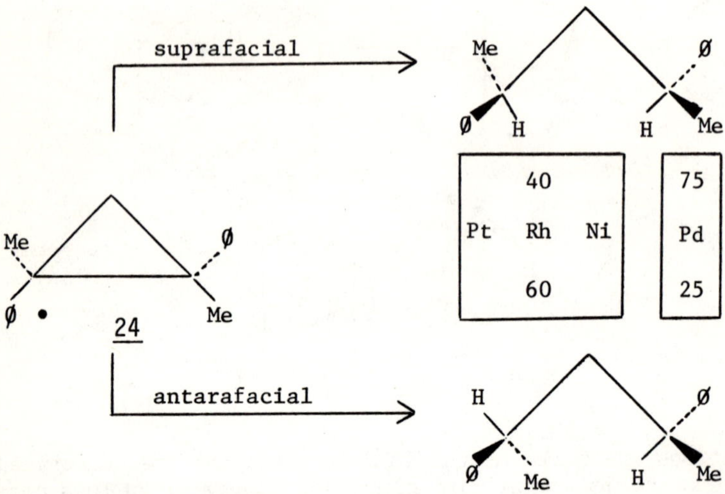

26

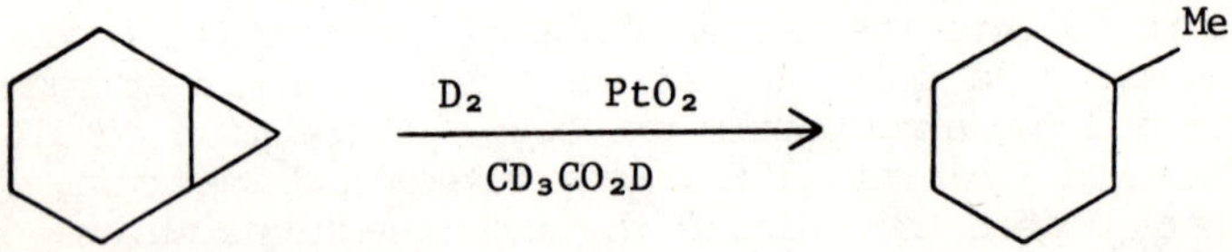

27 28

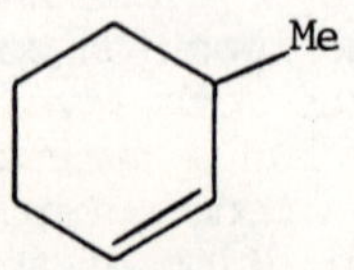

29

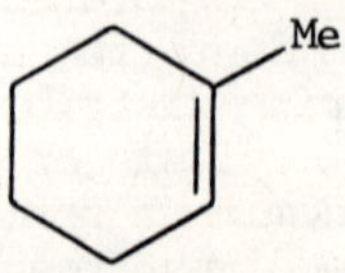

30

Since it has been established that no exchange occurred with either the starting material or the product and the solvent was completely deuterated, this hydrogen must be obtained from only one source, the reacting molecule. Thus, some type of 1,2-diadsorbed or olefinic intermediate is probably involved.

The deuterium distribution observed on deuterolysis of norcarane (27) was also compared with the deuterium distribution in the methylcyclohexane obtained from the deuteration of 3-methylcyclohexane (29) and 1-methylcyclohexene (30) using the same reaction conditions as those used to cleave norcarane. From the results pictured in Figure 1 and listed in the Table it is apparent that while the products from 29 and 30 have a reasonably similar deuterium distribution, this distribution is quite different from that found with the product obtained from the cleavage of 27. Even if one compensates for the dilution of the deuterium pool in the reaction of 27 by comparing the d_0, d_1, d_2 etc., of the product from 27 with the d_1, d_2, d_3 etc., of the products from 29 and 30, there is still no reasonable correlation of these data.

It has been established, however, that the primary mass spectral fragmentation pattern of methylcyclohexane is simply the loss of the methyl group (Meyerson, et al, 1963). The primary difference in the deuteration of the olefins, 29 and 30, and the cleavage of 27, is the direct involvement of this methyl carbon atom in the mechanism of the latter reaction. Thus, analysis of the cyclohexyl radical cation region in the mass spectra of these products should permit a more direct comparison of the deuteration data by removing from consideration this anomolous methyl group. This comparison is presented in Figure 2 and listed in the Table. Here the deuterium distribution from each of the products is quite different from that of the others. If the deuterium dilution in the cleavage of norcarane is considered and the d_0, d_2, d_2, etc., data from the product from 27 is compared with the d_1, d_2, d_3, etc., data from the products from 29 and 30 (Figure 3) a reasonably good correlation between the data from 27 with that from 29 is observed. The correlation between the data from 27 and 30 is not quite as good.

It appears, then, that these data do indicate that an olefinic or 1,2-diadsorbed intermediate is involved in the hydrogenolysis of polyalkylcyclopropanes. The intermediacy of a diadsorbed species can account for the deuterium distribution and even for the inversion of configuration if a "rollover" type of mechanism (Burwell, 1972) is occurring but it seems more likely that a fully desorbed olefin is present in solution reactions under low hydrogen availability conditions.

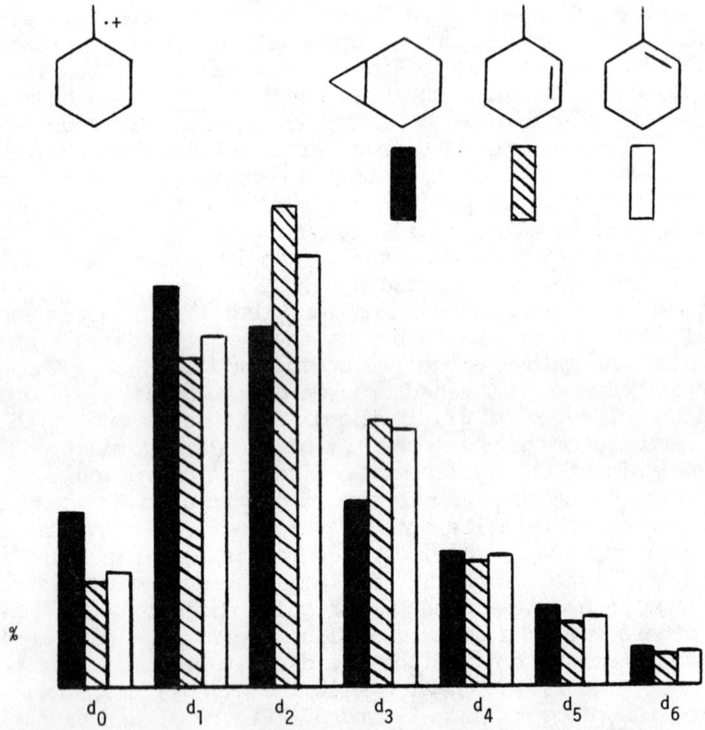

Fig. 1. Deuterium distribution in the molecular ion peaks of the methylcyclohexane obtained on deuterolysis of norcarane (27) and the deuteration of 3-methylcyclohexene (29) and 1-methylcyclohexene (30).

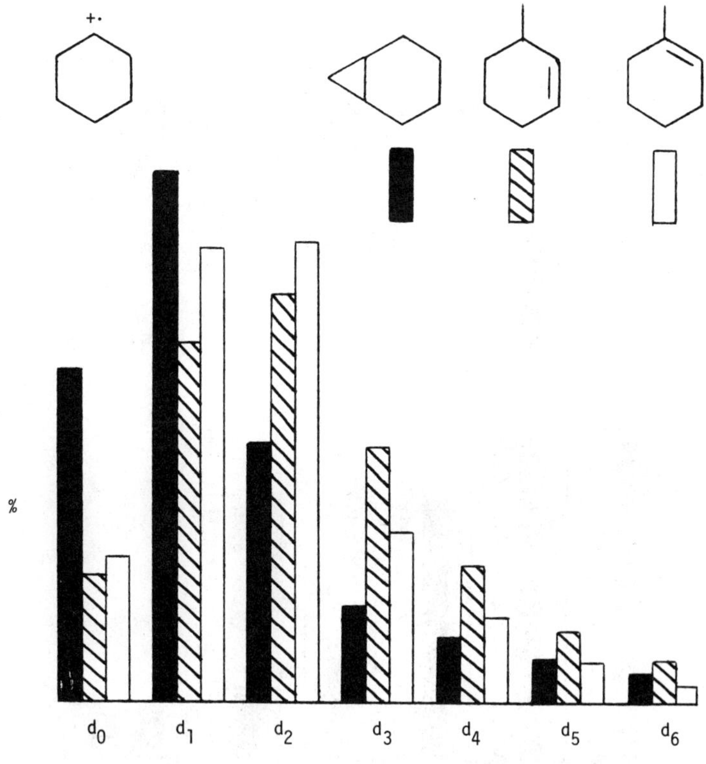

Fig. 2. *Deuterium distribution in the cyclohexyl radical cation peaks of the methylcyclohexane obtained from* 27, 29 *and* 30.

The difference in product stereochemistry obtained on cleavage of 21 and hydrogenation of 23 (Ghatak, et al, 1973) could possibly be due to the fact that the amount of intermediate olefin present in the reaction of 21 is expected to be quite small and thus, the olefin:catalyst ratio, a factor which influences product stereochemistry (Augustine, 1963), is different in the reaction of 21 than in the hydrogenation of 23. Further work is underway to attempt to clarify this problem more fully.

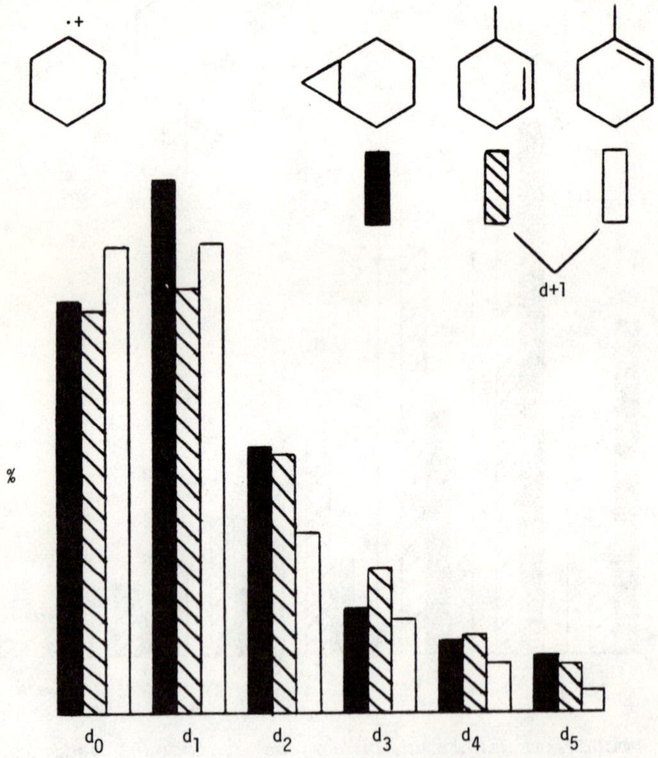

Fig. 3. Comparison of the deuterium distribution in the cyclohexyl radical cation peaks of the methylcyclohexane obtained from 27 and the d + 1 peaks from the products obtained from 29 and 30.

TABLE. *Deuterium Distribution in Methylcyclohexane Obtained from 27, 29 and 30*

mass	d	Percent of Total Deuterated Molecules or Fragments from the Products from		
		27	29	30
83	d_0 C*	24.3	9.2	10.1
84	d_1 "	38.9	24.2	33.6
85	d_2 "	19.1	30.2	33.8
86	d_3 "	6.8	18.8	12.5
87	d_4 "	4.4	9.9	6.0
87	d_5 "	2.9	5.1	2.8
88	d_6 "	1.8	2.5	1.1
98	d_0 M*	12.5	7.4	7.8
99	d_1 "	29.5	24.2	25.7
100	d_2 "	25.8	35.3	31.4
101	d_3 "	13.6	19.4	19.0
102	d_4 "	9.4	9.1	9.3
103	d_5 "	5.4	4.6	4.7
104	d_6 "	2.7		2.3

*C = cyclohexyl radical cation; M = molecular ion

REFERENCES

1. Akhtar, M. N. and Jackson, W. R., Chem. Commun. 813 (1972).
2. Augustine, R. L., J. Org. Chem., 28, 152 (1963).
3. Augustine, R. L., "Catalytic Hydrogenation", Marcel Dekker Inc., New York, N. Y., 1965, p 133.
4. Augustine, R. L. and Rearden, Jr., E. J., J. Org. Chem., 39, 1627 (1974).
5. Bond, G. C., and Newham, J., Trans. Faraday Soc., 56, 1501 (1960).
6. Bond, G. C., "Catalysis by Metals", Academic Press, New York, N. Y., 1962, p 275.
7. Burwell, R. L., Intra-Science Chem. Rept., 6, 135 (1972).
8. Fraisse-Jullien, R., Frejaville, C., Toure, V. and Derieux, M., Bull. Soc. Chim. Fr., 4444 (1968).
9. Frei, J., Ganter, C., Kagi, D., Kocsis, K., Miljkovic, M., Siewinski, A., Wenger, R., Schaffner, K. and Jeger, O., Helv. Chim Acta, 49, 1049 (1966).
10. Ghatak, U. R., Chakraborti, P. C., Ranu, B. C., Sanyal, B., Chem. Comm., 548 (1973).
11. Hendricksen, J. B. and Boeckman, Jr., R. K., J. Am. Chem. Soc., 93, 4491 (1971).
12. Horiuti, I. and Polanyi, M., Trans. Faraday Soc., 30, 1164 (1934).
13. Jorgenson, M. J., Tetrahedron Lett., 4577 (1968).
14. Kieboom, A. P. G. and Breijer, A. J. and Van Bekkum, H., Rec. Trav. Chim., 93, 186 (1974).
15. Laing, S. B. and Sykes, P. J., J. Chem. Soc., C, 937 (1968).
16. Lukina, M. Yu., Zotuva, S. V. and Kazanskii, B. A. Doklady Akad. Nauk S.S.S.R., 114, 792 (1957); Chem. Abstr., 52, 1930 (1958).
17. Lukina, M. Yu., Zotava, S. V. and Kazanskii, B. A., Izvest. Akad. Nauk S.S.S.R., Otdel. Khim. Nauk, 300 (1958); Chem. Abstr., 52, 12774 (1958).
18. Majerski, Z. and Schleyer, Von R., Tetrahedron Lett., 6195 (1968).
19. Meyerson, S., Nevitt, T. D., and Rylander, P. N., in "Advances in Mass Spectrometry" ed. by Elliott, R. M., Pergamon Press, Oxford, 1963, Vol. 2, p 313-336.
20. Mil'vitskaya, E. M., Plate, A. F., Sterin, Kh. E. and Ivannikova, M. P., Zh. Org. Khim., 4, 271 (1968); Chem. Abstr., 68, 104356 (1968).
21. Mitsui, S., et al, Bull. Chem. Soc. Japan, 34, 774 (1961).
22. Mitsui, S., et al, Chem. Ind. (London), 1354 (1963), 381 (1965).

23. Newham, J., Chem. Revs., 63, 123 (1963).
24. Pines, H., and Noguerisa, L., J. Catal., 27, 89 (1972).
25. Prudhomme, J. C. and Gault, F. G., Bull. Soc. Chem. Fr., 832 (1966).
26. Roth, J. A., J. Am. Chem. Soc., 92, 6658 (1970).
27. Roth, J. A., J. Catal., 26, 97 (1972).
28. Rylander, P. N., "Catalytic Hydrogenation over Platinum Metals", Academic Press, New York, N. Y., 1967, p 468.
29. Schultz, A. L., J. Org. Chem., 36, 383 (1971).
30. Smith, G. V. and Burwell, Jr., R. L., J. Am. Chem. Soc., 84, 925 (1962).

PREPARATION OF N-ALICYCLIC AMIDES BY NUCLEAR HYDROGENATION OF N-ARYL AMIDES

RUSSELL E. MALZ, JR. AND HAROLD GREENFIELD

Uniroyal Chemical, Division of Uniroyal, Inc.
Naugatuck, Connecticut 06770

The heterogeneous hydrogenation of N-aryl amides has been investigated with several catalysts. The rhodium-catalyzed reaction is excellent for the laboratory preparation of N-alicyclic amides. Promotion of the catalyst by acid and reactivation of used catalyst by acid treatment makes the process economical and of commercial interest. The hydrogenation of N-aryl amides followed by hydrolysis to the corresponding amines may be superior to the direct hydrogenation of aryl amines to alicyclic amines, particularly in the case of diamines, with respect to product purity and economics. Rhodium is by far the most active catalyst for this nuclear hydrogenation. Ruthenium also has useful activity.

The purpose of this investigation was to develop a suitable process for the laboratory preparation of N-alicyclic amides of high purity, and for their commercial manufacture. These amides can be hydrolyzed to the corresponding amines which, particularly in the case of diamines, are difficult to obtain by direct hydrogenation of the aryl amine in satisfactory purity because of side reactions (Rylander, 1967; Freifelder, 1971) and economically because of catalyst poisoning (Greenfield, 1973).

The nuclear hydrogenation of the aromatic ring in N-arylmonocarboxamides has been reported with nickel (Adkins and Cramer, 1930; Hartmann et al, 1939; Billman and Buehler, 1953; Della and Jefferies, 1961; Fraser and Swingle, 1970), platinum (Skita and Rolfes, 1920; Ferber and Bendix, 1939; Ferber and Bruckner, 1939; Pletneva et al, 1964, 1965; Fraser and Swingle, 1970; Knowles, 1972), platinum-rhodium (Pletneva et al, 1965; Muromova et al, 1965) and rhodium (Abraham and Lamb, 1966) catalysts. Freifelder (1971) has referred to unpublished nickel, rhodium and ruthenium catalyzed nuclear hydrogenations of 1,2-bisacetamidobenzene.

Side reactions to be anticipated in the nuclear hydrogenation of N-aryl amides are those involving the amido group itself, before, during, or after hydrogenation of the aromatic ring. Hydrogenolyses of carbon-oxygen and carbon-nitrogen bonds in amides have been reported with copper chromite (Adkins and Wojcik, 1934; Wojcik and Adkins, 1934; Adkins, 1937; Guyer et al, 1955), rhenium (Broadbent et al, 1959; Broadbent and Bartley, 1963), nickel (Guyer et al, 1955), and cobalt (Guyer et al, 1955) catalysts.

Acetanilide, for example, can produce N-ethylaniline by carbon-oxygen cleavage, whereas carbon-nitrogen cleavage can lead to aniline or ethylamine. Similarly, N-ethylcyclohexylamine and cyclohexylamine could be produced from the saturated N-cyclohexylacetamide, or by nuclear hydrogenation of N-ethylaniline and aniline. Such side reactions, even if present to a very limited extent, might have very important catalyst-inhibition effects. Poisoning of nuclear hydrogenation by amines has been reported with platinum (Devereux et al, 1957; Peeling and Shipley, 1958; Greenfield, 1964), palladium (Greenfield, 1964), and rhodium (Greenfield, 1964) catalysts.

EXPERIMENTAL AND RESULTS

All temperatures are in °C. Melting points are uncorrected.

Materials

Formanilide (distilled before use), acetanilide, butyranilide, stearanilide, benzanilide, 4,4'-methylenedianiline (practical grade), p-phenylenediamine (practical grade), and 2,4-toluenediamine (practical grade) were obtained from Eastman Kodak. The aminonaphthalenes were obtained from Aldrich Chemical Co. The m-phenylenediamine was prepared by the Pd-catalyzed hydrogenation of m-dinitrobenzene (Dovell et al, 1970). The following amides were prepared by reaction of the appropriate amine with acetic anhydride in glacial acetic acid (Butler and Adams, 1925); bis(4-acetamidophenyl)methane, mp 231-233° (Butler and Adams, 1925); 2-acetamidonaphthalene, mp 131-132° after recrystallization from methanol-water (Leonard and Boyd, 1946); 1,3-bisacetamidobenzene, mp 187-188° after recrystallization from methanol-water (Sasa, 1954; Morgan and Turner, 1966); 1,4-bisacetamidobenzene, mp 309-312° (Biedermann and Ledoux, 1874; Tröger and Westerkamp, 1909; Adams and Anderson, 1950; Tu et al, 1956; Sladkov and Vett, 1956; Mamlok, 1956; Tsatsas and Delaby, 1956; de la Mare and Hassan, 1958; Guyader, 1962); 2,4-bisacetamidotoluene, mp 223-224° (Landenburg, 1874; Domnin and Cherkasova, 1956; Ipatieff and Schmerling, 1937; Jansen, 1963); 1,5-bisacetamidonaphthalene, mp

>350° (Knuckell and Schneider, 1912; Buu-Hoi, 1945; Leonard and Hyson, 1949). The 5% Pt, Pd, Rh and Ru on carbon catalysts were obtained from Engelhard Industries. The 50% Ni on kieselguhr (Girdler G-49B) and 50% Co on kieselguhr (Girdler G-67RS) were obtained from Chemetron Corp. The Raney Ni (Grace 28) was obtained from Davison Chemical Co.

Hydrogenation of Acetanilide

A comparison of several catalysts for the nuclear hydrogenation of acetanilide to N-cyclohexylacetamide is shown in Table I. In each experiment, only traces of residual aromaticity were detected by UV analysis and only traces of or no side products detected by glpc analysis.

TABLE I. *Hydrogenation of Acetanilide*[a]

Metal	Catalyst[b] Wt. g	Catalyst[b] Conc. g/l	Temp. °C	Pressure psig	Time at Temp., hr
Rh	1.1	4.9	35-40	400-600	1.0[c]
Ru	1.1	4.9	75-80	400-600	1.5[c]
Pd	4.4	20	120-130	500-700	7.0[d]
Ni	4.4	20	170-180	900-1200	4.1[c]
Co	4.4	20	190-200	900-1200	1.5[e]

[a] Each experiment was run in a 600-ml Magne Drive autoclave with 27.0 g (0.20 mole) of acetanilide and 200 ml of 2-propanol.

[b] The Rh, Ru, and Pd catalysts were 5% metal on carbon. The Ni and Co catalysts were 50% metal on kieselguhr.

[c] Little or no gas absorption in additional 1.0 hr.

[d] Little or no gas absorption in additional 1.5 hr.

[e] Little or no gas absorption in additional 0.5 hr.

The procedure is illustrated in the following description of one experiment:

A mixture of 27.0 g (0.20 mole) of acetanilide, 200 ml of 2-propanol and 1.1 g of 5% Rh on carbon was added to a 600-ml

Magne Drive autoclave. The vessel was sealed, purged first with nitrogen and then with hydrogen, and pressured with hydrogen to 500 psig. The autoclave was heated with agitation at 35-40° and 400-600 psig for 2 hr with little or no gas absorption in the last 1 hr. The autoclave was cooled and depressurized. The reaction product was removed and filtered through Celite filter aid to remove the catalyst. The solvent was removed from a portion of the filtrate in a rotary evaporator under vacuum. The white crystalline residue consisted of N-cyclohexylacetamide mp 105-106° (Baeyer, 1894). No side products were detected by glpc analysis, and UV analysis showed only traces of residual aromaticity.

This experiment was repeated with purified acetanilide and purified 2-propanol. The acetanilide in solution in 2-propanol was treated with Raney Ni, the Ni removed by filtration on Celite filter aid, the 2-propanol evaporated on a rotary evaporator, and the acetanilide distilled at 202° at 43 mm. The 2-propanol solvent was treated with Raney Ni, distilled, and dried over magnesium sulfate.

There was no discernible difference in the rates of hydrogenation in the two experiments.

Hydrogenation of bis(4-acetamidophenyl)methane

A mixture of 70.5 g (0.25 mole) of bis(4-acetamidophenyl)-methane, 200 ml of 2-propanol, and 5.6 g of 5% Rh on carbon was added to a 600-ml Magne Drive autoclave. The vessel was sealed, purged first with nitrogen and then with hydrogen, and pressured with hydrogen to 1600 psig. The autoclave was heated with agitation at 95-110° and 1600-1900 psig for 1 hr when gas absorption ceased. Agitation was continued for an additional 0.5 hr at 105° . The autoclave was cooled and depressurized. The reaction product was removed and filtered through Celite filter aid to remove the catalyst. The solvent was removed in a rotary evaporator under reduced pressure. The residue consisted of 68.5 g (93% yield) of white bis(4-acetamidocyclohexyl)methane, mp 207-221° (Barkdoll et al, 1951), and was shown by UV analysis to contain 0.12 g (0.2% yield) of aromatic starting material. The product, a mixture of isomers (Barkdoll et al, 1951), was characterized by elemental analysis.
Anal. Calcd. for $C_{17}H_{30}N_2O_2$: C, 69.35; H, 10.27; N, 9.51. Found: C, 69.10; H, 10.31; N, 9.76. The trans-trans isomer, mp 271-273°, and the cis-trans isomer, mp 231-232° (Barkdoll et al, 1951), were isolated by fractional crystallization from 2-propanol.

A mixture of 8.5 g (0.030 mole) of bis(4-acetamidophenyl) methane, 150 ml of 2-propanol, and 4.8 g of 5% Rh on carbon was added to a Paar shaker type pressure reaction apparatus

(Paar Instrument Co., series 3910). The vessel was sealed, purged first with nitrogen and then with hydrogen, and pressured with hydrogen to 50 psig. The reaction mixture was agitated for 18 hr at room temperature. The apparatus was depressurized and the reaction product removed and filtered through Celite filter aid to remove the catalyst. The solvent was removed in a rotary evaporator under reduced pressure. The residue consisted of 8.5 g (96% yield) of white bis(4-acetamidocyclohexyl)methane, mp 208-228°, and was shown by UV analysis to contain 0.1% of aromatic starting material.

A mixture of 16.94 g (0.060 mole) of bis(4-acetamidophenyl)methane, 220 ml of 2-propanol and 4.8 g of 5% Rh on carbon was added to a 600-ml Magne Drive autoclave. The vessel was sealed, purged first with nitrogen and then with hydrogen, and pressured with hydrogen to 1050 psig. The autoclave was agitated at 15-18° and 890-1050 psig for 4.7 hr at which point the gas absorption was about 50% of the theoretical amount required for the hydrogenation of bis(4-acetamidophenyl)methane to bis(4-acetamidocyclohexyl)methane. The autoclave was depressurized and the reaction product removed and filtered through Celite filter aid to remove the catalyst. The solvent was removed in a rotary evaporator under reduced pressure. The residue consisted of 16.6 g that had a UV absorption maximum in methanol at 251 mm, identical with that of the starting material, bis(4-acetamidophenyl)methane. A TLC of the residue on silica gel with benzene-methanol (4:1) gave two spots. One, detected by acid charring (5% potassium dichromate in 40% sulfuric acid), had the same R_f value as bis(4-acetamidocyclohexyl)methane. The other, detected by fluorescent quenching as well as by acid charring, had the same R_f value as bis(4-acetamidophenyl)methane. There was no indication of the presence of 4-acetamidocyclohexyl-4-acetamidophenylmethane from either UV or TLC analyses.

The results of experiments on the nuclear hydrogenation of bis(4-acetamidophenyl)methane with various catalysts are summarized in Table II.

Effect of purity in hydrogenation of bis(4-acetamidophenyl)-methane

Each experiment was run in a 300-ml Magne Drive autoclave with 8.47 g (0.03 mole) of bis(4-acetamidophenyl)methane, 110 ml of 2-propanol, and 0.60 g (5 g/l) of 5% Rh on carbon at 65-70° and 900-1200 psig.

Using bis(4-acetamidophenyl)methane prepared from practical grade 4,4'-methylenedianiline, from distilled 4,4'-methylenedianiline, and from the latter material further purified by one recrystallization from methanol-water, the times for complete reaction were 5.0, 3,5, and 1.8 hr, respectively.

TABLE II. *Hydrogenation of Bis (4-acetamidophenyl)methane*[a]

Catalyst	Temp., °C	Time, hr	Residue Product mp, °C[b]	Residue Product % aromatic[c]
Rh	20-25	4.3	209-230	0.2
Ru	80	4.5	203-233	0.6
Pd	150-155	7.0	203-243	1.6
Pt	195-200[d]	0.8	-	ca 100[e]

[a]Each experiment was run in a 300-ml Magne Drive autoclave with 8.47 g (0.03 mole) of bis(4-acetamidophenyl)methane, 110 ml of 2-propanol, and 2.4 g (20 g/l) of a 5% metal on carbon catalyst at 900-1200 psig.

[b]Melting point of bis(4-acetamidophenyl)methane is 236-237°.

[c]Determined by UV analysis based on starting material; one mole of 4-acetamidocyclohexyl-4-acetamidophenylmethane would be calculated as 0.5 mole of bis(4-acetamidophenyl) methane.

[d]Pressure was 1400-1450 psig.

[e]Determined by gas absorption.

Hydrogenation of various aromatic amides

Table III summarizes the results of the nuclear hydrogenation of a number of aromatic mono and diamides.

Rhodium- and ruthenium-catalyzed hydrogenation of acetanilide

The results of a detailed investigation of the Rh- and the Ru-catalyzed nuclear hydrogenation of acetanilide are summarized in Table IV.

Effect of acid in hydrogenation of acetanilide

The effect of acid on the Rh-, Pd-, and Ru-catalyzed hydrogenation of acetanilide is summarized in Table V.

TABLE III. *Miscellaneous Hydrogenations*[a]

Starting Material			Wt. Catalyst	Vol. Solvent	Temp.	Pressure	Time	Product		
Compound	Wt. g	Mole	g	ml	°C	psig	hr	Compound	mp	Mole% Yield[b]
Formanilide	36.3	0.30	5.0	215	25-30 60	940-1060	9.3 2.5[c]	N-cyclohexylformamide	--	5[d]
Butyranilide	16.3	0.10	5.4	250	20-23	900-1200	0.2[e]	N-cyclohexylbutyramide	61-2[f]	100[g]
Stearanilide	35.9	0.10	5.4	250	65 50-65	900-1000	0.1 0.5[c]	N-cyclohexylstearamide	193.5-196	96.5[h]
Benzanilide	19.7	0.10	5.4	250	21-23	900-1200	2.3[i]	N-cyclohexyl-cyclohexylcarboxamide	170-1[j]	99
1,3-bisacetamidobenzene	19.2	0.10	5.5	250	20-60 60	900-1200	0.7 1.5[c]	N,N'-1,3-cyclohexylene-bisacetamide	237-251[k]	97[l]
1,4-bisacetamidobenzene	19.2	0.10	5.5	250	35-40 50-60	900-1200	2.0 2.2[m]	N,N'-1,4-cyclohexylene-bisacetamide	203-243[k]	94[n]
2,4-bisacetamidotoluene	8.2	0.040	5.5	250	25-60 60-65	900-1000	0.7 1.0[c]	N,N'-1,3-(4-methylcyclohexylene)bisacetamide	198-212	83[o]
2-acetamidonaphthalene	10.8	0.058	5.5	250	20-50 45-50	900-1200	0.4 0.9[c]	2-acetamidodecalin	70-6[p]	97[q]
1,5-bisacetamidonaphthalene	12.0	0.050	4.0	250	75 95	900-1200 1140-1180	1.8 3.5[r]	1,5-bisacetamidodecalin	250-270(dec)	85[s]

[a]Each experiment was run in a 600 ml Magne Drive autoclave with a 5% Rh on carbon catalyst and 2-propanol as solvent. [b]Yield of isolated product unless otherwise specified. [c]Little or no gas absorption. [d]Determined by glpc analysis; 90% formanilide and no detectable aniline or cyclohexylamine. [e]Little of no gas absorption in additional 0.5 hr. [f]Reported mp 68° (Harvill et al., 1950). [g]UV analysis indicated 0.2% aromatic starting amide. Anal. Calcd. for $C_{10}H_{19}NO$: C, 70.94; H, 11.34; N, 8.27. Found: C, 70.45; H, 11.32; N, 8.70. [h]UV analysis indicated 0.6% aromatic starting amide. Anal. Calcd. for $C_{24}H_{47}NO$: C, 78.84; H, 12.96; N, 3.83. Found: C, 78.52; H, 12.93; N, 3.61. [i]Little or no gas absorption in additional 1.9 hr. [j]Harvill et al., 1950. [k]Nielsen, 1962. [l]UV analysis indicated no aromaticity. Anal. Calcd. for $C_{10}H_{18}N_2O_2$: C, 60.56; H, 9.18; N, 14.12. Found: C, 60.73; H, 9.21; N, 13.86. [m]Little or no gas absorption in additional 1 hr. [n]UV analysis indicated no aromaticity. Anal. Calcd. for $C_{10}H_{18}N_2O_2$: C, 60.56; H, 9.18; N, 14.12. Found: C, 60.63; H, 9.22; N, 14.05. [o]UV analysis indicated 0.3% aromatic starting amide. Anal. Calcd. for $C_{11}H_{20}N_2O_2$: C, 62.06; H, 9.54; N, 13.26. Found: C, 62.03; H, 9.31; N, 13.20. [p]Dauben et al., 1954. [q]UV analysis indicated no aromaticity. Anal. Calcd. for $C_{12}H_{21}NO$: C, 73.78; H, 10.86; N, 7.17. Found: C, 73.79; H, 10.94; N, 6.85. [r]Little or no gas absorption in additional 3 hr. [s]UV analysis indicated no aromaticity. Anal. Calcd. for $C_{14}H_{24}N_2O_2$: C, 66.63; H. 9.59; N, 11.10. Found: C, 66.25; H, 9.78; N, 10.88.

TABLE IV. *Rhodium and Ruthenium Catalyzed Hydrogenation of Acetanilide*[a]

Catalyst							Yield, mole %[c]		
Metal	Wt. g	Conc. g/l	Temp. °C	Pressure psig	Added base[b]	Time hr	acetanilide	N-cyclohexyl-acetamide	Trace products[d]
Rh	0.03	0.15	75	580-740	No	21.5	59	41	CHA
Rh	0.03	0.15	125	520-770	No	6.0[e]	37	63	CHA
Rh	0.03	0.15	100-175	700-800	No	0.3[f]	66	34	CHA
Rh	0.20	1.0	100	500-800	No	0.3	0	100	-
Rh	0.10	0.50	100 100-165	500-800	No	1.7 0.1	0	100	-
Rh	0.10	0.50	165	740-800	No	3.2	0	100	-
Rh	0.10	0.50	100	500-800	Yes[g]	12.0	29	71	-
Ru	0.10	0.50	100	500-800	No	13.0	33	67	CHA
Ru	0.10	0.50	160	500-800	No	5.5[h]	32	68	CHA, DCHA
Ru	0.10	0.50	180	665-930	No	7.3	36	64	CHA, DCHA
Ru	0.30	1.5	100	500-800	No	5.0	0	100	CHA
Ru	0.30	1.5	100	500-800	Yes[g]	3.5[i]	0	100	-
Ru	0.30	1.5	100	750	Yes[j]	5.3	No reaction[k]		

[a]Each experiment was run in a 600 ml Magne Drive autoclave with 27.0 g (0.20 mole) of acetanilide, 175 ml of 2-propanol, and a 5% metal on carbon catalyst. [b]0.5 mole % based on acetanilide. [c]Determined by quantitative glpc and UV analyses. Material balances were in the high nineties, and yields are normalized to 100% because only traces of other products were indicated by glpc. [d]Determined by glpc. CHA = cyclohexylamine, DCHA = dicyclohexylamine. [e]Little or no gas absorption in additional 1.0 hr. [f]Little or no gas absorption in additional 1.0 hr. at 175-180°. [g]N-ethylaniline. [h]Little or no gas absorption in additional 1.5 hr. [i]There was less of an induction period or a faster initial rate than in the preceding experiment. [j]Cyclohexylamine. [k]Determined by lack of gas absorption.

TABLE V. *Effect of Acid in Hydrogenation of Acetanilide*[a]

Expt. No.	Catalyst Metal	Catalyst Wt., g	Catalyst Conc., g/l	Added Acid[b]	Temp., °C	Pressure psig	Time, hr
1	Rh	0.10	0.83	No	100	800-1000	1.9
2	Rh	0.10	0.83	Yes	100	800-1000	1.0
3	Pd	2.4	20	No	110-120	800-1200	13[c]
4	Pd	2.4	20	Yes	110-120	800-1200	11[d]
5	Ru	0.2	1.7	No	100	800-1000	3.0[e]
6	Ru	0.2	1.7	Yes	100	800-1000	3.5[f]
7	Ru	0.4	3.3	No	100	800-1000	0.9[g]
8	Ru	0.4	3.3	Yes	100	800-1000	1.2[h]

[a]Each experiment was run in a 300-ml Magne Drive autoclave with 27.0 g (0.20 mole) of acetanilide, 90 ml of 2-propanol, and a 5% metal on carbon catalyst.

[b]0.10 ml (3 x 10^{-4} mole) 6N H_2SO_4.

[c]Reaction about 90% complete based on gas absorption.

[d]Little or no gas absorption in additional 1.8 hr.

[e]Little or no gas absorption in additional 1.0 hr.

[f]Little or no gas absorption in additional 0.5 hr.

[g]Little or no gas absorption in additional 0.7 hr.

[h]Little or no gas absorption in additional 0.6 hr.

Acid-promotion in hydrogenation of bis(4-acetamidophenyl)methane

Results with a Rh catalyst and 2-propanol are summarized in Table VI. Experiments with methanol and acetic acid as solvents did not go to completion, apparently due to formation of an inhibitor. Incomplete reaction still resulted after addition of 0.20 ml of 6N H_2SO_4 to the methanol, and after the addition of 0.10 ml of 6N H_2SO_4 to the acetic acid.

TABLE VI. *Acid-Promotion of Fresh Rhodium Catalyst*[a]

	Acid Added			
Type	Quantity	Mole	Equiv.	Time, hr
None	–	–	–	2.7 ± 0.2[b]
6N H_2SO_4	0.20 ml	6×10^{-4}	12×10^{-4}	1.8
6N H_2SO_4	0.10 ml	3×10^{-4}	6×10^{-4}	1.8 ± 0.1[b]
6N HCl	0.10 ml	6×10^{-4}	6×10^{-4}	1.3
6N H_3PO_4	0.10 ml	2×10^{-4}	6×10^{-4}	1.2
6N acetic acid	0.10 ml	6×10^{-4}	6×10^{-4}	2.7 ± 0.1[b]
succinic acid	0.035 g	3×10^{-4}	6×10^{-4}	2.3

[a]Each experiment was run in a 300-ml Magne Drive autoclave with 28.2 g (0.10 mole) of bis(4-acetamidophenyl)methane, 87 ml of 2-propanol, and 1.1 g of fresh 5% Rh on carbon at 120° and 800-1000 psig.

[b]Average of 2 experiments.

A Pt catalyst had little or no activity in the absence of acid (Table II). The addition of two drops of 6N HCl produced only slight activity and the addition of 0.5 ml of acetic acid had little or no effect.

Reactivation of used rhodium catalyst in hydrogenation of acetanilide

Experiment 1 in Table V was repeated with the used catalyst from that experiment after it had been separated from the reaction product by centrifugation and then washed with one 400-ml portion of 2-propanol followed by washings with two 200-ml portions of 2-propanol. The reaction with this recovered catalyst was completed in 6.8 hr.

The used catalyst from the above experiment was separated from the reaction product by centrifugation, washed with one 400-ml portion followed by two 200-ml portions of 2-propanol, washed once with 200-ml of 6N H_2SO_4, and then twice with 200-ml portions of 2-propanol. When this recovered and acid-treated catalyst was used with the addition of 0.10 ml

(3×10^{-4} mole) of 6N H_2SO_4, the reaction was completed in 3.8 hr.

Reactivation of used rhodium catalyst in hydrogenation of bis(4-acetamidophenyl)methane

A mixture of 282 g (1.0 mole) of bis(4-acetamidophenyl)-methane, 870 ml of 2-propanol, and 11 g of 5% Rh on carbon was added to a 1-gallon stirred autoclave. The vessel was sealed, purged first with nitrogen and then with hydrogen, and pressured with hydrogen to 900 psig. The autoclave was heated with agitation at 120° and 800-1000 psig for 5.3 hr with little or no gas absorption in the last 0.8 hr. The autoclave was cooled and depressurized. The reaction product was removed using a large volume of methanol to wash all material out of the vessel. The catalyst was separated by centrifugation, washed with a large volume of methanol and then with about 200 ml of water. After a final centrifugation, the catalyst paste was divided into equal portions, each containing 1.1 g of 5% Rh on carbon on a dry basis. Each portion was then washed with two 200-ml volumes of 2-propanol.

The samples of used catalyst from above were washed with two 200 ml portions of a specified solvent or solution, followed in each case by a final wash with 200 ml portions of 2-propanol. The results of subsequent hydrogenations are summarized in Table VII. An experiment with fresh catalyst is included for comparison. The spent catalyst from experiment 11 (Table VII) was washed with methanol, two 200-ml portions of 6N H_2SO_4, and one 200 ml portion of 2-propanol. Experiment 11 was then repeated after the addition of 0.10 ml of 6N H_2SO_4 and resulted in complete conversion in 3.5 hr.

Catalyst promotion and reuse in hydrogenation of 2,4-bisacetamidotoluene to N,N'-1,3-(4-methylcyclohexylene)bisacetamide

Each experiment was run in a 300-ml Magne Drive autoclave with 10.3 g (0.05 mole) of 2,4-bisacetamidotoluene, 106 ml of 2-propanol, and 1.1 g of 5% Rh on carbon at 120° and 800-1000 psig. The experiment with fresh catalyst was completed in 2.3 ± 0.3 hr (average of 2 runs).

A repetition of these experiments with the addition of 0.10 ml of 6N H_2SO_4 resulted in complete conversion in 1.2 hr.

The used catalyst from one of the experiments with fresh catalysts without added acid was washed with methanol and then two 200-ml portions of 2-propanol. This catalyst resulted in about 90% conversion in 7.5 hr.

TABLE VII. *Reactivation of Used Rhodium Catalyst*[a]

Expt. No.	Wash Medium	Time, hr	Comments
1	None	2.7 ± 0.2[b]	Base run with fresh catalyst
2	2-propanol	5.8	ca. 90% complete[c]
3	dimethylformamide	11.0	ca. 72% complete[c]
4	tetrahydrofuran	10.8	ca. 92% complete[c]
5	6N acetic acid	4.1	
6	6N H_3PO_4	2.2	
7	6N HCl	4.3	
8	6N H_2SO_4	3.8	
9	6N H_2SO_4	4.1	Used catalyst from expt. 8
10	6N H_2SO_4	1.8	Used catalyst from expt. 9
11	methanol	6.1	Used catalyst from expt. 10 ca. 90% complete[c]

[a]Each experiment was run in a 300-ml Magne Drive autoclave with 28.2 g (0.10 mole) of bis(4-acetamidophenyl)methane, 87 ml of 2-propanol, and 1.1 g of 5% Rh on carbon at 120° and 800-1000 psig.

[b]Average of 2 experiments.

[c]Determined by gas absorption.

The used catalyst from the last experiment was washed with methanol, two 200-ml portions of 6N H_2SO_4, and one 200-ml portion of 2-propanol. After the addition of 0.10 ml of 6N H_2SO_4, this catalyst resulted in complete conversion in 2.8 hr.

DISCUSSION

The amido group in N-aryl amides should influence the nuclear hydrogenation of the aromatic ring by virtue of its electronic and steric effects on the absorption of the ring and the rate of hydrogenation of the absorbed species, by its competitive absorption on the catalyst, and by the competitive absorption of amine by-products formed from hydrogenolysis of the amido group.

The electronic effect of substituents on the hydrogenation of a benzene ring (Mochida and Yoneda, 1968) and of olefinic bonds (Kieboom and Van Bekkum, 1972) appears to be small. The electronic effect of the amido substituent on the hydrogenation behavior of the aromatic ring probably is also of minor importance.

The steric effect of substituents on the hydrogenation of an aromatic ring is considerable and presumably due to steric interference with absorption of the ring (Smith, 1967). The effect of the amido group can be anticipated from the fact that the hydrogenation of benzene was several times faster than that of isobutylbenzene with each catalyst used in this study (Greenfield, 1973).

A comparison of several catalysts for the nuclear hydrogenations of acetanilide (Table I) and bis(4-acetamidophenyl)-methane (Table II) shows that rhodium and ruthenium are the most active of the catalysts, with rhodium much more active than ruthenium. There was no detectable hydrogenation with the platinum catalyst (Table II). Acetanilide is hydrogenated much more rapidly than bis(4-acetamidophenyl)methane. The rhodium-catalyzed hydrogenation of isobutylbenzene (Greenfield, 1973) is much more rapid than that of acetanilide under the same conditions (Malz and Greenfield, 1969). This suggests that the steric effect of the polar amido group is less important than its competitive absorption and/or the absorption of amine by-products.

An attempt to detect 4-acetamidocyclohexyl-4-acetamidophenylmethane from a rhodium-catalyzed hydrogenation of bis(4-acetamidophenyl)methane was unsuccessful. Only the fully hydrogenated product, bis(4-acetamidocyclohexyl)methane, and the starting material were found.

Significant decreases in reaction time were effected by the use of better quality bis(4-acetamidophenyl)methane in the rhodium-catalyzed hydrogenation.

The general synthetic applicability of the nuclear hydrogenation of N-aryl amides is demonstrated by the examples in Table III. The products from the bisamides and the naphthalene compounds are mixtures of geometric isomers.

The difficulties experienced with formanilide were unusual, as often is the case with the first member of a series (Table III). Similar results were obtained with a ruthenium catalyst (not shown in Table III). The very low conversions to N-cyclohexylformamide probably is due to poisoning of the catalyst by the hydrogenolysis products of the formanilide. A similar poisoning has been reported for the hydrogenolysis of the formamides formed during the palladium and platinum-catalyzed hydrogenations of isocyanates (Knopf, 1970).

The results of a detailed investigation of the rhodium and the ruthenium-catalyzed nuclear hydrogenation of acetanilide are summarized in Table IV. The use of either catalyst is limited to a maximum temperature range above which poisoning of the catalyst seems to occur. This maximum temperature, as expected, decreases as the amount of catalyst decreases and the resulting susceptibility to poisoning increases. A similar effect was found in the rhodium-catalyzed hydrogenation of bis(4-acetamidophenyl)methane.

The addition of N-ethylaniline greatly inhibits the rhodium catalyst, but increases the initial activity and decreases the reaction time for the ruthenium catalyst. Induction periods have been reported for ruthenium in other systems (Rylander, 1967), and the added N-ethylaniline may have decreased such an induction period. It has been reported that added dicyclohexylamine retards the rate of the rhodium-catalyzed hydrogenation of aniline, although the rate with a ruthenium catalyst is little effected (Greenfield, 1964). Thus there is a precedent for the marked difference in behavior of the two catalysts.

Cyclohexylamine completely poisoned the ruthenium catalyst, showing that cyclohexylamine, in contrast to secondary amines, is an inhibitor for ruthenium as well as rhodium. Traces of cyclohexylamine were detected from experiments with each catalyst, particularly in reactions that were incomplete. The cyclohexylamine could have resulted from carbon-nitrogen cleavage of N-cyclohexylacetamide, or nuclear hydrogenation of aniline formed by carbon-nitrogen cleavage of acetanilide. The dicyclohexylamine detected in some experiments with ruthenium probably is a coupling product formed from nuclear-hydrogenation intermediates and/or cyclohexylamine.

The effect of added acid in the rhodium, palladium, and ruthenium-catalyzed hydrogenations of acetanilide is described in Table V. Small amounts of sulfuric acid (0.15 mole % based on acetanilide) markedly promote the rhodium and palladium-catalyzed reactions, but have no such effect on the ruthenium catalyst.

Experiments on the acid-promotion of a rhodium catalyst in the hydrogenation of bis(4-acetamidophenyl)methane are summarized in Table VI. The catalyst is promoted by the mineral acids, perhaps slightly promoted by succinic acid, and not effected by acetic acid. The hydrogenation did not go to completion when the 2-propanol solvent was replaced by methanol or acetic acid. Incomplete reaction still resulted after addition of sulfuric acid to either the methanol or acetic acid.

A platinum catalyst without activity in the absence of acid (Table II) showed only slight activity after the addition of hydrochloric acid and little or no activity after the addition of acetic acid.

The sulfuric acid-promotion of a rhodium catalyst was also demonstrated in the nuclear hydrogenation of 2,4-bisacetamidotoluene.

Extensive reuse of a rhodium catalyst for the hydrogenation of bis(4-acetamidophenyl)methane is possible when the used catalyst is washed with an aqueous solution of a mineral acid or acetic acid (Table VII). Even a catalyst that has been completely poisoned can be reactivated in this manner. Similar reactivations of rhodium catalysts by a combination of aqueous sulfuric acid washes and the addition of sulfuric acid have been demonstrated in the hydrogenation of acetanilide, bis(4-acetamidophenyl)methane, and 2,4-bisacetamidotoluene.

SUMMARY

The rhodium-catalyzed nuclear hydrogenation of N-aryl amides is an excellent laboratory method for the preparation of N-alicyclic amides. The acid-promotion and acid-reactivation of the catalyst makes the reaction of commercial interest (Malz and Greenfield, 1975).

REFERENCES

1. Abraham, E.G., and Lamb, G. (to Amer. Cyanamid), U.S. Patent 3,228,975 (Jan. 11, 1966).
2. Adams, R., and Anderson, J.L., J. Amer. Chem. Soc., 72, 5154 (1950).
3. Adkins, H., Reactions of Hydrogen with Organic Compounds over Copper-Chromium Oxide and Nickel Catalysts, The University of Wisconsin Press, Madison, Wis., 1937.
4. Adkins, H., and Cramer, H.I., J. Amer. Chem. Soc., 52, 4354 (1930).
5. Adkins, H., and Wojcik, B., J. Amer. Chem. Soc., 56, 247, (1934).
6. Baeyer, A., Ann., 278, 104 (1894).
7. Barkdoll, A.E., Gray, H.W., and Kirk, Jr., W., J. Amer. Chem. Soc., 73, 741, (1951).
8. Biedermann and Ledoux, Chem. Ber., 7, 1531 (1874).
9. Billman, J.H., and Buehler, J.A., J. Amer. Chem. Soc., 75, 1345 (1953).

10. Broadbent, H.S., and Bartley, W.J., J. Org. Chem., 28, 2345 (1963).
11. Broadbent, H.S., Campbell, G.C., Bartley, W.J., and Johnson, J.H., J. Org. Chem., 24, 1847 (1959).
12. Butler, C.L., and Adams, R., J. Amer. Chem. Soc., 47, 2610 (1925).
13. Buu-Hoi, Bull soc. chim., 12, 587 (1945).
14. Dauben, W.G., Tweit, R.C., and Mannerskantz, C., J. Amer. Chem. Soc., 76, 4420 (1954).
15. De la Mare, P.B.D., and Hassan, M., J. Chem. Soc., 1519 (1958).
16. Della, E.W., and Jefferies, P.R., Austr. J. Chem., 14, 610 (1961)
17. Devereux, J.M., Payne, K.R., and Peeling, E.R.A., J. Chem. Soc., 2845 (1957).
18. Domnin, N.A., and Cherkasova, V.A., Zhur. Obshchei Khim., 26, 1616 (1956); Chem. Abstr., 51, 1870a (1957).
19. Dovell, F.S., Ferguson, W.E., and Greenfield, H., Ind. Eng. Chem. Product Res. and Develop., 9, 224 (1970).
20. Ferber, E., and Bendix, H., Chem. Ber., 72, 839 (1939).
21. Ferber, E., and Bruckner, H., Chem. Ber., 72, 995 (1939).
22. Fraser, R.R., and Swingle, R.B., Can. J. Chem., 48, 2065 (1970).
23. Freifelder, M., Practical Catalytic Hydrogenation, Chapter 24, Wiley-Interscience, New York, N.Y., 1971.
24. Greenfield, H., J. Org. Chem., 29, 3082 (1964).
25. Greenfield, H., Ann. N.Y. Acad. Sci., 214, 233 (1973).
26. Hartmann, M., Ensslin, H., and Panizzon, L. (to Soc. Chem. Ind., Basle, Switzerland), U.S. Patent 2,152,960 (April 4, 1939).
27. Guyer, A., Bieler, A., and Gerliczy, G., Helv. Chim. Acta, 38, 1649 (1955).
28. Harvill, E.K., Herbst, R.M., Schreiner, E.C., and Roberts, C.W., J. Org. Chem., 15, 662 (1950).
29. Ipatieff, V.N., and Schmerling, L., J. Amer. Chem. Soc., 59, 1056 (1937).
30. Jansen, E.F., J. Biol. Chem., 238, 1552 (1963).
31. Kieboom, A.P.G., and Van Bekkum, H., J. Catal., 25, 342 (1972).
32. Knopf, R.J., Chem. and Eng. Data, 15, 196 (1970).
33. Knowles, R.N. (to du Pont), U.S. Patent 3,683,022 (August 8, 1972).
34. Knuckell, F., and Schneider, H., Chem. Ztg., 36, 1201 (1912).
35. Landenburg, Chem. Ber., 8, 1211 (1874).
36. Le Guyader, M., Compt. Rend., 254, 4182 (1962).
37. Leonard, N.J., and Boyd, Jr., S.N., J. Org. Chem., 11, 405 (1946).

38. Leonard, N.J., and Hyson, A.M., J. Amer. Chem. Soc., 71, 1961 (1949).
39. Malz, Jr., R.E., and Greenfield, H., unpublished work, 1969.
40. Malz, Jr., R.E., and Greenfield, H., (to Uniroyal) U.S. Patent 3,867,443 (Feb. 18, 1975).
41. Mamlok, L., Bull. soc. chim. France, 1182 (1956).
42. Mochida, I., and Yoneda, Y., J. Catal., 11, 183 (1968) and references therein.
43. Morgan, K.J., and Turner, A.M., Tetrahedron, 22, 1175 (1966).
44. Muromova, R.S., Pletneva, I.D., Demidova, T.V., and Shkhiyants, I.V., Tokareva, G.A., Vysokomolekul. Soedin., 7, 1354 (1965); Chem. Abstr., 63, 18269f (1965).
45. Nielsen, A.T., J. Org. Chem., 27, 1998 (1962).
46. Peeling, E.R.A., and Shipley, D.K., Chem. and Ind., 362 (1958).
47. Pletneva, I.D., Muromova, R.S., and Pervukhima, I.V., Zh. Obshch. Khim., 34, 1815 (1964); Chem. Abstr., 61, 8201d (1964).
48. Pletneva, I.D., Muromova, R.S., Pervukhima, I.V., and Shkhiyants, I.V., Zh. Organ. Khim, 1, 1981 (1965); Chem. Abstr., 64, 9607f (1966).
49. Rylander, P.N., Catalytic Hydrogenation over Platinum Metals, Academic Press, New York, N.Y., 1967.
50. Sasa, T., J. Soc. Org. Synthet. Chem. (Japan), 12, 185 (1954); Chem. Abstr., 51, 2780f (1957).
51. Skita, A., Rolfes, H., Chem. Ber., 53B, 1242 (1920).
52. Sladkov, A.M., and Vett, S.V., Zh. Obshch. Khim., 26, 1130 (1956); Chem. Abstr., 50, 16704e (1956).
53. Smith, H.A., Ann. N.Y. Acad. Sci., 145, 108 (1967).
54. Tröger, Westerkamp, Arch. Pharm., 247, 663 (1909).
55. Tsatsas, G., and Delaby, R., Ann. pharm. franc., 14, 621 (1956); Chem. Abstr., 51, 8670d (1957).
56. Tu, C-T., Meng, C-H., and Ho, Y-T., Hua Hsüeh Hsüeh Pao, 22, 134 (1956); Chem. Abstr., 52, 7179h (1958).
57. Wojcik, B., and Adkins, H., J. Amer. Chem. Soc., 56, 2419 (1934).

Subject Index

A

Asymmetric Hydrogenation, 203-230
Auger Electron Spectroscopy, 49-73

C

Carbon Monoxide, 123-135, 181-200
Carbon Monoxide Insertion, 181-200
Carbonylation,
  Allylic Halides, 193-200
  Olefins, 183
  Polyhalogenated Compounds, 198
  Vinylic Halides, 184-193
Catalysts,
  Al, Co Organic Complex, 305-323
  Ag (CH_3COO), 261
  Ag NO_3, 261
  Cobalt, 36
  Co-on-Kieslguhr, 345
  $Co_2(CO)_8$, 257, 297
  CoH $(CO)_2(Bu_3P)_2$, 300, 302
  Copper, 62, 65
  Cu (CH_3COO), 261
  Cu $(CH_3COO)_2$, 261
  CuCl, 259, 260, 261
  Copper Chromite, 117, 119-126, 344
  Cu, Raney, 244, 245
  Fe $(CO)_5$, 257, 264
  $Fe_2(C_5H_5)_2(CO)_4$, 238, 251, 264
  Hg Cl_2, 261
  Ir $(\phi_3P)_2$ (CO)(Cl), 166-172
  Lanthanum manganates, 113
  Li, Co Organic Complex, 305-323
  Mo $(CO)_6$, 242, 251, 253
  $Mo_2(C_5H_5)_2(CO)_6$, 238, 251
  Mo $(acac)_3$, 242, 251
  $MoO_2(acac)_2$, 242, 251
  Ni-on-Kieselguhr, 306, 345
  Ni, Raney, 80, 244, 245, 246, 247, 283, 306
  Ni $(\phi_3P)_2(CO)_2$, 165, 166, 174-178
  Pd $(CH_3CO_2)_2$, 189
  Pd Black, 244, 245, 250
  $PdCl_2$, 183, 189, 196
  Pd-on-Alumina, 59, 66, 82, 144, 184, 189-193, 200, 250
  Pd-on-Barium carbonate, 82
  Pd-on-Barium sulfate, 89
  Pd-on-Calcium carbonate, 82, 84, 144
  Pd-on-Carbon, 32, 80, 81, 82, 84, 86, 87, 89, 144, 145, 147, 149, 246, 274, 283, 306, 345
  Pd-on-Silk Fibroin, 206
  Pd, Au-on-Alumina, 185
  Pd, Au-on-Silica, 185
  Pd, Pt-on-Carbon, 147, 149
  Pt Black, 244, 245, 247
  PtO_2, 334
  Pt-on-Alumina, 144
  Pt-on-Carbon, 146-148, 150, 283, 331, 345
  Pt-on-Corderite, 118
  Rh-on-Calcium carbonate, 84
  Rh-on-Carbon, 283, 345
  $RhCl(CO)(\phi_3P)_2$, 248
  RhCl $(\phi_3P)_3$, 80, 91, 153, 157, 165, 174
  RhCl(DIOP), 212
    DIOP≡2, 3–O–Isopropylidene–2, 3–dihydroxy–1, 4–bis(diphenylphosphino)butane
  RhCl $(NMDPP)_3$, 211
    NMDPP≡Neomenthyl–diphenylphosphine
  $RhCl_3[P(CH_3)(C_3H_7)(\phi)]_3$, 208
  Rh (diene) X^- (ACMP), 214, 218
    ACMP≡o-Anisylcyclohexyl-methylphosphine
  $RhH(CO)(\phi_3P)_3$, 248, 300

Ruthenium Oxide, 144
Ru-on-Alumina, 150, 283
Ru-on-Carbon, 147, 275, 283, 345
$RuClH(\phi_3P)_3$, 264, 300, 302
$RuClH(CO)(\phi_3P)_3$, 292
$RuClH(CO)_2(\phi_3P)_2$, 287, 291
$RuCl_2(CO)_2(\phi_3P)_2$, 161, 164, 166, 174-178 264, 287-303
$[RuCl_2(CO)_2\phi_3P]_2$, 297
$RuCl_2(CO)_3$, 264
$RuCl_2(olefin)_2$, 209
$RuCl_2(\phi_3As)_3$, 264
$RuCl_2(\phi_3P)_3$, 248, 258, 263, 264, 267, 287-303
$RuCl_2(\phi_3Sb)_3$, 264
$RuH(OCOCF_3)(\phi_3P)_3$, 300
$V(acac)_3$, 238, 242, 251
$V(C_5H_5)(CO)_4$, 237, 238, 251, 252
$VO(acac)_2$, 238, 242, 251
$W(CO)_6$, 242, 251
WO_3-on-Al_2O_3, 288
Catalyst Activity, 12, 113-136, 137-151
Catalyst Life, 20
Catalyst Reactivation, 357
Chiral Catalysts, 203-230

D

Dehalogenation, 82-87
Dehydrogenation, 249
Deuterium, 89, 161, 334
Differential Scanning Calorimetry, 137-151
Differential Thermal Analysis, 113-136
Diffusion, 1-47

E

Enzymes, 89, 90, 101-109
Epoxidation, 101-109, 235-253
Exchange Reactions, 77

F

Fick's Law, 6, 7
Fourier's Law, 8

H

Heat Transfer, 1-47
Hydrazine Reductions, 273-285
Hydroformylation, 298
Hydrogenation,
Aldehydes, 297
Aromatic Amides, 343-357
Aromatics, 218
Epoxides, 244
Ketones, 81
Nitro Compounds, 33, 34, 257-271 273-285, 298
Olefins, 80, 81, 89, 153-159, 161-179, 206-216, 223-230, 287-296, 299-303, 305-324
Sulfur Compounds, 36
Hydrogenolysis Cyclopropanes, 325-339
Hydrogen transfer, 249
Hydrosilylation, 223-225
Hydroxylation, 101-109, 235-253

I

Isomerization, 247

L

Labeling, 75-98, 161

M

Mass Transfer, 1-47, 154

O

Oligomerization, 163-166, 172-179

P

Peclet Number, 37
Perovskites, 113-
o-Phenylenediamines, 273-285

Polymer Attached Catalysts, 153-179, 223
Polymer Hydrogenation, 305-324
Pore Diffusion, 41

R

Rubredoxin, 101

S

Stokes' Law, 40
Synergism, 147-149

T

Temperature Gradients, 9, 11, 19, 35
Tritium, 75-98

A 6
B 7
C 8
D 9
E 0
F 1
G 2
H 3
I 4
J 5